The Educated Eye

Interfaces: Studies in Visual Culture

EDITORS Mark J. Williams and Adrian W. B. Randolph, Dartmouth College

This series, sponsored by Dartmouth College Press, develops and promotes the study of visual culture from a variety of critical and methodological perspectives. Its impetus derives from the increasing importance of visual signs in everyday life, and from the rapid expansion of what are termed "new media." The broad cultural and social dynamics attendant to these developments present new challenges and opportunities across and within the disciplines. These have resulted in a trans-disciplinary fascination with all things visual, from "high" to "low," and from esoteric to popular. This series brings together approaches to visual culture—broadly conceived—that assess these dynamics critically and that break new ground in understanding their effects and implications.

For a complete list of books that are available in the series, visit www.upne.com

Nancy Anderson and Michael R. Dietrich, eds., *The Educated Eye: Visual Culture and Pedagogy in the Life Sciences*

Shannon Clute and Richard L. Edwards, *The Maltese Touch of Evil: Film Noir and Potential Criticism*

Steve F. Anderson, *Technologies of History: Visual Media and the Eccentricity of the Past*

Dorothée Brill, *Shock and the Senseless in Dada and Fluxus*

Janine Mileaf, *Please Touch: Dada and Surrealist Objects after the Readymade*

J. Hoberman, *Bridge of Light: Yiddish Film between Two Worlds*, updated and expanded edition

Erina Duganne, *The Self in Black and White: Race and Subjectivity in Postwar American Photography*

Eric Gordon, *The Urban Spectator: American Concept-Cities from Kodak to Google*

Barbara Larson and Fae Brauer, eds., *The Art of Evolution: Darwin, Darwinisms, and Visual Culture*

Jeffrey Middents, *Writing National Cinema: Film Journals and Film Culture in Peru*

Michael Golec, *The Brillo Box Archive: Aesthetics, Design, and Art*

Rob Kroes, *Photographic Memories: Private Pictures, Public Images, and American History*

Jonathan Beller, *The Cinematic Mode of Production: Attention Economy and the Society of the Spectacle*

Ann B. Shteir and Bernard Lightman, eds., *Figuring It Out: Science, Gender, and Visual Culture*

Anna Munster, *Materializing New Media: Body, Image, and Interface in Information Aesthetics*

Luc Pauwels, ed., *Visual Cultures of Science: Rethinking Representational Practices in Knowledge Building and Science Communication*

Lisa Saltzman and Eric Rosenberg, eds., *Trauma and Visuality in Modernity*

The Educated Eye

Visual Culture and Pedagogy in the Life Sciences

Nancy Anderson and
Michael R. Dietrich, Editors

Dartmouth College Press | *Hanover, New Hampshire*

DARTMOUTH COLLEGE PRESS
An imprint of University Press of New England
www.upne.com
© 2012 Trustees of Dartmouth College
All rights reserved
Manufactured in the United States of America
Designed by Katherine B. Kimball
Typeset in Electra, Bureau Empire, and Myriad Pro
by Passumpsic Publishing

5 4 3 2 1

University Press of New England is a member of the
Green Press Initiative. The paper used in this book meets
their minimum requirement for recycled paper.

For permission to reproduce any of the material in this
book, contact Permissions, University Press of New England,
One Court Street, Suite 250, Lebanon NH 03766;
or visit www.upne.com

Library of Congress Cataloging-in-Publication Data
The educated eye : visual culture and pedagogy in the life sciences /
Nancy Anderson and Michael R. Dietrich, editors. — 1st ed.
    p. cm.
Includes bibliographical references and index.
ISBN 978-1-61168-043-0 (cloth : alk. paper) —
ISBN 978-1-61168-044-7 (pbk. : alk. paper) —
ISBN 978-1-61168-212-0 (e-book)
1. Life sciences — Study and teaching. 2. Life sciences —
Technique. 3. Visual communication. 4. Communication and
culture. I. Anderson, Nancy, 1958– II. Dietrich, Michael R.
QH315.E326   2012
570.76 — dc23   2011030415

Contents

Preface ... vii

Introduction: Visual Lessons and the Life Sciences ... 1
NANCY ANDERSON AND MICHAEL R. DIETRICH

1. Trained Judgment, Intervention, and the Biological Gaze: How Charles Sedgwick Minot Saw Senescence ... 14
MARA MILLS

2. Facing Animals in the Laboratory: Lessons of Nineteenth-Century Medical School Microscopy Manuals ... 44
NANCY ANDERSON

3. Photography and Medical Observation ... 68
SCOTT CURTIS

4. Cinematography without Film: Architectures and Technologies of Visual Instruction in Biology around 1900 ... 94
HENNING SCHMIDGEN

5. Cinema as Universal Language of Health Education: Translating Science in *Unhooking the Hookworm* (1920) ... 121
KIRSTEN OSTHERR

6. Screening Science: Pedagogy and Practice in William Dieterle's Film Biographies of Scientists ... 141
T. HUGH CRAWFORD

7. Optical Constancy, Discontinuity, and Nondiscontinuity   162
   in the Eameses' *Rough Sketch*
   MICHAEL J. GOLEC

8. Educating the High-Speed Eye: Harold E. Edgerton's   186
   Early Visual Conventions
   RICHARD L. KREMER

9. On Fate and Specification: Images and Models of   213
   Developmental Biology
   SABINE BRAUCKMANN

10. Form and Function: A Semiotic Analysis of Figures   235
    in Biology Textbooks
    LAURA PERINI

11. Neuroimages, Pedagogy, and Society   255
    ADINA L. ROSKIES

12. The Anatomy of a Surgical Simulation: The Mutual   277
    Articulation of Bodies in and through the Machine
    RACHEL PRENTICE

    Contributors   311
    Index   315

Preface

The Educated Eye: Visual Culture and Pedagogy in the Life Sciences arose from the 2006 Dartmouth College Humanities Institute on Visual Culture and Pedagogy in the Life Sciences. Many of the contributors to this volume were participants in the Institute. Some presented their work at public lectures and others met weekly over the course of a term to develop and discuss the intersecting themes and individual contributions present in this anthology. These essays engage in a relatively new conversation concerning the roles of visual pedagogy in science, the impact of different media, different audiences, and different historical circumstances. We hope that *The Educated Eye* will foster scholarship on the important roles that visual representation plays in scientific communication and education in a broad sense.

As the Institute's codirectors and now editors of this volume, we would like to express our thanks to the Leslie Humanities Center at Dartmouth College for their generous support of the 2006 Humanities Institute and this publication. Specifically, Jonathan Crewe, Adrian Randolph, and Isabel Weatherdon provided invaluable support that made this project possible. We would also like to express our thanks to Brian Kennedy, Katherine Hart, and the staff of the Hood Museum of Art at Dartmouth College for their expert assistance and support for the exhibit *Life Forms: Visual Lessons from Biology*, which offered a public face for the 2006 Institute.

We would also like to acknowledge our thanks to *Common Knowledge* and *Social Studies of Science* for permitting us to reprint essays first published in these journals. An earlier, and very similar, version of "Screening Science: Pedagogy and Practice in William Dieterle's Film Biographies of Scientists" by T. Hugh Crawford appeared in *Common Knowledge* in 1997 (volume 6, no. 2: 52–68). Rachel Prentice's "The Anatomy of a Surgical Simulation: The Mutual Articulation of Bodies in and through the Machine" was first published in *Social Studies of Science* in December 2005 (volume 35, no. 6: 837–866).

Nancy Anderson
Buffalo, NY

Michael R. Dietrich
Amherst, MA

The Educated Eye

Introduction

Visual Lessons and the Life Sciences

Nancy Anderson and Michael R. Dietrich

The late nineteenth century embraced what media theorist Jean-Louis Comolli described as a "frenzy of the visible," in that by the 1880s the world seemed to become entirely visually accessible as a result of increased mobility combined with new visual media, such as advanced printing techniques, photography, and then cinema.[1] The life sciences and medicine, one might say, participated in this frenzy with an astonishing increase in the production, dissemination, and consumption of visual images across medical practice and biological disciplines. Thus, sometime before Comolli's assertion, Michel Foucault, in *The Birth of a Clinic*, a work he claimed in the introduction was "about . . . the act of seeing," presented a history of nineteenth-century medicine as defined by the eye, or vision, taking a dominant role over the other senses. His study of the rise of the medical clinic describes a shift from a medical glance to the medical gaze, from a multisensory engagement with the diseased patient, where touching and listening perform principal functions in the identification of pathological symptoms, to the promotion of the eye as the sensate purveyor of a final (visual) confirmation of disease, specifically through the autopsy. Foucault described this as a *suzerainty of the visual*. By this he means the triumph of the "absolute eye of knowledge" over the other senses, one fully associated with power and death, especially as presented by the cadaver, as the eye has "confiscated, and re-absorbed into its geometry of lines, surfaces, and volumes, raucous voices, whistlings, palpitations, rough, tender skin, cries."[2] Importantly, it is at this same point in time, Foucault postulated, that the clinic emerged as a dual domain of practice and teaching, a space in which physicians were trained in and utilized the medical gaze from the dead to the living, from the living to the dead.

As we move into the twentieth century it is abundantly clear that visual study and visual learning in science and medicine continued to expand, with the implementation of an array of new imaging technologies from X-rays and ultraviolet rays to electronics and, finally, digital computers coming into use. In some ways, the eye becomes more and more an isolated sense when technologies can open up the body to the scientist's or physician's eyes without opening up the body, and we can observe the opened body from immense distances. It is obvious that sight does not work in isolation, but it does seem that the eye remains the privileged sense within the sciences of medicine and biology, and perhaps there are reasons for the continued emphasis on sight at the expense of the other senses. Even as mathematical and theoretical representations fill the field, there is still the need for description and evidence, in image and text, of the necessity to depict all forms of life from full organism to individual organs, then cells, and even macromolecules as best one can in order to convey that knowledge to others.

It has been documented that by the mid-twentieth century, after World War II, the life sciences, widely armed with new visualizing technologies from electron microscopes to video and, finally digital imaging, rapidly upped the ante for pictures, including their role in training students. One biologist who had studied the large increase of illustrations in biology textbooks in the decades following the war, concluded: "In the forty years following the end of World War II the average number of pages in biology textbooks almost doubled, the number of photographs tripled, and the pages that had no illustration at all dwindled to only 22 percent," and that this trend would seem to imply that (1) "good artwork does sell textbooks," (2) "research has shown that illustrations are effective cognitive devices," and (3) "illustrations keep the length of textbooks down by presenting concepts in less space than text alone."[3] The textbook had become a compendium of visual representations knit together with narrative prose, problems sets, and lists of further reading. One result could only be a demand for greater visual literacy among contemporary students. For the most part, however, textbooks are considered to be limited to generally accepted knowledge, and not necessarily a means for presenting new and emerging scientific work, or even providing some plausible experience of creating new knowledge.

One point that we hope to make here, however, is that textbooks, and scientific education more broadly, contribute to the creation of the practical and epistemic foundations for future research. Scientific training and scientific knowledge production are inextricably linked. As many of the essays in this volume will suggest, textbook pictures, projections of biological processes, and movies (scientific films as well as popular science films) can simultaneously teach and convey the experience of research. An important point becomes, then, that in so far as visual representations form an essential element of biological education, they have a profound impact on approaches to and the understanding of scientific research.

This volume, *The Educated Eye: Visual Culture and Pedagogy in the Life Sciences*, brings together twelve essays on the role of scientific images and media as pedagogical objects or tools within professional circles (disciplines) as well as within the wider public sphere, covering the time period from the last decades of the nineteenth century to the present. The authors here all start with the premise that the medium or object they are analyzing and critiquing is involved in some pedagogical mission, whether that be in the laboratory or medical clinic, in a textbook or a classroom, or in public arenas like movie theaters. These papers, then, investigate how images or a visual medium—aside from, but often along with, text—convey lessons related to medicine or the life sciences, lessons regarding biological theory, experimentation and research, observation and human vision, instrumentation, and human health. Readers will confront the problems of employing cinema to convey a globally relevant public health lesson as well as how varying pictorial conventions in contemporary biology textbooks serve very different levels of pedagogical intentions.

But beyond this, many of the authors here also ask how these images or media might enmesh their specific disciplinary lessons within a broader cultural, perhaps ideological, world picture. For example, in this volume, Kirsten Ostherr shows the problems of developing a film for global consumption, one that must effectively convey the dangers of a disease, hookworm, from the Deep South in the United States to Nicaragua and Java. Readers will also encounter Harold Edgerton and his "high-speed eye." Richard Kremer shows that a retraining of the eye is put in play with the technological wizardry of a stroboscope and shutterless camera, whereby viewers encounter the accelerated pace of vision in the twentieth century and beyond—making the point that the human eye must "get up to speed," as it were. Another essay explores how illustrations found in nineteenth-century physiology and histology textbooks might be grappling with the issue of a human-nonhuman animal divide in the teaching of laboratory research methods, even as the animals (rabbits, newts, and frogs) stand in for the human. Whether the authors focus on pictures in an undergraduate or medical school textbook or projected on the cinema or computer screen, whether they discuss drawings, prints, photographs or motion pictures, the intention for many of the scholars included here is to think about content and medium as contributing to a seminar, as it were, that can take on scientific, technological, aesthetic, social, and cultural issues.

The scholarship offered here has concentrated almost exclusively on the use of visualizing technologies in laboratory work or images presented for the purpose of offering scientific knowledge and data or expressing scientific authority to a lay audience. All address the subject matter of images as pedagogical devices. This particular theme has been chosen because it seemed curious to us that so little attention has been given specifically to scientific images and media in the service of pedagogy, as specifically "teaching tools." *The Educated Eye: Visual Culture and Pedagogy in the Life Sciences* intends to draw attention to and encourage wider research in this aspect of the field of visual studies of science. Each essay questions the potential or limitations of the visual to serve as a teaching tool, how the role or meaning or visual form or medium aids or interferes with, or even misrepresents, the specific educational mission.

Over the course of the past thirty years, of course, scholars in science studies and disciplines that are committed to the study of visual images have increasingly looked at the role of visual representation and visual media in scientific practice, having built what is now the foundation for a subdiscipline one might call a "visual studies of science."[4] Drawing scholars from history and philosophy of science; science, technology, and science studies; art history; film/media studies; visual studies and visual culture; and science communication, this literature has been engaging critically with issues such as image production, processing, and consumption; the diverse uses of images in science and medicine; medium specificity; and specific visualizing technologies and techniques.

In the mid-1980s, sociologists of science such as Bruno Latour and Michael Lynch began introducing their ethnographic studies on laboratory work and image production, bringing together an international group of scholars in sociology and beyond around this topic. *Representation in Scientific Practice*, edited by Lynch and Steve Woolgar and published in 1988, marks a watershed moment for interest in this area.[5] By the mid-1970s interest in medical photography had appeared, with Sander Gilman introducing the work of Hugh Welch Diamond, a nineteenth-century doctor who used photography in psychiatric therapy, and art historian Georges Didi-

Huberman taking on the photographs of patients directed by early French neurologist Jean-Martin Charcot. In their intriguing volume, *Photographing Medicine: Images and Power in Britain and America since 1840*, Daniel Fox and Christopher Lawrence looked at various uses of photography in medical practice, except for teaching, but they did provide a detailed list of sources for studying the history of medicine and the photograph in order to encourage the opening up of the field to further work.[6]

Even if none of these publications concerned themselves particularly with the role of images in pedagogical work, the following decades saw a growing interest in analyzing the history of the laboratory classroom; developments of university botanical gardens; natural history museums and morphological museums; the increased production of atlases, graphs, charts, and three-dimensional models; projections of living working organs or model organisms; and the institution of assignments whereby students described and explained class material through the making of drawings after specimens seen with the microscope. The list of publications would have to be too long to do justice, but it is important to note that, still, so much of this scholarship concerns itself with the important issues of evidence, authenticity, and authority within a professional realm or discipline, or tackles the problem of using images to assert scientific authority within the wider public sphere. The problem of images in teaching information and values, that is, as tools of instruction, is rarely acknowledged.

Fairly recently, however, two anthologies have appeared that take their focus from issues of pedagogy and the work of acquiring disciplinary skills, visual skills, as a rite of passage to professionalization. Published in 2005, David Kaiser's *Pedagogy and the Practice of Science* makes important scholarly headway into the analysis of the intersection of science and education. Kaiser's volume, however, focuses exclusively on the physical sciences and only in part on the importance of visual representation in science education. Two years later, Cristina Grasseni's *Skilled Visions: Between Apprenticeship and Standards* appeared with a section of three essays devoted to "the social schooling of the eye in scientific and medical settings," including a marvelous piece on training the naturalist's eye in the eighteenth century.[7]

Pedagogy practiced inside the classroom or in public spaces seems like an obvious topic of study for scholars of visual and material culture in medicine and the sciences, and indeed interest is growing. Nick Hopwood used the exhibition galleries of the Whipple Museum of the History of Science at Cambridge and an exhibition catalog to introduce a contemporary audience to the wax embryos created by the Zeigler family from the 1870s through the early twentieth century. While these beautiful colored wax models were first created as a form of "plastic publishing" to support various pieces of scientific research, Hopwood also discussed how they developed into mainstays of classrooms in Europe, the United States, and elsewhere. Together with Soraya de Chadarevian, Hopwood has edited a group of essays under the title of *Models: The Third Dimension of Science*, in which certain scholars did directly and specifically consider models as teaching tools.[8] Quite recently one can find current sociological studies of current practices in medical school, particularly on the implementation of virtual reality simulators for training future physicians and surgeons and the use of digital modeling to teach structural biology at the molecular scale.[9] Finally, we note the highly influential essay from 1992 by Lorraine Daston and Peter Galison, "The Image of Objectivity" that was then expanded into the book *Objectivity*, published in 2007. The authors focus on images from atlases, often enough publications used for teaching students and professionals (continuing education, as it were), and thus touch upon pedagogical issues.[10]

The contributions to *The Educated Eye: Visual Culture and Pedagogy in the Life Sciences* are drawn from areas as diverse as developmental biology, neuroimaging, medical school physiology and microscopic histology, public health film, and science exposition for the public, and they consider a variety of media—including drawing, photography, cinema (including animation), stroboscopes, shadow projection, graphic methods, textbook illustration, functioning magnetic resonance images (fMRIs), and computer graphics. Because many of these media contributed specifically to the production of new knowledge and new practices (X-rays, fMRIs, stroboscopes, digital graphic modeling), one motif running through *The Educated Eye*, as noted, is the interplay of the crafting of innovative scientific research work and medical technique with the role of teaching and edification, informing and communicating to others. But we want to emphasize that the focus of the essays in this volume will not be how the visual image was used to persuade peers and colleagues in the research science arena, but rather how the visual image is a teaching tool that has worked to convey information or train the eye and the mind in the ways of seeing, researching, organizing, and acting in the world.

In the first essay, Mara Mills offers what she calls a "didactography" of nineteenth-century Harvard anatomy professor Charles Minot, whose research career focused on senescence and the organism, especially at the cellular level. It may seem odd to begin a book on images in the life sciences with a history of the study of death, but, as she points out, Minot's emphatic conclusion was that "death was the condition that made life possible." Minot also placed at the core of his work as researcher and teacher the development of skills of observation, suggesting to Mills that in Minot's mind scientific study was equivalent to educating the eye. The development of visual expertise was imperative, for, as Minot would announce in class, "cells furnish our text." In this mode, Minot's pedagogical contributions emerged in his laboratory, in his public lectures, and in textbooks he authored. Most importantly, as is Mills's focus, Minot visualized the temporal aspects of embryological growth through senescence of the cells in mature organisms, primarily through histological techniques, to present the physiological body in time, from conception to aged maturity.

In "Facing Animals in the Laboratory," Nancy Anderson also addresses nineteenth-century physiology, looking specifically at textbooks instructing medical students on the preparation and observation of specimens under the microscope. The question here is how illustrations in the books—what is presented pictorially and what is not—provide surplus lessons related to the scientist's and the physician's attitude toward the body under investigation, especially when it is a nonhuman body. Her focus is a consideration of the rarity of images of intact animals in physiology and microscopic histology textbooks, and the even rarer inclusion of a face on the animal when illustrated. Are there plain values regarding human and nonhuman, subject and other, being taught or reinforced here when the acts undertaken are to kill and dismember in order to fix, stain, and observe bits of the animal's tissue? Anderson suggests this may be the case and, beyond this, the absent or faceless creatures (if shown) contribute to the message that a certain desensitizing is a crucial goal in undertaking these lessons.

By concentrating on the relationship between image and observer, instead of object and camera, in considering the use of photography in scientific and medical work, Scott Curtis suggests that certain salient values and skills of research and analysis important in the nineteenth century become more clearly understood. Certainly the advantage of stopping biological movement, stopping time, in the photographic still is the luxury of careful attention to detail. And the photographic medium, introduced in the mid-nineteenth century, soon was able to deliver detail in abundance, giving what Curtis calls a "complete description," a key value and valued skill for scientific and medical observation. In addition, Curtis emphasizes photography's capacity for repeatability and facilitating the important work of making correlations and finding patterns among a series of comparative examples. Photography, when considered in the relationship of the image and observer, then, moves beyond the argument that it is an objective document and becomes a tool for training the eye of the observer.

Anschauung—visuality, visibility, the act of visual perception—is a term that held great pedagogical significance in German physiological programs of the nineteenth century, in particular, manifesting itself in lesson plans and even in the architectural design of classroom spaces. In his paper, Henning Schmidgen discusses the rise of the lecture and demonstration room called the spectatorium, the viewing hall, as initially a response to calls for *unmittelbarer Anschauung* (immediate perception). While the professor spoke to the audience, a large screen would be available behind and above him for projecting "images, preparations, and living processes," perhaps photographs of embryos but also an "isolated, still contracting frog heart." As the twentieth century dawns, though, a more rational conception of physiology education shifts the value of the spectatorium from this experience of immediate perception to a more modern need of conveying a rapidly growing body of knowledge to larger and larger groups of students with ever more efficiency and speed. Schmidgen also points to the dual use of the spectatorium as a teaching apparatus and an instrument for research. The enlarged projection both delivers the lesson and creates new knowledge by revealing details of the phenomenon on display, which would be invisible without this technology.

The next three papers are concerned with the potential of cinema, as a mass medium, to convey scientific information, either as a tool for teaching public health or as entertainment, that might engage the viewer in the wonders and experiences of scientific visualization. First, Kirsten Ostherr analyzes the public health film, *Unhooking the Hookworm*, to consider various difficulties arising in the use of cinema in the 1920s to inform the public worldwide of health risks. Finding a way to engage the rural poor in a public health film was seen as the initial problem, and this was followed by broader social and cultural issues that, in the final analysis, dictated the creation of several versions of the film in order to engage audiences of diverse cultures, ethnicities, and races. With problems of creating an effective lesson across diverse viewerships, there was also the vexing challenge of making the subject matter entertaining enough to maintain viewers' attention long enough to convey the facts, the film's concrete educational message. Here Ostherr gives an in-depth analysis of the ideas exchanged on developing effective pedagogical techniques and the combining of live action, cinemicroscopy and stop-action animation to visualize various aspects of an information-laden story to diverse lay audiences.

Next, Hugh Crawford analyzes public entertainment in the form of scientific biographic cinema as scientific practice and as scientific pedagogy. "Science and film," he writes "require a heavy investment in teaching their respective audiences how to see, how to read, and to warrant the accuracy of their perceptions." He insists that the production of scientific knowledge is always pedagogical, and in two films by William Dieterle, *The Story of Louis Pasteur* (1936) and *Dr. Ehrlich's Magic Bullet* (1940), he contends that audiences find a public demonstration allowing them to partake in the validation of the facts produced by the scientist. In the film on Pasteur, it is the demonstration of proof through the presentation of the natural objects themselves, the microbes. In the film on Ehrlich, inscriptions focus the data, the proof, including biological staining of bacilli as well as images of immunological theory drawn on the table linen of a potential benefactor with whom the scientist is dining. Both, Crawford argues, link scientific practice and cinema to the role of pedagogy in the creation of new regimes of vision in science. Laboratories, public presentations, and movie houses are all "theaters of proof," a term he borrows from Bruno Latour.

Finally, Michael Golec considers the work of the American designers Charles and Ray Eames. Most well known is their 1977 film *Powers of Ten*, but Golec has chosen to analyze an earlier version of the project, the *Rough Sketch* from 1968. The sprint through magnitudes of scale, up to the galactic heavens of far-off stars and planets followed by the rapid descent to the very elemental components of matter, molecules, and atoms, makes up most viewers' memories of the *Powers of Ten* experience, and it is the makeup of *Rough Sketch* as well. Golec's essay reminds us of that rush-like feeling the filmmakers intentionally effected. There is more to this film, though, according to Golec, as he analyzes the way that the far and the small are positioned vis-à-vis the human scale, noting that the experience for the audience begins with a most familiar setting: a picnic, and a shot that will place the domestic sphere at the center, as it were, of scientific work, or visualization of the world. Here, at the start, viewers experience the domestic world meeting scientific discovery. But his larger point about the film is the science fiction ride from the very minute scale of cells and chromosomes to the far reaches of the outer galaxy, images that make up a sequence arranged by powers of ten. It is with the medium of film itself, its cuts and splices, that the filmmakers offer this heterogeneous collection of images, physical and biological nature at its smallest, its largest, and its farthest, as an unlikely continuum, or, as Golec emphasizes, a visual nondiscontinuity.

Richard Kremer explores the idea of technology and the training of a "high-speed" eye. His topic is Harold Edgerton, engineering professor at MIT, and his development in the 1930s of a photographic method using stroboscopic light, a technology to assist in the capturing of crisp images of a drop of milk splashing or a bullet smashing through an apple. These images are as much about seeing "time" as the cinema is—not with motion, but with the capturing of a small increment of time, allowing the human eye to experience in an image a fraction of a moment, an event that would be entirely inaccessible without the aid of Edgerton's apparatus. Edgerton was a man after the "facts," Kremer emphasizes, much in the vein of an earlier scientist, Etienne-Jules Marey, who employed photography to capture the smallest increments of time in studies of motion. But Edgerton was not satisfied with merely displaying his visual data for the edification of diverse audiences, professional and lay. Curiously, Kremer notes, Edgerton seemed determined to expose the process to his viewers. Kremer discusses how the photographer often published images of himself creating these high-speed photographs, showing himself as the director of the mechanical process, thus revealing, as part of the lesson, the technological setup that allowed the human eye to experience the views of the "high-speed" eye.

As Kremer introduces Edgerton's photography of extreme speed, Sabine Brauckmann draws our attention to the diagrammatic innovation of the fate map in twentieth-century embryology, which also sought to capture changes through time. Understanding how embryos develop from undifferentiated single cells to highly specified bodies composed of hundreds of thousands of cells is the central problem of embryology. Biologists, faced with communicating their observations of the transformation of cells and embryonic regions over time, created a visual convention known as the fate map and the specification map, which represented a temporal process of development in a single diagram. In effect, these maps imposed the future fate of the embryo back on its originating form. As Brauckmann shows, these maps became iconic representations of a complex dynamic process that was very difficult to convey otherwise in the static pages of a textbook.

The next two essays, penned by analytic philosophers, provide a look at how representations bear on issues of abstraction and authority, as objects for teaching as well as images representing science to the wider public. Contemporary college biology textbooks are rich in images, with various conventions mixed to convey information in either more or less detail, to visually clarify relations, or to present composite information. Laura Perini analyzes and compares various conventions used to convey data to students—from what she calls "replete pictorial representations" to "schematic diagrams"—and shows how each type of image plays its particular productive pedagogical role. Perini's philosophical interest, however, is in determining how one can determine or distinguish the uses of different sorts of pictures, and she will proceed on her analysis with the conviction that it is not "the distinction between whether or not an image is conventional or resembles its referent that explains the use of different kinds of images, but the different *kinds* of conventions and resemblance relations involved." She then discusses how one can use these different kinds of conventions to determine form-content relations and create meaningful categories for these images in biology textbooks as teaching tools, as pedagogical props.

Adina L. Roskies then meets head-on the problem of unleashing complex images, especially images produced through highly theorized technologies, to a public that does not properly understand the technology. In this case, she addresses functional magnetic resonance imaging (fMRI) of the brain and what she identifies as an ignorant acceptance of what the medium is capable of depicting. Certainly, there is an age-old problem of the lay person being convinced of the image's authority as scientific knowledge, and this still can have dire consequences, she insists, as when it might occur that fMRI is employed to expose a lying witness in court by showing "activity" in the proposed "deception area" of the brain. She emphasizes what she calls the unintuitiveness of this technology and insists that the public needs to understand both causal and counterfactual relations between image and the phenomenon represented. She concludes that education is the key, and that by introducing a scientific visualizing technology like the fMRI into high school curricula, a generation develops not just an understanding of this particular technology, but a stronger awareness of scientific methods, values, and, yes, limitations.

Finally, the last essay addresses the early development of virtual reality systems to teach surgical technique, Rachel Prentice begins with the premise that acknowledgment of the visuality of the lessons should not overshadow other senses (touch) and the importance of embodiment. Prentice spent time in a medical informatics laboratory at Stanford

University where she observed and participated in practicing anatomy, which would be translated into a computer program for a digital program to teach and practice a minimally invasive procedure, something done with cameras and small incisions, such as, in this case, the removal of an ovary. She discusses how various participants in the building of this tool—computer programmers and gynecologists/surgeons, for example—worked together to bring a model of the female pelvis to an interactive screen. It is certainly a visual lesson, but visuality working with the sense of touch, in particular, and with the awareness and training of one's own working body engaging with another, even if it is a virtual one. In creating this digital experience between the real body of the surgical student and the patient as virtual body, Prentice asserts, there is a "mutual articulation"; bodies and machines are mutually constructed during the design of the surgical simulation. It seemed fitting to end with an essay that speaks to the development of digital tools in the teaching of biology and medicine as well as acknowledges explicitly that such learning is always multisensory and embodied.

We have allowed each essay to stand on its own, although it is obvious that there are connections to be made between individual texts in terms of the analysis of medium, their role in a professional or public sphere, and the way in which visual images can always convey more than the obvious intention, carrying their lessons, for example, into the cultural, the aesthetic, the social. Thus, it is the diversity of these essays that is the strong point of this volume as a whole. It is hoped that the works will provide, individually and in cross-consideration, pedagogical opportunities for exploring various productions and disseminations of scientific, medical, and public health information from a multidisciplinary standpoint.

Notes

1. Jean-Louis Comolli, "Machines of the Visible," in *The Cinematic Apparatus*, ed. Theresa De Lauretis and Stephen Heath (New York, St. Martins, 1980), 122.

2. Michel Foucault, *The Birth of the Clinic* (London: Routledge, 1989), ix, 204.

3. Robert J. Blystone, "Biology Learning Based on Illustrations," in *High School Biology: Today and Tomorrow* (Washington, DC: National Academy Press, 1989), 155–164.

4. Nancy Anderson, "Eye and Image: Looking at a Visual Studies of Science," *Historical Studies in the Natural Sciences* 39 (2009): 115–125. Also, Norton Wise, "Making Visible," *Isis* 97 (2006): 75–82.

Introduction · 13

5. Michael Lynch and Steve Woolgar, eds., *Representation in Scientific Practice* (New York: Kluwer Academic, 1988). Here Greg Myers begins his work on textbook illustrations, but all in all the volume is not concerned with the teaching role of scientific visual documents.

6. Sander Gilman, ed., *The Face of Madness: Hugh W. Diamond and the Origin of Psychiatric Photography* (New York: Brunner/Mazel, 1976); Georges Didi-Huberman, *Invention of Hysteria: Charcot and the Photographic Iconography of the Salpetriere*, trans. Alisa Hartz (Cambridge, MA: MIT Press, 2003) (first published in Paris in 1982); Daniel M. Fox and Christopher Lawrence, *Photographing Medicine: Images and Power in Britain and America since 1840* (New York: Greenwood Press, 1988).

7. David Kaiser, ed., *Pedagogy and the Practice of Science: Historical and Contemporary Perspectives* (Cambridge, MA: MIT Press, 2005); Cristina Grasseni, ed., *Skilled Visions: Between Apprenticeship and Standard* (New York: Berghahn Books, 2005), including Daniela Bleichmar, "Training the Naturalist's Eye in the Eighteenth Century: Perfect Global Visions and Local Blind Spots," 166–90. Also, the editors of this book series, Interfaces, published Luc Pauwels, ed., *Visual Cultures of Science: Rethinking Representations in Knowledge Building and Science Communication* (Hanover, NH: University Press of New England, 2006), in which, although issues of pedagogy and the visual did not play a direct role in the contributed essays, Massimiano Bucchi tackled wall charts and science education, and Jean Trumbo examined visual learning in science communication.

8. Nick Hopwood, *Embryos in Wax: Models from the Ziegler Studio*, Whipple Museum of Science (Cambridge: Cambridge University Press, 2002); Nick Hopwood and Soraya de Chadarevian, eds., *Models: The Third Dimension of Science* (Stanford, CA: Stanford University Press, 2004). Soraya de Chadarevian's book *Designs for Life* (Cambridge: Cambridge University Press) offers a chapter on models, discussing their role in scientific discovery and then as objects for public relations of science, but she does not really reach into the area of pedagogy.

9. Rachel Prentice, "The Anatomy of Surgical Simulation: The Mutual Articulation of Bodies in and through the Machine," *Social Studies of Science* 35 (December 2005): 837–66 (reprinted in this volume). Natasha Myers, "Modeling Proteins, Making Scientists: An Ethnography of Pedagogy and Visual Cultures in Contemporary Structural Biology," PhD dissertation (Massachusetts Institute of Technology, Cambridge, MA, 2007).

10. Lorraine Daston and Peter Galison, "The Image of Objectivity," *Representations* 40 (1992): 81–123. Lorraine Daston and Peter Galison, *Objectivity* (Cambridge, MA: MIT Press, 2007).

chapter one

Trained Judgment, Intervention, and the Biological Gaze
How Charles Sedgwick Minot Saw Senescence

Mara Mills

> What a man saw when he looked through a microscope often depended more on what was behind his eye than on what was in front of it.
> —HENRY H. DONALDSON, "Charles Sedgwick Minot" (1915)

1.

In an 1897 letter to the editor of *Science*, acidly titled "Literary Embryology," Harvard biologist Charles Sedgwick Minot quoted at length from a piece in the *Atlantic* on the application of embryology to teacher training. The article's author, Frederic Burk, had tried to demonstrate that "embryology throws some suggestive light upon the radical difference of childhood from maturity"—with embryos and children undergoing a tortuous, even regressive, development.[1] Minot italicized every claim he found erroneous, leaving very little of Burk's original argument standing. "It appears singular that in an article on *teaching*, severely criticizing prevalent methods," Minot commented, "there should occur a very striking example of inaccurate *learning*."[2]

Minot considered development to be progressive and teleological, beginning at the moment of fertilization.[3] His contemporaries credited him with being "the first among us systematically to examine the phenomenon of growth after birth."[4] According the *Oxford English Dictionary*, as a graduate student in 1879 Minot gave the first biological definition of senescence—one that paradoxically lashed growth to decline: "With each successive generation of cells the power of growth diminishes ... this loss of power I term senescence."[5] His statistical studies of maturation in guinea pigs upended the medical notion of old age as "the chief period of decline," instead characterizing it as "the period of slowest decline."[6] Minot later turned to the question of growth *before* birth, chasing his curves back to the first embryonic cell divisions. He was among the first to theorize that death might be "biological," a prerequisite of most somatic cells.[7]

Charles Minot was not a pleasant man. His obituaries unsparingly portrayed the bodily manifestations of his intellectual severity. According to Frederick Lewis, who worked in his histology lab, Minot had "a small rather prim mouth with something about it of the unmated and no longer youthful female."[8] He did not have any close friends. Edward Morse recalled that "his walk was alert and in a straight line," moreover, "the nicety and precision which marked all his endeavors" often graded into intolerance and a "fearless" critique of his peers.[9] Lewis believed that Minot's work had in fact been hampered by his personality; his "characteristic intensity of conviction was frequently vigorously expressed, and not always in such a way as to facilitate his progress."[10]

Even more than demeanor, the happenstance of birth date and education left Minot vulnerable to the rebuffs of history. He appears here and there in recent scholarship as a member of that first generation of U.S.-trained physiologists who had such difficulty finding placements in universities.[11] He followed an undergraduate degree in chemistry from MIT with a PhD in Natural History from Harvard. He began his dissertation under Louis Agassiz, but soon became the first student in the first American physiology laboratory—that of Henry Pickering Bowditch, under whom he studied the cardiovascular effects of anesthesia at the medical school. Minot considered himself a biologist, but a graduate program in biology was not available in the United States until 1876, at Johns Hopkins.[12]

On the fault plane of the discipline, his novel training made employment difficult.[13] Two years after his graduation in 1880, Charles Eliot "procured for him with some difficulty" a lectureship split between the Harvard Medical and Dental schools; the medical school, in particular, was hesitant to accept a faculty member whose primary interest was not doctoring.[14] Despite these uncertain beginnings, he remained at the medical school for the length of his career, eventually chairing the Department of Histology and Embryology. This institutional setting—and Minot's untimely turn *from* working with live animals and tissues *to* embryonic sections—may have obscured his role in the history of biology. Nick Hopwood observes that narratives of a "rising" experimental biology often exclude embryology: "Historians of biology have investigated embryology as a paradigm of the life sciences in transformation around 1900. Between the 1870s and World War I, comparative research addressing grand evolutionary themes gave way to often-experimental analyses of focused problems in simpler systems. To historians in search of the new biology, studies of the relatively inaccessible and experimentally intractable human embryo seemed irremediably descriptive and unappealingly medical."[15] Bowditch founded the American Physiological Society in 1887; the field's focus on "medical" experiment also seemed to separate it from the "new biology." Accordingly, Lynn Morgan cites Minot as an example of traditional morphology's limitation to the analysis of structure.[16] Elsewhere, Minot can be found writing notes to Franklin P. Mall on the finer points of embryo collecting, in the small hours before the experimentalization of life prevailed.[17]

Donna Haraway has argued that developmental biology underwent something akin to a paradigm shift between the 1920s and the 1930s, years that saw the rise of the gene and the molecule in biology at large. The generation that followed Minot adopted organicism—"systems and their transformations in time"—as a central metaphor.[18] At the leading edge of organicism, "directed" by an ascendant constellation of scientific analogies and images, Ross Harrison devised the technique of tissue culture. Haraway credits tissue culture with turning attention to cell movement, "inaugurating study of parts of the organism outside the body of the animal," and making embryology more experimental.[19]

Biologists and historians have subsequently disputed the relevance of Kuhn to the field of biology, and have focused more wholly on the topic of technological innovation. A 1998 *Nature* editorial explained, "Non-Kuhnian revolutions in biology are often, even usually, sparked by innovations in technique."[20] In this vein, Susan Squier diagnoses the emergence of a "tissue culture point of view" in the first decades of the twentieth century. With image and metaphor now following technique, tissue culture "resituated the 'body of science' from the realm of the static, graphic, and dead to that of the dynamic, photographic, and living."[21] In *Culturing Life*, Hannah Landecker suggests that tissue culture was the very source of time's biological definition; the "new cytology" of Alexis Carrel and others was "a science that studied cells as dynamic, temporal beings rather than as static, killed entities of histological staining."[22] Elsewhere, Landecker argues that the technique prompted embryologists to "think in terms of cells" and provided the context in which death became cellular.[23]

Without disputing the magnitude of culture techniques for the understanding of cell behavior, mutability, autonomy, and immortality (this last being the opposite of time), I argue that Minot's "senescence" was part of an earlier restructuring of life — when death, and hence time, lodged itself within the living cell.[24] Here, visual techniques governed points of view. Minot's stained sections, as well as the graphs he derived from living bodies, revealed signs of decline where previously there had been no visible symptoms; here the graphic was not necessarily static. He framed his work as a refutation of the "medical" tendency to pathologize senescence; the portraits of aging as atrophy and disease given by physicians "belong only to extreme old age, and do not represent or illustrate, so far as we can see at present, any general law or operation going on throughout the entire period of life."[25] Subsequently, tissue culture would rework Minot's thesis, suggesting that senescence — if a biological imperative — could be delayed, deferred, even overcome. Landecker has documented the technique's contributions to "the engineering ideal" in biology.[26] Tissue culture resulted in an unprecedented appreciation of cellular plasticity — and manipulation. "Techniques of plasticity," Landecker explains, "are modes of operationalizing biological time, making things endure according to human intention."[27]

Michel Foucault identifies a "great break in the history of Western medicine" at the end of the eighteenth century with the introduction of the medical, or clinical, gaze.[28] By making death a "technical instrument" through the practice of autopsy, Xavier Bichat presented the tissues beneath the body's skin for study. The medical gaze traveled "along a path that had not so far been opened to it: vertically from the symptomatic surface to the tissual surface, plunging from the manifest to the hidden."[29] Bichat traced disease symptoms to pathological tissues, defining "life" as the resistance to death.

The "biological gaze," on the other hand, preceded the possibility for "technologies of living substance."[30] Evelyn Fox Keller defines the biological gaze as interventionist, always requiring the manipulation (and frequently — even after tissue culture — the outright destruction) of its object. From natural history collections to electron microscopy, this gaze "has become increasingly and seemingly inevitably enmeshed in actual touching, in taking the object into hand, in trespassing on and transforming the very thing we look at."[31] For some early biologists, intervention was limited to the "preparation" of specimens for observation and display. By the end of the nineteenth century, Keller argues, "for the experimental biologist, the function of the microscope is not simply to enhance looking, or even to validate the tangibility of that which we are gazing at, but to employ that gaze as a probe in anticipation of action."[32] If life could not in fact resist death of its own accord, then for many biologists after Minot the action to be taken was life-extension.

Minot, too, theorized biological looking, and his "gaze" frequently depended upon intervention—as in the staging of life for the microscope. Nevertheless, he did not make the transition to the emerging culture of enhancement, or transformative intervention. By his account, the biological gaze included the recording of "variable relations," both *changes across time* and *differences between individuals*. In his 1910 vice presidential speech before the experimental medicine division of the AAAS, Minot proposed that "the method of science" was "the art of making durable trustworthy records of natural phenomena,"[33] records that could be used to register time or identify norms.[34] The "graphic method" was one means to this end—for Minot, it included automatic recording devices as well as diagrams of mathematical data.[35] Specimen collecting, in the case of his embryo series, became another form of temporal recording, a set of samples that could be used to simulate development.

Although the interventionist principle in biology does not match up cleanly with the history of objectivity in other fields, Minot nonetheless exemplifies the standard of "trained judgment" that emerged during his lifetime. Lorraine Daston and Peter Galison have anchored this variety of objectivity in education: a "late nineteenth-century explosion in pedagogical innovation blazed a path to scientific formation."[36] New departments and training methods led to new ideals for seeing; scientists *interpreted* machine-made images, and in particular they learned to judge the range of the normal.[37] "The trained expert possessed and conveyed to apprentices the means (through the 'trained' or 'seeing' eye) to classify and manipulate."[38] For Minot, *making* records and building recording devices—rather than discovering their output in a library—was essential to becoming-scientific. To see and to interpret records required "training by actual use," similar to the playing of a musical instrument.[39] If error could stem from "the variation of the phenomena, the imperfections of the methods and the inaccuracy of the observer"—and if, as Minot believed, the personal equation could rarely be determined—then practice lay at the heart of the method of science.[40]

According to Minot's own teaching philosophy—"we must teach how to learn, and how to learn from the unknown"—every possible technical method should be called to the service of the eye, for "to observe correctly and reason correctly are the most difficult accomplishments a man can strive for."[41] The laboratory was the ideal space for this teaching and self-teaching. Minot often opened his own classes with the proclamation, "cells furnish our text."[42] In an 1899 commencement lecture to the medical students at Yale, on "Knowledge and Practice," he urged his audience to pair clinical observation with structured laboratory investigation—namely because "experimental physiology, bacteriology and pathology offer far better discipline of the observational power than anatomy alone."[43] (This lecture cannot have gone over very well, barbed as it was with comments such as "medicine is one department of applied biology, just as dyeing is one department of applied chemistry."[44]) Minot's preoccupation with purposeful development in the classroom obtained from his theory of life: "The most striking distinction of the processes in living bodies, as compared with those in inanimate bodies, is that living processes have an object— they are teleological."[45] What follows, then, is a "didactography"—Minot's history as it was impressed upon that of the discipline; the theories he derived through lecturing, textbook-writing, and embryo-collecting; the visual aids to his developmental research, from graphs to microtomes; his equation of science with educating the eye.

2.

Garland Allen, for one, places Minot with Jacques Loeb and Thomas Morgan, as a lead experimentalist at the turn of the twentieth century: "These younger men were more concerned with biological mechanisms—the way in which an organism developed, or how cell differentiation was triggered in early embryonic development. They were dissatisfied with the basically descriptive, taxonomic, and frequently speculative viewpoint of the nineteenth century naturalists."[46] Yet the better part of Minot's career was devoted to embryo collecting, institution-building, and administration— despite his sermons on laboratory investigation and the scientific method. He helped found the Association of American Anatomists, the American Society of Naturalists, and the laboratory at Woods Hole; he served as president of the Boston Society of Natural History and the American Association for the Advancement of Science; he edited *Science*, among other journals. A paradox of Minot's career, and perhaps another explanation for his relative absence from the secondary literature, is that he spent all of his professional years training doctors rather than the next generation of biologists. In that capacity, he designed the Harvard Medical School laboratories, from the sizes of the rooms to the placement of the windows and workstations. Tidy as the periodic table, he wrote articles on scholarly information management—down to the level of pamphlet storage on bookshelves.[47] His pedagogical manifestos include one on lengthening the school year, another on lowering the age for college admission (given the drift of senescence), and very many on making medicine "scientific."[48] His engrossment with architecture and infrastructure often inhibited experimental research. The remains of Minot's career are indeed mostly to be found in laboratory implements and methods.

Minot spent three years of his own graduate study abroad, primarily in Leipzig at Carl Ludwig's laboratory. Returning in 1877, he published a series of articles on "The Study of Zoology in Germany." He found much to recommend in the German laboratory system—the proximity to notable faculty, the self-guided activity of the students, the availability of museum collections. Laboratories, Minot explained, were "at once training-schools and the scene of active original research."[49] He took pains to describe the central technique of the zoological laboratory—microscopy—and the creative labor required "to render the objects suitable for investigation."[50] Tissues had to be hardened in order to be sectioned thinly enough for mounting on slides. Material could be made transparent through immersion in glycerin, which allowed the passage of light for better viewing. Staining, in Minot's account, was a sort of ocular prosthesis:

> Preparations for the microscope cannot be felt or dissected, but only seen; therefore, the differential coloring produced by the carmine, for example, is an assistance to the eye, comparable to the raised alphabets of the blind. In both cases, the conditions under which the special sense, whether sight or feeling, has to act are greatly exaggerated, so to speak, thus producing magnified or strengthened perceptions.[51]

When Minot began his career at the Harvard Medical School, his laboratory had eighteen Hartnack microscopes, which magnified an object no more than fifty times; hence the assemblage required for the biological gaze included "preparations," or biological objects that had been altered for compatibility with the eye.[52]

Beyond the rendering of tissues into new perspectives, the human eye and mind required preparation. Minot urged life scientists to study the laws of optics, for instance, to learn how to minimize the refraction of light through microscope specimens. More importantly, he believed that experience was the surest solution to such problems as the microscope's inversion of the field of vision. "A beginner," he explained, "finds it almost impossible to move a preparation under the microscope in the way he wishes, but with practice the coordination of sight and movement becomes so perfect that the adjustment is unconscious."[53] He believed that vision itself was largely learned—it was a "psychological and not a sensory phenomenon."[54] Thus he urged a tactical enlargement of the eye's capabilities.

Throughout his career, Minot dedicated himself to the philosophical topics of consciousness and temporality, appropriating them for the field of biology with the conviction that "the biologist must necessarily become more and more the supreme arbiter of all science and philosophy, for human knowledge is itself a biological function."[55] In 1879, still a graduate student, he authored "On the Conditions to be Filled by a Theory of Life," which demanded that all aspects of life—including stimulus-response, ingestion, sex, senescence and memory—be explained on the order of cells.[56] In another article that year, Minot specifically attempted to define growth as cellular (boasting, "the discussion of growth from this stand-point has not, as far as I am aware, been hitherto attempted"[57]). Minot was not alone in his enthusiasm for cells, however, or in his conviction that they were the ciphers of life, death, and development. In the late 1870s, as Abigail Lustig has meticulously explained, the study of single-celled organisms prompted scientists to question whether such milestones as death were actually common to all living things.[58] August Weismann most famously theorized that protozoans were immortal, therefore death was merely an "adaptation" of multicellular organisms—not a defining characteristic of all life. Weismann saw death as a "secondary adaptation," Lustig explains, "necessary to clear away older, more maladaptive individuals."[59] Contemporaries Emile Maupus and Otto Bütschli insisted that protozoans senesced.

Minot weighed in on this debate, at first disbelieving that any organisms could escape natural death, but by the end of his career conceding that protozoans may well do. He consistently argued for a focus on cellular rather than organismic death; single-celled protozoans (with their capacity for conjugation, or the exchange of nuclei), could not be compared to multicellular individuals.[60] The somatic cells of the latter were fated to die—even if they escaped trauma—as a consequence of senescence or developmental imperatives. Minot tried to defend, in print, his title to "the conception of the biological problem of death." His hypothesis (he insisted) "was formed several years before Weismann's first publication"[61] Regardless, the details of "natural" cell death would not begin to be worked out until the mid-1880s.[62]

Minot had observed Ludwig's work on muscle physiology and conducted his own experiments on the capacity of muscles to function outside of the body. Just as the heart beats for a time after its dissection from a living creature, so do other body parts exhibit what Minot called "over-life": an independent capacity for living and dying. Death was not organismic, but occurred in pieces.[63] At the Harvard Medical School, Minot turned to "experiment" to test his theoretical claims.

An early line of his research on senescence was conducted using guinea pigs. They were not yet model organisms; Minot appreciated that they were inexpensive, and easily penned-in to limit experimental conditions. (He also remarked that he found them "so unintelligent that I have been unable to feel any interest except scientific in them."[64]) His intention was to analyze the weight, chemical composition, and cell structure of a population from birth until death.[65] He cared meticulously for 100–400 guinea pigs—depending on birth and death rates—over the course of five years. Minot cleaned their pens himself, visited twice each day, gave them names such as "Snout" and "Hypocrite," and took upwards of 8,040 weight measurements. In addition, he induced abortions in pregnant guinea pigs so as to weigh their embryos at different stages and examine those early cells under the microscope. He described this project as one of "experimental biology."[66]

A medical tradition already existed for measuring growth—Bowditch, Minot's advisor, conducted a famous study of schoolchildren in Boston and concluded that growth increases just before puberty.[67] Rather than directly measure growth, as had most comparative anatomists, Minot decided to calculate the average *rates of change* for his guinea pig population. The results, he believed, would be most readily apparent through graphic representation. "For our accuracy," Minot theorized, "it is necessary often to have a number of data in their correct mutual relations presented to our consciousness at the same time, and this we accomplish by the visual image, which is far more efficient for this service than any other means of which we dispose."[68] Minot's hand-drawn "charts of senescence" showed curves that plummeted and then seemed to flatten out over the x-axis (figure 1.1). He was surprised by his own conclusion—namely, that growth increasingly weakens after the *prenatal* stage.[69]

Under the microscope, the cells of his animals revealed a seemingly related phenomenon—"the proportion of protoplasm to the nucleus increases with the age of the organism."[70] However, the guinea pig experiments came to a premature end, a fact Minot detailed with great pain:

1.1 Bowditch's growth chart compared to Minot's chart of senescence (below). Reprinted from Charles Sedgwick Minot, *The Problem of Age, Growth and Death* (New York: The Knickerbocker Press, 1908), 91, 96.

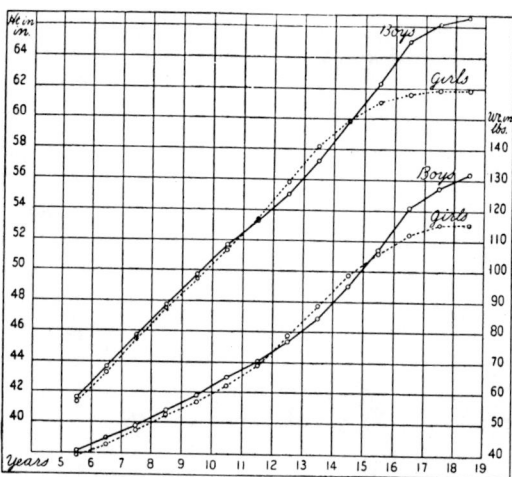

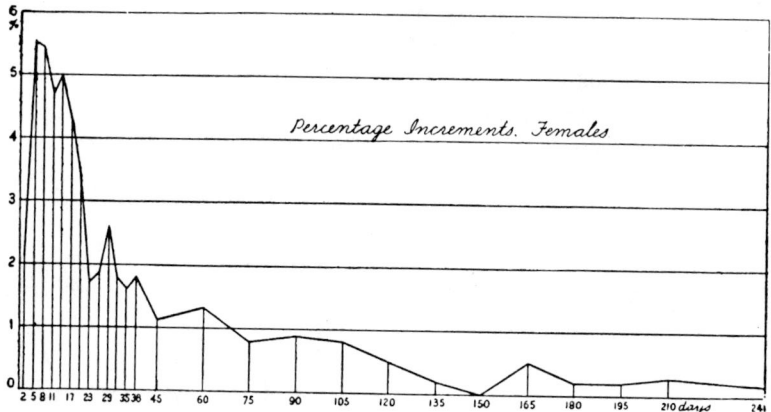

"Three times in a row, dogs got in and killed the animals leaving me only four alive of all those I had kept and weighted for a long period, making as I went along a careful biological record of each individual."[71] He nonetheless attempted several partial explanations for why cells senesced as they grew, one of which seemed to be their inexplicable teleological striving, or plan.[72] Surveying the work of Oskar Hertwig and others, he surmised that this plan resided in the "nuclear substance" or "chromatine."[73] Susan Squier remarks upon the prescience of these insights (still relegating his techniques and his times to the outdated): "Although working in the pregenetic era, Minot nevertheless glimpsed a relationship between the beginning of life and its end that scientists are now rearticulating as the relationship between telomere length and rate of aging."[74] The shortening

1.2 Minot's microtome. From Charles Sedgwick Minot, "On Two Forms of Automatic Microtomes," *Science* 5, 127 (1897): 859.

of the telomeres (chromosome tips) with each cell division is now believed to lead to cell senescence, often followed by apoptosis, in many (but not all) multicellular organisms.[75]

During his first ten years at the Harvard Medical School, Minot simultaneously undertook the massive project of authoring a textbook on embryology, with medical students as his primary audience. He explained this turn in his research as a happenstance of the classroom—"Necessity early led me into teaching embryology."[76] He began a small collection of human embryos, obtained from Boston and New York physicians, which he sketched throughout the textbook to illustrate the stages of embryonic and fetal development.[77] Even at the end of his career, his laboratory did not have the resources for photography, so most of the examples in his publications were drawings of preserved embryos—whole, or divided into tissue sections. With sectioning, Minot argued, "every portion . . . may be subjected to minute examination, and, further, the sections once made and mounted they may be stowed away and investigated at any leisure moment. For example, during a few weeks at the seaside, material for a winter's occupation may be very easily procured. Neither is there so much hurry in drawing, as when an animal is living we are afraid it may die."[78] Sketching this material required "a special degree of skill and a considerable faculty of plastic imagination."[79]

Minot's fussiness lent itself to the invention of a precision microtome, which yielded extremely fine sections of .002 mm or less (figure 1.2). He collaborated with Edward Bausch on this project in the 1890s—a man with whom he had previously quarreled in *Science* regarding the quality of American lenses.[80] According to a 1947 review in the *Transactions of the American Microscopical Society*, their microtome "rightly received wide attention" and became something of an American standard.[81] More importantly, Minot's instrument work helped move more bodies into wax and under glass, prior to the possibility of life *in vitro*. Nick Hopwood, writing on Wilhelm His as the initiator of the mechanist perspective in embryology, has contextualized these new ways of seeing thus: "The microtome symbolizes a transformation in the practice of microscopy, a sea change in the experience of laboratory work in the life sciences, and a reorientation of the objects of research from living organisms in their environments to the internal topography of fixed and sectioned specimens."[82]

In 1892, Minot finally published *Human Embryology*, an 800-page volume that was generally noted as the most important American text on the topic.[83] Reviewers primarily recommended the book for its exhaustive in-line citations and notes; one described feeling "indebted . . . for the compilation of such of valuable résumé of the work of other embryologists."[84] This comprehensive pedagogical endeavor undoubtedly stimulated much of Minot's own, largely theoretical, work.

After the decimation of the guinea pig study, Minot began to pursue embryology almost exclusively. If old age "is merely the culmination of changes which have been going on from the very first stage of the germ up to the adult," Minot postulated, "We must expect from the study of the very young stages to find a more favorable occasion for analysis of the factors which bring about the loss in the power of growth and of change."[85] Thus, Minot began what he called his "unremitting" development of the Harvard Embryological Collection in January 1896.[86] Less than two decades later, at the time of his death, he had supervised the preservation and sectioning of 2,000 vertebrate embryos in his medical school laboratory. Minot's collection was designed to serve as a "cyclopedia" for researchers worldwide, with at least 3 embryos each for 22 species, including human, cat, pig, chick and opossum. "So far as I am aware," he announced in *The Journal of Medical Research* in 1905, "no other collection of this kind has yet been attempted."[87]

In Minot's own research, comparative embryology was an avenue toward the broad biological definition of senescence. His collection of rabbit embryos revealed a connection between "growing power" and differentiation (figure 1.3). Obtaining these embryos at half-day age intervals, he counted the number of cells in each and estimated the percent that looked to be actively dividing. After eight days or so of relative uniformity and rapid growth, the cells began to change and at the same time their number of divisions decreased.[88] Minot ventured to estimate that vertebrates "have lost at the time of their birth 99 percent of their original growth capacity."[89] This mitotic slowing, or decrease in cell reproduction, meant not only a

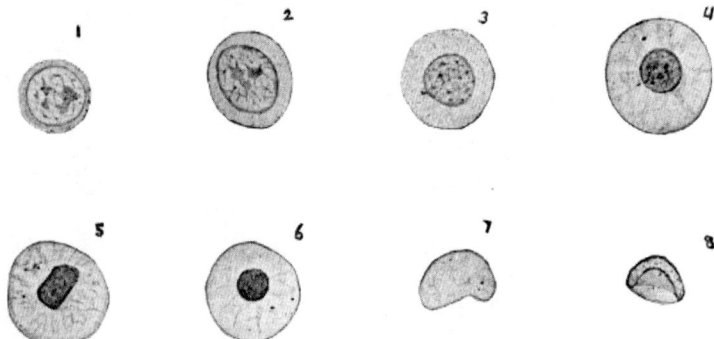

1.3 Minot's representation of the life history of blood corpuscles from rabbit embryos. From Charles Sedgwick Minot, *The Problem of Age, Growth and Death* (New York: G.P. Putnam's Sons, 1908), 78.

loss in growth but in repair. Minot compared his straightforward observations to the experimental work of Hans Driesch and Wilhelm Roux, whose shaking and pricking and centrifuging of embryos had revealed their undifferentiated cells to be capable of transforming into any tissue type (what is now described as totipotence).[90] He began to see differentiation as "the central problem of all biological research . . . and if we understood fully the nature of differentiation and the cause of it, we should have probably got far along towards the solution of the final problem of the nature of life itself."[91]

Minot was equally interested in applying his human series to the question of biological time, through the careful study of tissue metamorphoses. This method resembled the chronophotographic, with time and development emerging from a sequence of slides or preserved bodies. Unlike the graphic representations from his guinea pig research, each embryological stage derived from a different individual. Nick Hopwood has detailed the techniques applied by Wilhelm His to "visualize embryos and make them comparable" in the process of assembling his "normal" series of human embryogenesis.[92] The dearth of examples, the ambiguities of pathology, and the variation of individual growth persistently dragged on these typological endeavors. There were no observations, at all, for several early stages of human embryonic development. One four-millimeter embryo, in particular, represented an enormous quandary for Minot. "So far as the condition of the embryo itself indicates, the specimen is normal," he wrote, "but it differs in many respects from the few human embryos of this size which have hitherto been described. . . . The question arises whether this embryo is really not more normal than the others."[93] Staging raised the additional question of how time should be measured—if cuts should be made according to the calendar (i.e., development measured by day) or according to the embryos themselves (i.e., development measured by length or by landmark).

In 1903, Minot published the *Laboratory Textbook of Embryology* as an activity-guide and companion volume to *Human Embryology*. Minot often marveled at the recency of the laboratory phenomenon; here was his own somewhat paradoxical effort to entrench lab work into the life sciences curriculum.[94] "Knowledge lives in the laboratory," Minot declaimed, "when it is dead we bury it, decently, in a book."[95] His *Laboratory Textbook* introduced students to research methods: how to collect embryos from fishermen and use a chick incubator (the Harvard labs themselves had only just switched to mechanization, after years of keeping hens); how to remove, measure, preserve, stain, dissect, and section embryos; how to make camera lucida drawings. This lab manual made the fetal pig an embryological standard.[96] It also described the procedure for enlarging, and then reconstructing, a sectioned embryo in a series of wax plates. As Hopwood has demonstrated, His and other embryologists *saw* development through the practice of building these wax models.[97] The plastic reconstruction of embryos was equally a scientific method and a pedagogical tool, instructing makers and students on matters of growth and differentiation.

Minot did not miss the opportunity to discuss cell death in his lab book, urging readers to recognize that without death "on a large scale the normal round of human life would be impossible. The student should free himself from the unfortunate tradition that these processes are exclusively pathological."[98] Normal, or intrinsic, cell death accounted for the continuous shedding of cells by the intestines, the uterus, the hair, and the skin. Cells also died through accident or through attack by other cells (now called necrosis). Moreover, somatic cells seemed to have intrinsically limited life spans; they followed an arc of decelerating growth followed by decomposition.

Perhaps Minot's most influential work, *The Problem of Age, Growth and Death*, was based on his 1907 Lowell Lectures, a series sponsored by Harvard and open to the public. Published six years prior to his own death in 1914, the book was "the first comprehensive presentation" of his research, from theoretical speculation to physiological experiments to embryo collection.[99] By the time of the Lowell Lectures, he had incorporated his

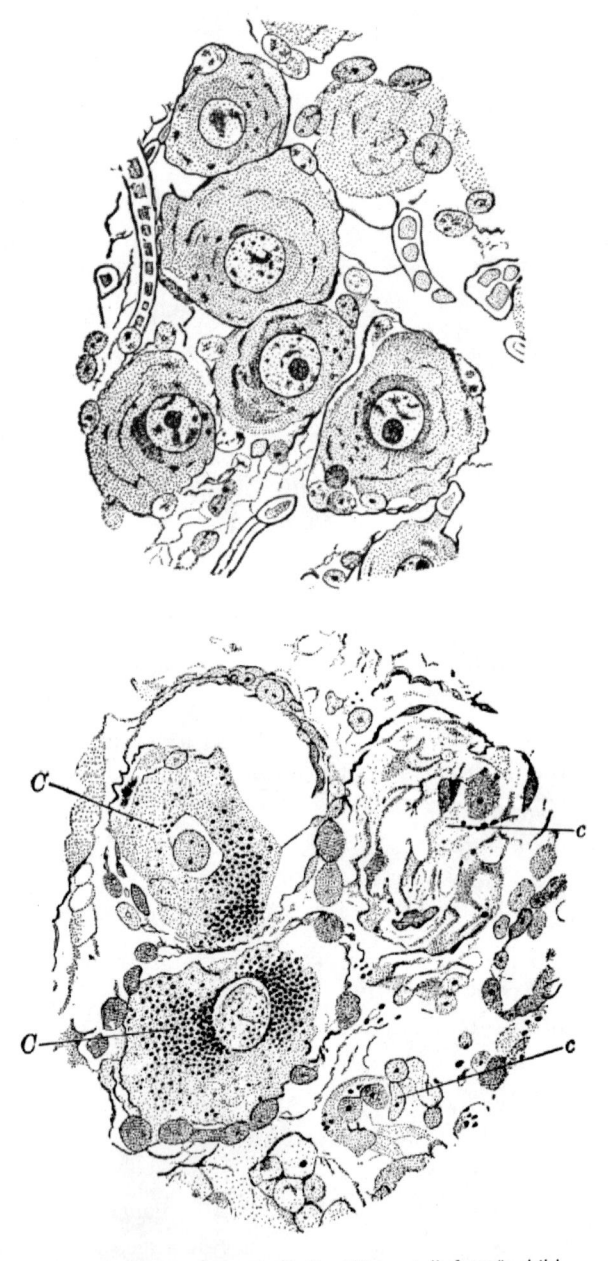

1.4 Minot's images of nerve cells. (top) Nerve cells from "a child at birth." (bottom) Nerve cells from "a man dying of old age at ninety-two years." From Charles Sedgwick Minot, *The Problem of Age, Growth and Death* (New York: G.P. Putnam's Sons, 1908), 68, 69.

ideas about death and senescence into a theory of "cytomorphosis."[100] This concept was meant to embrace the full development of the cell, parsed into four stages: undifferentiated, differentiated, degenerated, and decomposed or "removed." Senescence accrued across the lifetime, thus death should be thought of as a process rather than an "event." Minot consolidated his research into four hypotheses: the process of differentiation somehow strips cells of their regenerative powers and thus leads eventually to death; a "law of genetic restriction" precludes differentiated cells from returning to an undifferentiated state; the germ cells, including those stored in the marrow, slowed their production of new cells in aged bodies; and finally, "the period of most rapid decline is youth."[101]

Historians of gerontology have tended to assume that biologists simply miniaturized the symptoms of aging and transposed medical interpretations onto the cell cycle.[102] Yet Minot opened his Lowell Lectures with lantern slides that contrasted a newborn baby with the "shrinkage and shriveling" of an old man in a counterintuitive manner. "We commonly think of the old as those who have lost most, who have passed beyond the maximum of development and are now upon the path of decline . . . [but] the period of old age," he told his audience, "so far from being the chief period of decline, is in reality essentially the period in which the actual decline going on in each of us is the least."[103] He subsequently exhibited images of "old" and "young" cells, again reversing the expected vision of aging by focusing on growth rather than atrophy (figure 1.4). Minot offered his Lowell audience a bit of encouragement in the face of this declension: "The maintenance of the life of each individual of us depends partially on the continued death going on in minute fragments of our body here and there."[104] The final stage of cytomorphosis—"removal of the dead"— provided material and room for the generation of new cells. Senescence proper (as opposed to other processes leading to cell death), Minot defined as a "consequence" of differentiation—not an outcome of natural selection with an evolutionary advantage.[105]

In his last public lecture, the year before his death, Minot returned cell death to the human scale:

> To it we are indebted for our organization which makes us men; to it we owe the possibility of knowing our earth, its inhabitants, and ourselves; to it we owe all the advantages of our existence; to it we owe the possibility of carrying on our physiological work much better than the lower organisms; to it we owe the possibility of those human relations which are the most precious

of our experiences. These advantages and many others do we owe to differentiation, the price of which is death. The price is not too high.[106]

Rather than being essential, as in Weismann's "adaptive" interpretation, Minot had made death at once innate and incidental.[107]

3.

The publication of the Lowell Lectures as *The Problem of Age, Growth and Death* generated widespread response. In 1914, Dr. Ignatz Leo Nascher authored *Geriatrics: The Diseases of Old Age and Their Treatment*, the first medical textbook on the topic since that of Jean-Martin Charcot, a half-century prior. Nascher reprinted Minot's illustrations of young and old nerve cells and defined aging as an intrinsic, physiological process. The bulk of *Geriatrics* was nosological, however, yielding what Lawrence Cohen has termed a "geriatric paradox": old age simultaneously included and excluded from normalcy.[108] With precisely the medical approach to aging that Minot had rebuked, Nascher insisted that the manifestations of senility would be pathological at any other time of life. Moreover one became an entirely new person as the result of age: "In advanced life none of the early cells are left (except brain cells) . . . [and] . . . the aged individual is in fact an entirely different individual from the one who was formed from the ancestors of the late cells."[109]

In 1922, G. Stanley Hall published his interdisciplinary textbook *Senescence: The Last Half of Life*. Of Minot's theory, he wrote that it "might almost justify a kind of homesickness for the state of the ovum or the immortal germ plasm. . . . Life itself as we know it from this viewpoint seems a little falsetto."[110] Hall stopped short of accepting the inevitability of cell death, reassuring his readers that if Minot "gives us a haunting sense of loss [he] also reinforces the hope that the high potential with which we all started somehow, sometime, may be better conserved."[111] Popular authors on longevity turned more forcefully to this question of conservation. In *Adding Years to Your Life*, Dr. Henry Williams rebuffed Minot's claim that "the higher diversified life is purchased at the price of ultimate death." And even if Minot were correct, Williams argued, contemporary medical experiments held great therapeutic promise — especially those of Serge Voronoff, a surgeon whose grafting and transplantation fame rested on moving the glands of young criminals and chimpanzees into the testicles of wealthy old men.

Voronoff's teacher, Alexis Carrel, cited *The Problem of Age, Growth and Death* in a 1923 publication for having established the fact that cells and bodies lost "growth energy" over the course of their life spans.[112] Many of Carrel's own investigations were devoted to growth and senescence, which he tried to prove were media phenomena—the result of a cell's interactions with its environment. The medical tradition of longevity research had always focused on environment, especially as related to ingestion and habit, producing catalogs of the diets of super-centenarians from around the globe that recommended the miraculous powers of yogurt, turnips, or egg yolk. Keeping tissues alive *in vitro*, the technique Carrel greatly expanded, was also a matter of providing the appropriate medium—which included temperature, moisture, and chemical composition.[113] Perrin Selcer has demonstrated that Carrel turned to tissue culture through his concern, as a surgeon, with speeding the cicatrization (healing) of wounds.[114] Carrel soon became interested in extending the lives of cells, perhaps indefinitely.[115]

In a book titled *Biological Time*, Pierre Lecomte du Noüy outlined the trajectory of Carrel's early research. This included their collaboration on the healing-rates of soldiers' wounds during World War II. In 1914, they began their association at the Rockefeller Laboratories in Compiegne, in the converted Hotel du Rond Royal. Laboratory animals with geometric shapes cut into their abdomens roomed alongside wounded soldiers—all became part of a project to understand healing and, it was hoped, the laws of senescence.[116] This hotel never slept; du Noüy described the sounds of gunfire and the aftershocks of bombs that attended his research into less extraordinary forms of death. As a physical biologist, du Noüy hoped to "introduce a new concept of time . . . [which] we will indicate a possibility of deriving from our own organism" through the measurement of wounds. He and Carrel traced soldier's lesions using a fountain pen and cellophane. These drawings were transferred to paper and cut out, and then the papers themselves were weighed, and compared, every four days. Deeper wounds were filled with water to obtain a measurement of volume. Calculating an index of cicatrization, or healing, with the data from hundreds of wounds, du Noüy determined that the age of a patient negatively corresponded to the rate of healing, an unacknowledged corroboration of Minot's research.

Earlier, at the Rockefeller Institute, Carrel had attempted the "regeneration" of an elderly dog through removing, "cleansing," and re-injecting two-thirds of its blood. As du Noüy explained, "the logical conclusion which Carrel deduced from this remarkable experiment was that . . . the symptoms of senescence are the expression of profound physico-chemical and chemical changes occurring in the organism through the influence of time."[117] Carrel and du Noüy searched for a "chemical clock" by which toxicity caused cells to senesce—and which hygienic conditions might reverse.

In 1931, Carrel published an article on "physiological time." He contrasted it to the uniform ticks on the clock-face that marked "physical time." On the one hand, Carrel assailed traditional histology for failing to adequately represent or even *see* time. Merging Henri Bergson's philosophy of *duration* with the evidence of tissue culture, Carrel argued that "physiological age" was the result of accumulated *events* in the life of a tissue, its ongoing relationship with the environment: "The cause of duration seems to consist of the modification of their medium produced by living structures, and of the secondary changes undergone by these structures under the influence of the modified medium. Time is recorded by a cell community only when the metabolic products are allowed to remain around the tissue."[118] In *Creative Evolution*, moreover, Bergson had used Minot's article "On Certain Phenomena of Growing Old" as a counterexample to duration. The philosopher scorned the captivity of most biologists to "the image of the hourglass"—their conviction that aging was a process of "constant accumulation or loss," that time was external to the histories of individual bodies.[119]

Even as Carrel assailed traditional histology, he simply repeated many of its conclusions regarding development:

> Dead organs and histological sections are nothing but useful abstractions. The body really consists of a flux of structural and functional processes, that is, of an uninterrupted modification of tissues, humors, and consciousness. Such is physiological duration. The process of aging starts simultaneously with embryonic life. It is expressed by irreversible changes progressing during the entire span of our existence. The decrease in the rate of growth during infancy and youth, the occurrence of puberty and menopause, the lowering of basal metabolism and the modifications of the skin and hair, etc., appear as the stamp of time on the organism.[120]

Carrel, too, saw physiological time as something regular and quantifiable; the index of cicatrization, in fact, was one of its measures.[121]

In his widely translated *Man, the Unknown* of 1935, Carrel outlined a philosophy in which "physiological age" was the result of the accumulated events in an organism's life: "Our present does not drop into nothingness as does the present of a pendulum. It is recorded simultaneously in mind, tissues, and blood. We keep within ourselves the organic, humoral, and psychological marks of all the events of our life. . . . We are the result of a history. . . . Each thought, each action, each illness has definitive consequences, inasmuch as we never separate ourselves from our past. We may completely recover from a disease, or from a wrong deed. But we bear forever the scar of those events." At the same time, Carrel echoed Minot regarding the inevitable senescence of organisms with differentiated cells: "Death is the price he has to pay for his brain and his personality."[122] The life span of tissues could be controlled through careful culturing, and so might longevity be increased for humans, but Carrel did not ultimately believe this would translate into human immortality. "Where are we going?" he asked the readers of *Reflections on Life* (published posthumously).

> Towards death. The structure of our body makes death a necessity. From the very inception of its existence in the womb, the young organism begins to age. This process of senescence is much more rapid in the foetus and the young child than in the adult and still more than in the old. The progress towards death slows down considerably with advancing years but it never stops and never changes its direction. Whatever the future successes of science may be, every human being is condemned sooner or later to disappear from this world.[123]

Carrel, who later directed the Foundation for the Study of Human Problems for the Vichy government, also felt that heredity had a strong influence. Individual differences (between humans, or between tissue types) resided in structure as well as in history.[124] Thus he did not universally recommend the application of tissue culture, or any other technique, to increase human longevity: "It is imperative that the number of the diseased, the paralyzed, the weak, and the insane should not be augmented. Besides, it would not be wise to give everybody a long existence. The danger of increasing the quantity of human beings without regard to their quality is well known. Why should more years be added to the life of persons who are unhappy, selfish, stupid, and useless?"[125]

Many biologists contemporary to Carrel focused entirely on the ways the lived past accumulated in, "acted upon," the present. Aging, then, was the result of wear and tear, unhealthy environments, or the build-up of poisons in the body. Jacques Loeb, for instance, in his book *The Organism as a Whole: From a Physico-Chemical Viewpoint,* pitted Minot against Ilya

Metchnikoff, who won the Nobel Prize in 1908 for his discovery of phagocytosis. Metchnikoff proposed that aging resulted from "autointoxication" caused by intestinal bacteria. This, he believed, could be prevented through the regular consumption of sour milk. Siding with Metchnikoff, Loeb placed "natural" death in quotation marks.[126] As G. Stanley Hall later put it, Metchnikoff turned aging into an "infectious chronic disease."[127]

More damaging, Raymond Pearl took apart Minot's theory of cytomorphosis in his own Lowell Lectures of 1920, published as *The Biology of Death*. Earlier, Pearl had praised Minot for conducting the "the pioneer researches" into growth, but these Lowell Lectures set out to refute his predecessor's basic concepts.[128] After discussing the immortality of single-celled organisms, Pearl argued that differentiated cells, with their limited powers of reproduction, were simply sensitive to external injuries and unable to repair themselves.[129] Pearl revised cytomorphosis into the result rather than the cause of aging. He based these contentions on the evidence of life tables (death was medical, not biological—no one actually died of old age) and on that of tissue culture: "If cells of nearly every sort are capable, under appropriate conditions, of living indefinitely in undiminished vigor, and cytological normality, there is little ground for postulating that the observed senescent changes in these cells while in the body, such as those described by Minot and others, are expressive of specific and inherent mortal processes going on in the cells."[130]

Near the end of his life, Minot continued his attempts to expand the Harvard embryological collection. In 1912, he wrote a letter to James Jackson, soliciting money "to purchase for the school Dr. Mall's collection of human embryos, the best in the world, and which, if added to the collection which I have formed, already the finest in the world for the study of comparative vertebrate embryology, will assure Harvard the first place for the study of this important science."[131] But Mall's work at Johns Hopkins would soon be funded by the Carnegie Institute and "the Carnegie Stages" would become the standards of human development. Mall had long since trained Ross Harrison, whose tissue culture technique was rapidly circulating. Minot discussed this emergent method in his final book, *Modern Problems of Biology*, a transcription of his lectures at Jena in 1913, the year before his death.[132] Throughout his career, Minot had waffled on the relative importance of theory and technique to scientific progress.[133] But regarding Harrison's in vitro cultures, he announced, "Sometimes we find the paradox justified which says: 'New methods are more important for science than new thoughts.'"[134] Minot went on to explain that tissue culture confirmed many of his own arguments regarding the progressiveness of development, and the decreased growth of somatic cells as compared to embryonic.[135] Some aspects of cell death — trauma, infection — might be "curable diseases," but Minot continued to view decay as an essential characteristic of life.

Carrel had explanted cells from a chicken heart in 1912, and they were beating at the Rockefeller Center. They would "live" for decades, raising hopes about human life-extension. Hannah Landecker has shown that the promise of immortality made by tissue culture deferred research into cell death. Not until 1965 would Leonard Hayflick use culture techniques to corroborate Minot's theory of senescence.[136] Hayflick's cells proved to have finite life spans of approximately fifty cycles, and this "Hayflick limit" cast suspicion on Carrel's beating heart-tissue as having been contaminated, or cancerous.[137] Nonetheless, the biological gaze had by then expanded its purview: from intervening to prepare "the normal and the pathological" for observation to intervening for enhancement.

Notes

Henry H. Donaldson, "Charles Sedgwick Minot," *Proceedings of the Boston Society of Natural History* 35 (July 1915): 84. Donaldson was paraphrasing Thomas Henry Huxley.

1. Frederic Burk, "The Training of Teachers," *Atlantic* (October 1897). Quoted in Charles Sedgwick Minot, "Literary Embryology," *Science* 6, 146 (October 1897): 596.
2. Minot, "Literary Embryology," 595.
3. Charles Sedgwick Minot, *Modern Problems of Biology* (Philadelphia: P. Blakiston's Son & Co., 1913), 20.
4. Donaldson, "Charles," 927.
5. Charles Sedgwick Minot, "On the Conditions to be Filled by a Theory of Life," *Proceedings of the American Association for the Advancement of Science* 28 (August 1880): 413. An abstract from the 1879 meeting, printed in 1880.
6. Charles Sedgwick Minot, *The Problem of Age, Growth and Death* (New York: The Knickerbocker Press, 1908), 5.
7. Likewise historians of aging, from Simone de Beauvoir to Thomas Cole, name Minot as one of the earliest investigators of biological temporality. Simone de Beauvoir, *Old Age* (London: André Deutsch, 1972), 22; Thomas Cole, *The Journey of Life: A Cultural History of Aging in America* (New York: Cambridge University Press, 1992), 195.
8. Frederick Lewis, "Charles Sedgwick Minot," *The Anatomical Record* 10, no. 3 (January 1916): 134.

9. Edward Morse, *Biographical Memoir of Charles Sedgwick Minot* (Washington DC: National Academy of Sciences, 1920), 265, 274.

10. Lewis, "Charles," 149.

11. See W. B. Fye, "Growth of American Physiology," in *Physiology in the American Context: 1850–1940*, ed. Gerald Geison (Bethesda, MD: The American Physiological Society, 1987), 55.

12. For more on the history of biology education, see Philip J. Pauly, "The Appearance of Academic Biology in Late Nineteenth-Century America," *Journal of the History of Biology* 17 (1984): 369–97.

13. Despite his illustrious genealogy (which could be traced back to Jonathan Edwards), and the prominence of his Boston family.

14. Lewis, "Charles," 145.

15. Nick Hopwood, "Producing Development: The Anatomy of Human Embryos and the Norms of Wilhelm His," *Bulletin of the History of Medicine* 74 (2000): 34–35.

16. Lynn Morgan, "Embryo Tales," in *Remaking Life and Death*, ed. Sarah Franklin and Margaret Lock (Santa Fe, NM: School of American Research Press, 2003), 269.

17. See Adrianne Noe, "The Human Embryo Collection at the Department of Embryology of the Carnegie Institution of Washington," in *Centennial History of the Carnegie Institution of Washington, Volume 5: The Department of Embryology*, ed. Jane Maienschein (Cambridge: Cambridge University Press, 2005), 21–62.

18. Haraway's first book is newly subtitled "Metaphors That Shape Embryos." Donna Haraway, *Crystals, Fabrics, and Fields: Metaphors of Organicism in Twentieth-Century Developmental Biology* (New Haven, CT: Yale University Press, 1976), 17.

19. Quote from ibid., 71. The other arguments noted above can be found on 2, 199, 64.

20. "Biology versus physics?" *Nature* 391 (1998): 107.

21. Susan Merrill Squier, *Liminal Lives: Imagining the Human at the Frontiers of Biomedicine* (Durham, NC: Duke University Press, 2004), 66. Italics mine.

22. Ibid., 85. Carrel earned the Nobel Prize for Physiology in 1912, and was most famous for extending Harrison's method of tissue culture (a phrase Carrel coined).

23. Hannah Landecker, "The Lewis Films: Tissue Culture and 'Living Anatomy' at the Carnegie Institute for Embryology, 1919–1940," in *Centennial History of the Carnegie Institution of Washington, Volume 5: The Department of Embryology*, ed. Jane Maienschein (Cambridge: Cambridge University Press, 2005), 118. "Paradoxically, the practices of tissue culture created the highly inhospitable but also necessary conditions under which cell death became not just a biologically significant concept but a medically relevant one." Hannah Landecker, "On Beginning and Ending with Apoptosis: Cell Death and Biomedicine," in *Remaking Life and Death: Toward an Anthropology of the Biosciences*, ed. Sarah Franklin and Margaret Lock (Santa Fe, NM: School of American Research Press, 2005), 26, 41.

24. Haraway comments that the search for historical "continuities" can make "revolutions look ultimately conservative." Haraway, *Crystals*, 192. In a similar vein, Landecker warns readers that "making connections and discovering precursors does nothing to explain why tissue culture appeared to Harrison's contemporaries as something startlingly new." Landecker, *Culturing*, 58. My point is not to dispute the innovation of tissue *engineering*, but to examine the ways earlier optical techniques, often aimed at "dead" cells, brought about their own revolution in thinking about biological time.

25. Charles Minot, "On Certain Phenomena of Growing Old," *Address by Charles Minot before the Section of Biology*, AAAS (August 1890) (Salem, MA: Salem Press, 1891), 12. Reprinted in the *Proceedings of the American Association for the Advancement of Science* 39 (1891): 271–89. Jean-Martin Charcot's *Clinical Lectures on the Diseases of Old Age*, published in Paris in 1868 and translated into English in 1881, was one of the first medical texts devoted specifically to the physical aspects of aging itself. Indigent elderly patients at Salpetrière provided bodies for his research, and historians have suggested that the *physical* isolation of these patients from "healthy" society was reflected in Charcot's *textual* "disordering" of old age. Charcot's book focused on the topic of disease and "normalized" degeneration within the period of old age. Georges Guillain, *J. M. Charcot: His Life, His Work*. (New York: Paul B. Hoeber, 1959), 5.

26. A phrase Philip Pauly applied to the work of Jacques Loeb. Philip Pauly, *Controlling Life: Jacques Loeb and the Engineering Ideal in Biology* (New York: Oxford University Press, 1987).

27. Hannah Landecker, *Culturing Life* (Cambridge, MA: Harvard University Press, 2007), 11.

28. Michel Foucault, *The Birth of the Clinic: An Archaeology of Medical Perception*, trans. A. M. Sheridan Smith (New York: Vintage, 1994), 146. Reprint of the 1975 edition.

29. Ibid., 144.

30. Loeb, quoted in Landecker, *Culturing*, 1.

31. Evelyn Fox Keller, "The Biological Gaze," in *Future Natural: Nature, Science, Culture*, ed. George Robertson (London: Routledge, 1996), 108.

32. Ibid., 114.

33. Charles Sedgwick Minot, "The Method of Science," *Science* 33 (January 1911): 122.

34. Minot also discusses "the comparative method" in "Knowledge and Practice," *Science* 10 (July 1899): 6.

35. Here Minot was undoubtedly influenced by his years of study with Carl Ludwig, whose kymograph brought graphic inscription to physiology.

36. Lorraine Daston and Peter Galison, *Objectivity* (New York: Zone Books, 2007), 27.

37. Ibid., 337

38. Ibid., 322.

39. Minot, "Method," 121.

40. Ibid., 129.

41. Minot, "Knowledge," 5, 4.

42. "Lecture Notes and Fragments" (n.d.), Charles Sedgwick Minot Papers, Countway Library, Harvard University. Cambridge, MA.

43. Minot, "Knowledge," 6. In this same article, he explains, "The practical work is the instructive work. . . . [T]he actual direct contact with the objects and with the phenomena *is* knowledge." p. 10.

44. Ibid., 5.

45. Charles Sedgwick Minot, "The Problem of Consciousness in its Biological Aspects," *Science* 16 (July 1902): 4.

46. Garland Allen, "T. H. Morgan and the Emergence of a New American Biology," *The Quarterly Review of Biology* 44 (1969): 168, 185.

47. Charles S. Minot, "The Storing of Pamphlets," *Science* 8 (December 1898): 944–45.

48. He himself was the youngest to graduate from MIT in 1872. On vacations, see Charles S. Minot, "The Distribution of Vacations at American Universities," *Science* 15 (March 1902): 441–44.

49. Charles Sedgwick Minot, "The Study of Zoology in Germany," *The American Naturalist* 11 (June 1877): 330.

50. Charles Sedgwick Minot, "The Study of Zoology in Germany," *The American Naturalist* 11 (July 1877): 392. Elsewhere he expansively argued, "The applications of the invention of placing pieces of glass of particular shapes in the two ends of a brass tube have more profoundly influenced human thoughts and beliefs than any other single invention, excepting only printing. The telescope has revolutionized our conception of the universe, and the microscope our conception of life." Quoted in Lewis, "Charles," 155.

51. Minot, "Study of Zoology," 396.

52. Merriley Borell has detailed the equipment in Harvard's physiology lab at the end of the nineteenth century, in "Instruments and an Independent Physiology: The Harvard Physiological Laboratory, 1871–1906," in *Physiology in the American Context: 1850–1940*, ed. Gerald Geison (Bethesda, MD: American Physiological Society, 1987).

53. Charles S. Minot, "The Inverted Image on the Retina," *Science* 2 (November 1895): 693.

54. Ibid., 692–93. "Sight has long been acknowledged by science as the supreme sense," he insisted elsewhere. Minot, "Method," 127.

55. Minot, "Problem of Consciousness," 3.

56. Minot, "On the Conditions," 411–15.

57. Charles S. Minot, "Growth as a Function of Cells," *Proceedings of the Boston Society of Natural History* 20 (1879): 190. This article developed the themes of an 1877 publication.

58. A. J. Lustig, "Sex, Death, and Evolution in Proto- and Metazoa, 1876–1913," *Journal of the History of Biology* 33 (2000): 221–46. Lustig also discusses the contributions of Maupus, Bütschli, Herbert Spencer Jennings, and others to the immor-

tality controversy. In the *Quarterly Review of Biology* in 1931, W. W. Lepeschkin suggested that the science of death was initiated by protozoan research as well as *in vitro* culturing techniques. W. W. Lepeschkin, "Death and Its Causes," *Quarterly Review of Biology* 6, 2 (July 1931): 167–177. To these fields, I would add Minot's studies of growth (in the adult and in the embryo).

59. Ibid., 233.

60. C. S. Minot, "Death and Individuality," *Science* 4 (October 1884): 398–400.

61. Charles Sedgwick Minot, "On Heredity and Rejuvenation," *The American Naturalist* 30 (January 1896): 6.

62. Guido Majno and Isabelle Joris have argued that what is now known as apoptosis (one type of "programmed cell death") was "born with a bang in 1885," when Walther Flemming visualized its occurrence with the assistance of new staining techniques. Guido Majno and Isabelle Joris, "Apoptosis, Oncosis, and Necrosis: An Overview of Cell Death" *American Journal of Pathology* 146 (January 1995): 4. Other authors have looked to the 1840s, shortly after the elaboration of cell theory, for early examples of the recognition of programmed cell death in the context of metamorphosis and development. See Peter G.H. Clarke and Stephanie Clarke, "Nineteenth century research on naturally occurring cell death and related phenomena," *Anatomy and Embryology* 193 (February 1996): 81–99. For a discussion of the ways "senescence" has been replaced by "programmed cell death" (PCD), see Wouter G. van Doorn and Ernst J. Woltering, "Senescence and programmed cell death: Substance or Semantics?" *Journal of Experimental Botany* 55 (2004): 2147–53.

63. Minot, *Problem of Age*, 214.

64. Ibid., xix. He later repeated many of these measurements using chicks and rabbits.

65. As a side project, he intended to trace the inheritance of fur markings.

66. Minot, "Senescence and Rejuvenation," *Journal of Physiology* 12 (May 1891): 102.

67. Longevity researchers, too, were interested in discovering the source of growth, and the reasons for its cease. Jean Pierre Marie Flourens, at the College de France, looked to the epiphyses for the secret of life in the mid-nineteenth century; he theorized that one would live five times longer than the period of time that the long bones grew. Jean Pierre Flourens, *De la longévité humaine et de la quantité de vie sur la globe* (Paris: Garnier Frères, 1854). Others examined intestine length or hair growth.

68. Minot, "Method," 128. Discussing both mathematical plots and Ludwig's kymograph in that same section, Minot announced, "The aim of science goes beyond the attainment of exact records to the attainment of accurate knowledge, and the accuracy of our knowledge depends chiefly on what we see."

69. Though Minot's method was critiqued by statisticians, he was quite proud of it at the time, and once commented, "If the absolute increments are constant the *rate* of growth diminishes, a point which so far as I am aware has been entirely overlooked hitherto," Minot, "Senescence," 147.

70. Minot, "On Certain Phenomena of Growing Old (Abstract)," *Science* 16 (August 1890): 123.

71. Minot, "Senescence," 102.

72. For one of his musings on this topic, as it related to regeneration, see Charles S. Minot, "The Formative Force of Organisms," *Science* 6 (July 1885): 4–6.

73. Charles Sedgwich Minot, "The Physical Basis of Heredity," *Science* 8 (August 1886): 125–30.

74. Squier, *Liminal*, 126.

75. However the pathway from telomere shortening to phenotypic aging is hardly understood.

76. Minot, *Problem of Age*, xvi.

77. He drew on Wilhelm His almost entirely for his discussion of the first two months, a developmental period that remained mysterious: few abortions occurred in those months, and miscarriages either went undetected, or involved abnormal embryos. Minot also relied on a rare, twenty-six-day embryo from Mall. In 1887, John S. Billings, curator of the Army Medical Museum, offered Minot $150 for "some specimens illustrating embryology for us." The following year, Billings invited Minot to exhibit part of his collection at the Museum. J. S. Billings to Charles S. Minot, October 22, 1887 and May 6, 1888, Outgoing Correspondence Letterbook 11, Walter Reed Archives.

78. Charles Sedgwick Minot, "The Sledge Microtome," *The American Naturalist* 11 (April 1877): 206–7.

79. Charles Sedgwick Minot, *A Laboratory Textbook of Embryology* (Philadelphia: P. Blakiston's Son & Co., 1903), 358.

80. T. Mitchell Prudden, Charles S. Minot, Edward Bausch, and John A. Ryder, "American Microscopes," *Science* 10 (December 1887): 310–12.

81. R. P. Cowles and Oscar W. Richards, "The Pfeifer and Minot Automatic Rotary Microtomes," *Transactions of the American Microscopical Society* 66 (October 1947): 379.

82. Nick Hopwood, "Giving Body to Embryos: Modeling, Mechanism, and the Microtome in Late Nineteenth-Century Anatomy," *Isis* 90 (1999): 476.

83. According to Charles Eliot, the textbook made Minot "famous throughout the learned world" and caused him to be awarded numerous honorary degrees. Charles W. Eliot, "Charles Sedgwick Minot," *Science* 41 (May 1915): 703.

84. "Review: Two Text-Books of Human Embryology," *American Naturalist* 27 (February 1893): 141.

85. Minot, *Problem of Age*, 130.

86. Charles Sedgwick Minot, "The Harvard Embryological Collection," *The Journal of Medical Research* 13 (August 1905): 499.

87. Ibid., 501.

88. He located the most massive amount of differentiation in the period between 7 and 16.5 days. Minot, *Problem of Age*, 160, 175.

89. "We are really old by the time we are born and the alterations which make us old have for the most part already occurred." Minot, *Modern Problems*, 33.

90. Minot discussed Driesch's work on the totipotency of early embryonic cells in "The Embryological Basis of Pathology" *Science* 13 (March 1901), 482. "The young type of cells really is physiologically and functionally important," he mused in *Problem of Age*, 213.

91. Minot, *Problem of Age*, 64.

92. Nick Hopwood, "Producing Development," 41.

93. Minot, *Human Embryology*, 510. J. M. Tanner has critiqued Minot's data on human embryos as resulting from circular logic rather than factual conditions. See *History of the Study of Human Growth* (Cambridge: Cambridge University Press, 1981), 262

94. To the Yale medical students in 1899, he pronounced regarding laboratories, "Seventy-five years ago there were none. There are but few laboratories which have stood for as much as twenty-five years." Minot, "Knowledge," 8.

95. Ibid., 10.

96. Minot recommended the pig for being cheaply procured from packinghouses, as well as for its optical advantages: "Owing to the enormous precocious development of the chorionic vesicle in pigs, it produces an enlargement of the uterus which is usually sufficient, by the time the embryo has attained a length of 6 mm, to be observable to the untrained eye. It is, therefore, only necessary to ask the man who removes the viscera from the pigs to lay aside for examination all of the uteri which appear distended." Minot, *Laboratory*, 157.

97. Nick Hopwood, "Giving Body."

98. Ibid., 31.

99. Minot, *Problem of Age*, 92.

100. He had publicly announced this notion of cytomorphosis earlier, in a 1901 lecture before the New York Pathological Society. See Minot, "The Embryological Basis," 481–98.

101. Minot, *Problem of Age*, 85.

102. Many works subsequent to Carole Haber's *Beyond Sixty-Five* cite her maxim: "General vital energy had, by the mid-nineteenth century, been replaced by a degeneration of tissues, and finally by an inexorable devolution of cells." Carole Haber, *Beyond Sixty-Five: The Dilemma of Old Age in America's Past* (New York: Cambridge University Press, 1983), 63.

103. Minot, *Problem of Age*, 5.

104. Ibid., 36.

105. Lustig has also discussed Minot's reduction of death to an "epiphenomenon of the essential phenomena of multicellularity." See Lustig, "Sex," 235.

106. Minot, *Modern Problems*, 80.

107. Ibid., 161.

108. Lawrence Cohen, *No Aging in India*, 62.

109. I. L. Nascher, *Geriatrics: The Diseases of Old Age and Their Treatment* (Philadelphia: P. Blakiston's Son & Co., 1914), 2.

110. G. Stanley Hall, *Senescence: The Last Half of Life* (New York: D. Appleton and Company, 1922), 273.

111. Ibid., 274.

112. Alexis Carrel, "Measurement of the Inherent Growth Energy of Tissues," *The Journal of Experimental Medicine* 38 (1923): 521–27.

113. Alexis Carrel and Montrose T. Burrows, "Cultivation of Tissues in Vitro and Its Technique," *The Journal of Experimental Medicine* 13, no. 3 (1911): 387–96.

114. Perrin Selcer, "Standardizing Wounds: Alexis Carrel and the Scientific Management of Life in the First World War," *The British Journal for the History of Science* 41 (March 2008): 79–80.

115. Alexis Carrel, "On the Permanent Life of Tissues Outside of the Organism," *The Journal of Experimental Medicine* 15 (1912): 516–28.

116. Carrel primarily researched the sterilization of wounds, in an attempt to reduce gangrenous infection and hence amputation, and return more soldiers to the battlefields. He published *The Treatment of Infected Wounds* in 1917.

117. Pierre Lecomte du Noüy, *Biological Time* (New York: The Macmillan Company, 1937), 115.

118. Alexis Carrel, "Physiological Time," *Science* 74 (December 1931): 621.

119. Henri Bergson, *Creative Evolution*, trans. Arthur Mitchell (New York: Dover, 1998; New York: Henry Holt, 1911), 17–18. Citations are to the Dover edition.

120. Carrel, "Physiological Time," 619.

121. Ibid. Carrel in fact believed that cells "show their true physiognomy" on film, rather than "live" under the microscope. Interestingly, Bergson had precisely critiqued the "cinematographical method" of the scientific mind: the cinematic interpretation of time as "a series of successive states"; the faith that an object could be "reproduced" through analysis and resynthesis. See Alexis Carrel, "The New Cytology," *Science* 73 (March 1931): 300. Bergson, *Creative*, 31, 329–32.

122. Alexis Carrel, *Man, the Unknown* (New York: Halcyon House, 1938), 183–184. Reflecting on the life spans of different peoples, he more directly left open the possibility of an innate mechanism of senescence, "To estimate true, or physiological, age, we must discover, either in the tissues or in the humors, a measurable phenomenon, which progresses without interruption during the whole lifetime."

123. Alexis Carrel, *Reflections on Life*, trans. Antonia White (London: Hamish Hamilton, 1952): 150–51.

124. As Selcer explains, "For Carrel, differentiation caused death because each cell type required its own particular environment, but the internal milieu of the organism could only provide a sort of physiochemical compromise." Selcer, "Standardizing," 79.

125. Carrel, *Man*, 180.

126. Jacques Loeb, *The Organism as a Whole: From a Physico-Chemical Viewpoint* (New York: G.P. Putnam's Sons, 1916), 362. In his own research, Loeb had found that he could prolong the viability of urchin eggs based on the chemical composition of their environments. Charles Manning Child, who worked with planaria, upheld Minot's thesis regarding the rapidity of decline during youth.

Titling his own book *Senescence and Rejuvenescence*, however, Child argued that regressive development regularly occurred, wherein old cells returned to young states. Charles Manning Child, *Senescence and Rejuvenescence* (Chicago: University of Chicago Press, 1915), 459.

127. Hall, *Senescence*, 249.

128. Pearl, "The Service and Importance of Statistics to Biology," *Publications of the American Statistical Association* 14 (March 1914): 46.

129. Ibid., 48–49.

130. Raymond Pearl, *The Biology of Death* (Philadelphia: J.B. Lippincott Company, 1922), 71.

131. Charles Sedgwick Minot to James Jackson, 1912, Charles Sedgwick Minot Papers, Countway Library Harvard University, Cambridge, MA.

132. In these lectures, he backed away from the signal importance of cells, focusing instead on chromatin and metabolism.

133. For his dismissal of staining, dissecting, and culturing (as opposed to habits of mind), see "Knowledge," 7. For his elevation of methodological innovation above all else, see *Problem of Age*, 9.

134. Minot, *Modern Problems*, 27.

135. Ibid., 32.

136. Hayflick discusses the work of Weismann and Minot in Leonard Hayflick, "Origins of Longevity," in *Modern Biological Theories of Aging*, ed. H. L. Warner (New York: Raven Press, 1987), 21–34. Other researchers subsequently set out to test "Minot's hypothesis that cellular aging and death of metazoan animals are the result of cell differentiation." See J. Miquel et al., "Review of Cell Aging in Drosophila and Mouse," *Age* 2 (July 1979): 78–88.

137. The Hayflick limit is now tied to the shrinking of the telomeres with each mitotic division—with cells also dying as a result of things like accumulated DNA damage from the environment. For a refutation of the consequences of telomere shortening, see Stanley Shostak, *The Evolution of Death: Why We Are Living Longer* (Albany: State University of New York Press, 2006). For a recent revision of Minot's argument, which argues that cell death serves as an adaptive advantage against mutations (i.e., those leading to cancer), see William Clarke, *Sex and the Origins of Death* (Oxford: Oxford University Press, 1996).

chapter two

Facing Animals in the Laboratory

Lessons of Nineteenth-Century Medical School Microscopy Manuals

Nancy Anderson

In the course of the nineteenth century we see the rise of various biological fields of research, including physiology, microscopic anatomy, histology, and as the cell theory gains momentum in the last decades of the century, cytology. In assessing this moment when such areas of study emerged as bona fide disciplines, gathering all the resources of such a status (university departments, professional societies, professional journals, and course textbooks), one debate in the history of science and medicine has been about how and how quickly these bodies of knowledge infiltrated and transformed medical education and practice. Was there a scientific medicine in the second half of the nineteenth century, and how did the biological sciences intersect, even influence, the medical profession? The answer seems to be, at least for Great Britain, that well into the century practicing physicians often resisted science. Supported by patients' expectations, many clung to the idea of the gentleman-doctor educated in the classics and practicing his "art" at the bedside.[1] It was thought widely enough that a scientist-physician—too narrowly trained and putting faith in instruments instead of his own skills—was of little use in the real world of the sick. Even in the wake of the success of vaccinations and breakthroughs in the area of bacteriology, many senior British physicians would take on "vocabulary, which routinely invoked science as the foundation of medicine but which prescribed for science only a limited role in clinical practice."[2] The antivivisectionist movement played a part in this as well, leading a crusade well into the later decades of the nineteenth century "against the professionalization and institutionalization of experimental medicine, which was seen as a trend of foreign origin."[3]

The biological sciences, however, had begun to influence medical education by midcentury. As professional research in physiology and histology expanded in continental Europe, by the 1820s forward-thinking British physicians and ambitious medical students had begun to travel to France and Germany to study in laboratories, returning with the belief that these disciplines held the key to their profession's future. In her novel *Middlemarch* (1871), George Eliot captured this trend of medical students embarking on "study abroad" trips in the first half of the century and their indoctrination into the new microscopic life sciences. Her character, the young doctor Tertius Lydgate, who comes to practice in Eliot's eponymous town in 1829, studied in Edinburgh and Paris as a student and owned a microscope, hoping in his spare time in Middlemarch to advance Xavier Bichat's work on webs of tissues so as to "demonstrate the more intimate relations of living structure, and help to define men's thought more accurately after the true order." Bichat himself had rejected the microscope as an untrustworthy tool of investigation, and Lydgate is then meant to exemplify the next generation who showed more interest and confidence in this instrument. Eliot, who invented this character of circa 1830 from the vantage point of the late 1860s (when she wrote *Middlemarch*) and who was the partner of physiologist George Lewes, explained this in her text, noting that Lydgate was "ambitious above all to contribute towards enlarging the scientific, rational basis of his profession."[4]

So, when these adventurous students, who traveled to laboratories on the Continent, returned home, they sought to spread their new knowledge, placing it within the context of modern means of understanding health and disease. They did so by offering classes and demonstrations for medical students in their own institutions, which would have an impact on the future of the medical curriculum. Classes in the basics of microscopic anatomy and histology were made available in medical schools in Great Britain in the 1840s, first as extramural offerings, and then later becoming curricular requirements.[5] Pedagogical innovations, occurring not just in England, but across Europe and in the United States, included instituting the seminar in which students, no longer subject to passive learning by lecture alone, were "actively inducted into the craft and standards of their specialty by repeating exercises that were already part of the repertoire of the discipline," and this included experimental procedures.[6] Students were even finding their research worthy of space in the emerging professional scientific journals. Joseph Lister, while undertaking the duties of a medical resident at King's College, for example, published his work on the muscular tissue of the skin in 1853, an experiment that replicated, confirmed, and expounded on data he found in *Kolliker's Microscopische Anatomie*, the textbook from which many a professor built his syllabus before the flood of English manuals beginning in the 1860s.[7] Then, in 1871, curricular reform for medical students arrived in the form of a decree by the Council of the Royal College of Surgeons making general anatomy and physiology as well as practical anatomy and histology prerequisites for earning a Diploma of Membership.[8]

With these new requirements came not only the need to build teaching laboratories but to publish and make available practical microscopy, histology, and physiology manuals. In 1875 the *British and Foreign Medico-Chirurgical Review* reviewed a group of new English texts, taking a moment to emphasize the importance of visual aids in the study of anatomy from organism to organ to cell. One of the books discussed was *The Handbook for the Physiological Laboratory* (1872), which had been put together with sections by four different authors, John Burdon-Sanderson, Edward Klein, Michael Foster, and Thomas Lauder Brunton. The review hailed the *Handbook* for its practical (actually, dogmatic) approach to teaching histology (Klein's contribution to the project), and particularly praiseworthy, it seemed, was that "the student is told what he will see as well as how he should set about seeing it." Plates and figures were crucial, although the most detailed and "realistic" image was not necessarily the most effective pedagogically, and so the article pointed out that "diagrams or rough sketches are sometimes better than true pictures for elementary instruction."[9]

Almost all of the new publications contained illustrations, an assortment of pictures showing prepared tissues and individual cells. Rarely, however, did these books offer images of the organisms, the beings serving as the origins of this microscopic matter. Of course, it was obvious that histology and physiology demanded animal bodies as sources of material to examine, but text and images treat the organism in its parts, as a dismembered thing. Tables of content reveal the primer of anatomy as an already analyzed organism—presented part by part—with the text then filled with descriptions of violent acts inflicted on animals whose particular attributes (an elongated kidney or transparent mesentery) provided the anatomy lesson, even as the student was encouraged to connect the particular view to organisms in general and, eventually, to the human body. The following section, then, will consider how manuals explained the preparation of specimens, with special attention given to the overt brutality toward animals these instructions and descriptions unambiguously admitted. But what of the individual animal and its intact body? In the third section of the paper I will address the problem of images not of cells and tissues but of intact animals. These were not often included in the manuals, but the examples that can be found offer an opportunity to think about how these publications acknowledged or deliberately ignored the life forms as thriving (breathing, heart-beating) wholes from which the isolated organs or slices of tissue or single cells necessarily originated.

In considering this last point, I will direct attention to one aspect of the organism, if not something we might actually consider a body part: the face. Philosophically, the meaning of "face" is poignantly rich. The face, certainly the human face, claim Gilles Deleuze and Felix Guattari, "is the location of signification and subjectification." As the surface upon which messages embodying feeling and emotion (joy, pain, fear, pleasure) are visually communicated, emanating as a "series of micromovements on an immobilized plate of nerve," the face is a field of immanence and point of transience.[10] There is, then, the intensified experience of subject-subject communication via the face: face-to-face encounters. There is a problem, though, in the engaging of the face of the nonhuman animal. The question for this essay is how to imagine these face-to-face encounters between the nonhuman and the human, the experimental animal and the experimenter (medical student) who might likely inflict pain, dismember, even kill a creature to get a glimpse at or an image of kidney cells or brain tissue. Or, at least, how are we to consider the body of a being described in a text or from which an abstract image showing a pattern of cells has derived? It is almost never that a textbook allows the student to meet an image of an animal's face directly on its pages, even as animals' bodies are mentioned explicitly in the text or (very rarely) presented in an image. Is there a reason, even unconscious, for such an absence, or *effacement*?

Deleuze and Guattari might offer an explanation for this in their description of the hierarchal relation between the one in power ("the despot") and the subjugated (e.g., "the one who is tortured"): The despot "has never hidden his face," they write, while the tortured "loses his or her face."[11] As will be described below, animals, the subjugated, were subjected to various degrees of what could be defined as torture in order to get the desired view of an organ or a mass of cells. It is, then, interesting and informative to consider what the philosophy of the face can reveal about relationships between human and nonhuman animals.

Emmanuel Levinas hinged a theory of ethics on the face-to-face meeting between humans in that he claimed that this encounter epitomized the inescapable, absolute responsibility: "Thou Shall Not Kill." However, when asked if (nonhuman) animals had faces, Levinas hedged, claiming "the priority here is not found in the animal, but in the human face." Continuing, he admitted, "I cannot say at what moment you have the right to be called 'face.' I don't know if a snake has a face. I can't answer that question."[12] In the past two decades, with growing attention to animal rights, concerted effort has been given to breaking down the stronghold of anthropocentrism in systems of philosophy and ethics and to understanding the "sentient" nonhuman animal, or as Tom Regan puts it, "Descartes' downfall."[13] Philosophers now consider with much more frequency questions of awareness, consciousness, and desire among nonhuman animals. Jacques Derrida, for one, famously rebuked Levinas's anthropocentrism when he took on the issue of the animal and proclaimed that the question philosophers must now address is: "How can an animal look you in the face?"[14] Matthew Calarco, who has considered the philosopher considering the animal (Heidegger, Agamben, Levinas, Derrida, Deleuze) writes that now "philosophy finds itself *faced by animals*, a sharp reversal of the classical philosophical gaze."[15]

It is the possibility or impossibility of the reversal of this cross-species gaze, the idea that "Other" animals can engage the human, in the world of nineteenth-century medical education that I want to explore by looking at physiology and microscopic histology textbooks of the time. And I want to do this by looking for faces, literally, in these publications. The fact is that images of intact animals are rarely presented in these texts, and the inclusion of a face is even rarer. My question here is what might be said about the presence, but mostly absence, of the animal's face among these textbook images.

Taking Animals Apart

In 1853, at age twenty-five, Lionel Beale, a petulant personality and vocal vitalist, who had studied zoology at King's College, was appointed professor and chair of physiology and general and morbid anatomy at King's College Medical School. The next year he would publish *The Microscope and Its Application to Clinical Medicine*, in which he promoted the microscope as an invaluable tool for medical work and medical education. Without this instrument, he consistently insisted, "how can he [the student] be expected to be able to investigate successfully the changes which take place in certain textures in diseases; or to make out the complicated structure of morbid growths?"[16] A year later, in 1855, he delivered an introductory lecture to open King's College's Medical Department session for that year entitled "The Medical Student, a Student in Science."[17] Beale also became the founding editor of *Archives of Medicine* in 1857, stating the journal's goal as a space for clinical observations, original research in physiology and pathology, as well as chemical and microscopic examinations of the body, particularly to serve medicine.[18] Beale stood at the forefront of advocating a "scientific" medicine for the nineteenth century. Through teaching, speaking, and publications, he sought to implement the scientific method into students' lessons and, then, into the work of healing.

In 1861 the first edition of Beale's textbook *How to Work with a Microscope* appeared. This manual would be frequently updated and expanded to remain state-of-the-art, with the fifth edition appearing in 1880. By this later edition Beale was providing detailed accounts of specimen preparation, and this meant gruesome descriptions of the demise of live animals in the pursuit of tissue specimens. Of the frog he wrote, "few creatures were better suited for minute anatomical investigation," and then went on to advise that "the animal is killed by being dashed suddenly upon the floor, but it must first be carefully folded up on the centre of a cloth, so that the tissues may not be bruised or injured in the least degree."[19] Beale's instructions as to what to do with the dead animal were quite detailed. Once the frog was dead, the sternum should be opened immediately, exposing the heart so that the arteries could be injected with Prussian blue fluid. Next, he directed the student to open the abdominal cavity and wash it in glycerin. Then, he was to remove the legs, slit open the mouth and wash the pharynx with glycerin as well. Following this, the entire torso and head of the frog should be placed in carmine solution. After up to eight hours in the stain, the tissue was washed in glycerin and placed in strong acetic acid, where it might be left for days. When "the injected vessels are a bright blue and the bioplasts of the tissues of a bright red," Beale writes, "the specimen is ready for minute examination."[20] At this point the student-microscopist could begin slicing thin sheets of tissue from various organs to be placed between glass slides for viewing through the microscope. The manual would then guide the student as to what to look for.

Beale understood the importance of providing visual images. When he founded the *Archives of Medicine*, he promised readers that articles would be "freely illustrated," making the claim that "drawings are really of much more use than long descriptions."[21] In the classroom, illustrations, Beale believed, were indispensible to teaching students how to see. He would stress the image over the word in his first bit of advice to users of his *How to Work with the Microscope,* instructing them that "this work may be 'read' by carefully studying the figures, and then referring to the text."[22] Although he did not offer images of an intact frog in *How to Work with the Microscope,* Beale did provide numerous plates to guide the student in viewing magnified details of the book's laboratory lessons, such as series of images showing nerves, feet, arteries, cells, and connective tissue. Some of the figures refer to dead tissue, while others claim to record the cells of living animals—frog, mouse, newt, cat. There is no single frog, or any animal at all, that can be seen as an origin once the images are set to show objects of this scale, and Beale made no attempt to have the individual pictures of frog fragments, and fragments of other animals for that matter, add up to a whole. In fact, the breaking down of the body, the analysis of its component parts, is the point. As for the once whole, thriving animal, it might still exist in the mental background of the reader or student, but it might just as well have given way to an understanding of the abstract views of cells and tissues standing in for organized life forms in general.

Michael Lynch reported some years ago on observations he made of work in a neuroscience laboratory, concluding that the intent of the research was to reduce live animals as objects of experimentation to bits and pieces that serve as "bearers of generalized knowledge." The gist of his explication was that "laboratory animals are progressively transformed from holistic 'naturalistic' creatures into 'analytic' objects of technical investigation." As he saw the work unfold, the "'analytic animal' . . . becomes the real animal in a scientific system of knowledge, while tacitly depending upon the 'naturalistic animal' for its foundation."[23] In this process, we can see not just loss of bodily integrity, but of species specificity as well.

Physiology and, certainly, the rise of the "analytic" breakdown of the body to organs and cells, the detail offered by microscopy as it ascended as a field of knowledge, a way of knowing, in the nineteenth century, seem to offer a foundation for this view of life in a laboratory. Animals, enjoying the discretion of "species," a form of distinctiveness in a world of natural history, lose their individuality when studied as body parts. Claude Bernard, explaining physiology in the nineteenth century, insisted, "physiologists take a quite different point of view: they deal with just one thing, the properties of living matter and the mechanism of life, in whatever form it shows itself. For them, genus, species and class no longer exist."[24] Comparative anatomy research and education, thus, presented itself as abandoning this classificatory differentiation. Animals brought into the laboratory were fre-

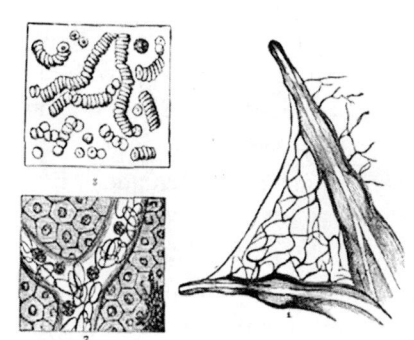

2.1 A view of blood vessels in the web of a frog's foot with a magnified view showing "blood discs" within a vessel. The top image is of "human blood discs." From Lionel Beale, *How to Work with the Microscope* (London: Harrison, 1880).

quently used to make points about human anatomy at the level of organs as well as that of the cell, and images in textbooks would help to drive this message home, as in this triad of images that move from the vessels in a frog's foot to a close-up of the cells in the amphibian blood vessel, and finally to isolated blood cells from not a frog but a human (figure 2.1). Explicitly put, though, this loss of integrity as species is a result of the destruction of individual life forms and, more specifically, individual lives.

To get to these slides of tissue was brutal work, and sometime ago Stewart Richards pondered the question of emotions and aesthetics in assessing nineteenth-century scientists' responses to antivivisectionists' condemnations of their practices. In particular, Richards wondered about a possible missed connection between aesthetics and anesthetics.[25] His essay questions whether British researchers may have faltered in neglecting the potential of anesthetics to persuade their opponents to their side. If physical anguish, the suffering of animals, was the concern of the animal rights activists, desensitization might be a persuasive argument. But my primary interest here is in the training of students, and so from that viewpoint one could imagine not only the advantage of anesthetizing the animals but the desire to numb students against such brutality through repeated descriptions in textbooks of experiments meant to be replicated in the classroom laboratory.

Could anesthetics used on animals also help to numb the student's emotions? Or is it that the medical students' sensibilities could be deadened from the sheer repetition of the shocking acts carried out in spaces of scientific research and education? As historian Richard French noted of one physiology textbook of the period, "it seemed to set no limits to callow youth being indoctrinated in animal experiment."[26] Paul White, in a recent essay, suggested that in the nineteenth century, "evolutionary continuities between humans and animals, useful in underwriting practices like vivisection and ethology, could also facilitate claims about the degeneration of 'gentlemen of science' whose natural sympathies for their human and animal fellows had been deadened by the abstract pursuit of truth."[27] Indeed, antivivisectionists saw it this way, as when in 1875 mathematician and author of *Alice's Adventures in Wonderland* Lewis Carroll described with regret that the time was coming "when successive generations of students, trained from their earliest years to the repression of all human sympathies, shall have developed a new and more hideous Frankenstein—a soulless being to whom science shall be all in all."[28] If this truly were to be the future, how might the education of the late nineteenth-century physician facilitate this?

How is it possible to turn brutal destruction into an experience under anesthesia? Such an act, actually, turns out to be one of modernity's innovations. In an essay addressing Walter Benjamin's own critique of aesthetics and politics and, in particular, the Fascist success of engaging the masses in the enjoyment of viewing their own destruction in the joy of perpetual war and killing, cultural historian and theorist Susan Buck-Morss introduced her thesis by reminding readers that "aesthetics," quoting Terry Eagleton, "is born as a discourse of the body." The main point is that the senses are biological, they are media of cognition, and they are the organs on the frontline of self-preservation. It was the eighteenth-century philosophical project of aesthetics that called for a training and refining—a repressive control—of the senses of taste, smell, touch, hearing, and sight, that is, the tempering of the corporeal reality of the sensorium. From there, Buck-Morss says, a sense-dead moral being, Kant's "autonomous, autotelic subject," rose up—and, no surprise, the warrior was its ideal.[29] But, physicians might fit this bill as well.

Buck-Morss emphasizes nineteenth-century modernity's anesthetizing of the senses through widespread presentations of art and mass culture as phantasmagoria and aesthetic overload, and she links this to the recorded escalation of violent casualties in modern life—in industry and in war. More interestingly, she makes a connection with the introduction of anesthetics into medical and surgical practice after 1846, and she claims that the professionalizing physician of the mid-nineteenth century is the exemplar of the modern subject anesthetized, or made sense-dead, by the technologies of bodily comfort, including ether and chloroform: "Anaesthesia was central to this development [of the medical profession and surgeons as technical experts]. For it was not only the patient who was relieved from pain by anaesthesia. The effect was profound on the surgeon. A deliberate effort to desensitize oneself from the experience of the pain of another was no longer necessary. Whereas surgeons earlier had to train themselves to repress empathic identification with the suffering patient, now they only had to confront an inert, insensate mass that they could tinker with without emotional involvement."[30] How, then, might the use of anesthetics on experimental animals have affected the medical student experimenter? Or was this trick for numbing the person causing pain as well as the body in pain limited to human-human interaction?

One of Beale's colleagues at King's College was William Rutherford, who in his own texts and classroom described acts as ruthless as those included in *How to Work with the Microscope*. But Rutherford was one who noted the use (even necessity) of anesthetics or, at least, something to still the body when working with living organisms. Rutherford, who had studied at Edinburgh under John Hughes Bennett (who in the 1840s had studied microscopic anatomy and histology with the eminent researcher Alfred Donné in Paris), served as chair of histology at King's College only from 1869 to 1875, when he would leave England to take up the position of chair of histology at the University of Edinburgh when Bennett retired. In 1876, after publishing his syllabus in both the medical journal *Lancet* as well as in the *Quarterly Journal of Microscopy*, a full textbook, *Outlines of Practical Histology*, appeared, based on the courses he led in London. The text wasted no time getting to the task of turning bits of animals into microscopic specimens. "Most of the requisite tissues and organs may be obtained from the cat and guinea pig," Rutherford explained to readers at the start of his manual, and then he jumped right into describing the procedures for procuring tissue from, in particular, a cat.[31]

Unlike Beale's frog, an animal to be dashed upon the floor, Rutherford's cat received a final meal before being knocked out with chloroform and summarily bludgeoned. Once the cat was dead, however, the feline body's fate was similar to that of the frog. Its chest was to be opened immediately in order to "speedily open the right ventricle, and allow the animal to bleed to death." Then, one should "divide the trachea immediately below the cricoid cartilage, and inject it with ¼ percent chromic acid fluid; tie it to prevent escape of the fluid and place the distended lungs in the same fluid." Next, "open up the linear incision of the stomach, small and large intestine, and wash the inner surface with ¾ percent salt solution. Place the tongue, divided transversely into a number of pieces, and portion of the small intestine in chromic and bichromate fluid." The directions continue, insisting that more and more organs be removed, sliced up, and placed into various solutions to soak:

> Open the bladder, wash it with salt solution. . . . Remove the kidneys, divide one transversely, the other longitudinally and place them in Muller's fluid. . . . Cut half of the liver into small pieces . . . Place the spleen, uterus, some thin muscles from the limbs or abdomen in ¼ percent chromic acid . . . Remove both eyes. Divide them transversely behind the crystalline lens. . . . Cautiously open the cranial and spinal cavities. Remove brain and spinal cord, and strip off arachnoid. . . . Place the brain in chromic and bichromate fluid. Change at the end of eighteen hours, and then once a week, until the brain is hard and tough.

Rutherford provided no images of prepared specimens, and this was certainly one of the criticisms of the book when it was reviewed in the *Monthly Microscopical Journal*.[32] What he does offer, however, are pictures of instruments (microscopes, microtomes, hot stages for maintaining temperatures, etc.) and in the first section diagrams related to the optics of microscope lenses. Words, then, are meant to conjure up the things in one's mind, and the text is organized explicitly as a registry of body parts and how they are to be prepared ("the tongue, when cut into small pieces, may be hardened in chromic acid and then in alcohol.") and then what should be viewed ("carefully examine the pulp close to the dentine for odontoblasts.")[33] The table of contents reveals, in a way found in most textbooks of the period, chapters ordered by cells, blood, glands, muscle nerve fibers, urinary deposits, sputum, organs, and so on. The book, an aggregate of knowledge arranged as a body in pieces, becomes a blazon, perhaps a perverse one, but still "a complete inventory" of a total body (now in pieces).[34] This pedagogical blazon, as laid out by the histology and physiology textbook, is, of course, a literary genre as well, one that we might suggest courts an anti-aesthetic intended to be a numbing anesthetic to violence against animal—nonhuman—bodies.[35]

Certainly the brutality of the laboratory lessons was a concern of the antivivisectionists. By the early 1870s the British Association of Advancement of Science (BAAS) felt sufficiently compelled by growing complaints to establish guidelines for animal testing, including the use of anesthetics on laboratory animals before beginning an experiment or if the creature was to be killed. In spite of the BAAS guidelines, however, the antivivisectionists continued their protests, and in 1876 Parliament enacted a law requiring all researchers who wished to work with animals to apply for a license. Restrictions were placed as well on the use of animals in teaching laboratories. Anesthetics absolutely had to be used when it was not a matter of creating new knowledge but merely an act of experimental replication for demonstration, which would certainly be the case in the classroom.[36]

In recognition of these guidelines and laws, Lionel Beale may have advocated taking out the lives of more and more frogs as a method of comparison, but he warned that to do this "the observer must . . . study in France or Germany, as in England all investigations upon living frogs are prevented by law, and the police have authority to seize any one detected in causing pain to a frog arranged for the purposes of scientific investigation."[37] In fact, this is just what Rutherford himself had done as a young scientist-investigator, not with frogs but with dogs. In undertaking research on the effects of drugs on bile secretion, work he performed in Edinburgh in 1875 and without providing anesthetics to the experimental dogs, the licensing board sanctioned him. His solution was to continue his experiments in France.[38] This was just a continuation of the differences, spanning the century, between British and Continental concerns over animals used for vivisection experiments. In *Middlemarch*, Eliot notes that British scientists and physicians understood the possibilities on the Continent, as opposed to on English soil, when she describes the actions of young Dr. Lydgate while a student of physiology in France: "One evening, tired with his experimenting, and not able to elicit the facts he needed, he left his frogs and rabbits to some repose under their trying and mysterious dispensation of unexplained shocks, and went to finish his evening at the theatre of the Porte Saint Martin."[39] That Lydgate so easily abandons his frogs and rabbits in their discomfort allows one to ponder the levels of sensitivity of both animals and experimenter. It does seem quite clear from the novel, however, that Lydgate does not repeat this work once he is back in Great Britain and practicing medicine in Middlemarch.

Of course, there was resistance in England to the antivivisectionist cause. This can be heard emphatically in Thomas Henry Huxley's 1877 speech to the Domestic Economy Congress in Birmingham. Huxley, comparative anatomist, professor of Natural History and fervent Darwinist, would play a key role in the formation of science education in England during the second half of the nineteenth century, but he himself never undertook any vivisectionist experimentation and admitted publically that he didn't have the stomach for such activity. He did, however, use the occasion of the Birmingham gathering to decry the 1876 Act of Parliament regarding the use of animals in science, denouncing such a law that would allow a boy to use live frog bait to troll for pike while punishing the boy's teacher for wishing to utilize a frog for "exhibiting one of the most beautiful and instructive physiological spectacles, the circulation in the web of the foot." Oh no, he goes on, "you must not inflict the least pain on a vertebrated animal for scientific purposes (though you may do a good deal in that way for gain or for sport) without due license of the Secretary for the Home Department."[40] As for defending such experimentation, it was often in vain, but researchers tried again and again to convince their detractors that such work advanced knowledge and medical progress in the service of the public's health, pointing to successes such as Louis Pasteur's work on rabies and Robert Koch's on anthrax and tuberculosis.[41]

There were certainly other challenges, not least of all ethical ones, in addressing the problem of anesthetics and the inert body of the experimental animal. Acknowledging BAAS rules, Rutherford is careful to point out when describing how best to set up a view of the blood vessels and circulation in the mesentery of a frog that "the student is not permitted to do the following experiment, but it may be shown to him by his teacher thus." He then explained to his readers "the curarised frog is first stunned by a blow on the head so that no pain can possibly be produced." Once this is done "a vertical incision is made on the left side of the abdomen . . . the small intestine is gently drawn out with forceps, and fixed with small pins to a crescent of cork close to a window in the cardboard."[42] It should be noted, anesthetically speaking, that it is deceptive to suggest that it is at all compassionate to inflict a blow on a curarized frog. Curare was known to paralyze, but it did not numb. Curare left the nerves alive and sensitive. Rutherford, less willing to provide pictures of his proposed educational activities, publishes only an empty frog plate — no frog included.

Images of Naturalistic Animals

In *How to Work with the Microscope* Lionel Beale included the image of a dissected animal with its body left fairly intact (figure 2.2). A rather small picture, the reader encounters a newt with its abdomen sliced and pulled open with the intention of examining cilia in motion on the upper portion of the kidney. Pinned by paws and tail, and the skin folded away, the animal is represented with simple hatching and shadowing. This splayed body is not intended to represent a whole intact animal so much as to suggest the body as background, or visceral context, for the highlighted ana-

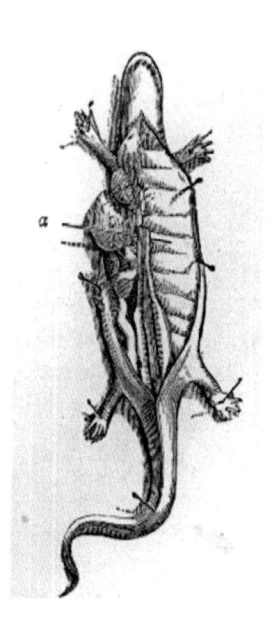

2.2 Newt dissected to show kidneys. From Lionel Beale, *How to Work with the Microscope* (London: Harrison, 1880).

tomical detail. In this case, Beale, who is less meticulous about discussing anesthetics, does note in the text that the creature should be decapitated before opening up its body. In the image, the head is meant to be missing, although this does not seem so clear pictorially. At the top of the newt's body, the reader finds a smooth lozenge or bullet-like shape just at the point where a face and head should be. What we see might represent the neck and underside of a chin. In any case, this space for a face is blank, merely the untouched white of the page framed by a bit of hatching.

Why remove a head, anyway? In his essay, "Does a Frog Have a Soul?" Thomas Henry Huxley discusses attempts to dismember the frog in search of a soul, and in doing so introduces evidence from various experiments in which a frog's head is first removed so that the experimenter can test the leg's ability to react by applying some sort of irritant. In these experiments, the leg on the headless body invariably contracted. In turn, Huxley emphasized, the separated head "will show signs of retaining all its nervous energies." He comes to comparing the acts of the headless frog body to what is known about human pathology and damage to the spinal cord for a moment, but then turns to philosophical debate regarding mechanism and animal behavior, leaving the animal-human connection behind. Thus, he declines to discuss "the question whether the soul of the frog possesses consciousness," insisting "this appears to me to be a totally insoluble problem."[43] Descartes's ghost hovers. So, what does this suggest about the sensing capabilities of the frog, in particular, its experience of pain in life and death? George Eliot's partner, the physiologist George Lewes believed "so far from Pain being common to all animals, it is, on the contrary, the consequence of a very high degree of specialization, and is only met with in animals of complex organizations. It is probable that reptiles have only a very slight capacity for pain, and animals lower than fish none at all." Lewes employed similar experimental evidence as Huxley, that is, that "a decapitated frog manifests the same movements of self-preservation as it manifested when its head was on."[44] It is a mechanical response. In the midst of tortuous experiments, then, he suggests "shrinking, struggling, crying, etc. are no certain indications of pain."[45] Charles Darwin, addressing the same experiments in his 1872 treatise on the expression of emotions in humans and animals, quite plainly asserts that the decapitated frog "cannot of course feel, and cannot consciously perform, any movement," even if the headless body will still react when an irritating acid is applied to its foot.[46]

It has been noted that bodies, and this means heads and faces as well, were not illustrated in the British manuals.[47] In the above discussion, of course, the sentient object in the head is the brain, but it is also to be noted that the sense organs—eyes, ears, nose, mouth—are situated on the front of the head and are the main features of what has come to be called the face. With this in mind, it is significant to note that physiology textbooks did, indeed, emphasize, as a group of lessons, the study of just these organs (eyes, ears, nose, tongue). In an 1887 edition of *A Manual of Physiology for the Use of Junior Students of Medicine* the physiology of face is addressed by dismantling one. "Cut off the head of a newt," the student is told, then "remove the lower jaw, cut off the nose by an incision carried just in front of the eyes, with a fine pair of scissors remove the roof of the nasal cavities, and place the remainder of the nose in osmic acid." This nose will be taken apart and put under the microscope, with preparations to view the nasal membranes emphasized. Included in the lesson on the senses of smell and taste is a series of experiments on a live organism, this being the student himself, who is instructed to pinch his own nose to cut off any odors and then attempt to distinguish two drastically different flavors, an apple and an onion, perhaps.[48] The intact human face is brought into pedagogical connection with the nonanimal face in pieces.

At the conclusion, the final section of material to be studied in this particular physiology textbook is the development of the human embryo's skull and face, with the description of a face unfolding, nose and ears arising, "in connection with ridges known as visceral folds or arches." Here

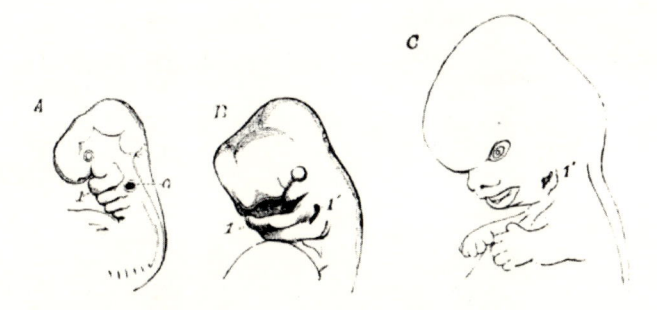

2.3 Different stages of development (four to nine weeks) of the head and face of a human embryo. In G. F. Yeo, *A Manual of Physiology for the Use of Junior Students of Medicine* (London: JA Churchill, 1887), 639, figure 300.

the student does encounter the image of a face. In presenting the "different stages of the development of the head and face of a human embryo" the image, exemplifying nine-weeks' gestation, clearly presents a familiar set of general facial characteristics—eye, forehead, nose, mouth—of a developing human (figure 2.3).[49] It is the only face found in the book, and it is there to illustrate the development of the human sensorium, the organs representing the promise of an ability to connect aesthetically (in its original sense noted above) to the world.

Now I would like to look outside England and consider a publication by French physiologist Louis-Antoine Ranvier to find slightly more explicit hints at a nonhuman animal face. Ranvier ran a private laboratory in Paris, along with a research associate, Victor Andre Cornil, during the 1850s, where they also taught histology courses to medical students. They also coauthored a histopathology textbook, published in 1869. In the mid-1870s Ranvier, who had since worked in the laboratory of the eminent physiologist Claude Bernard, moved on to assemble his own manual, *Traité technique d'histologie*. In this text the author provides various images of animals in the service of experiment, including a print of a frog with a slit in its abdomen out of which a bit of intestine has been removed (figure 2.4). (This is the same experiment that Rutherford described but did not fully illustrate, opting to show just the empty plate.) Here, the reader is offered both a view of the whole animal, as a background body "contextualizing" the organ on display, and then the display itself, a close-up of the cork plate upon which the intestine lies. This bit of interior organ will be placed under the microscope for examination, in this case, of blood vessels and circulation.

60 · *The Educated Eye*

2.4 Apparatus for observing mesentery in living frog. Louis-Antoine Ranvier, *Traité technique d'histologie* (Paris: F. Savy, 1875).

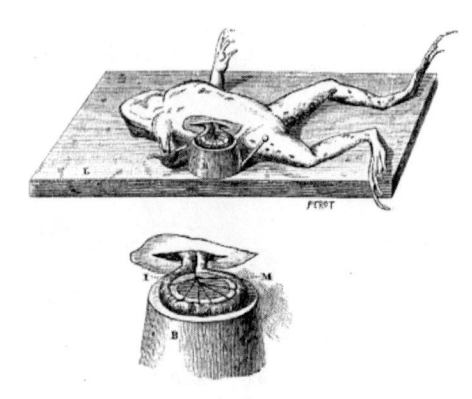

2.5 Apparatus for observing circulation in a frog's lung. Louis-Antoine Ranvier, *Traité technique d'histologie* (Paris: F. Savy, 1875).

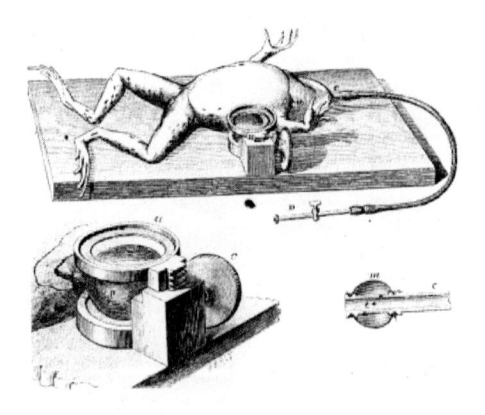

The image itself is composed of clean lines, strong contours, and just enough shadow to suggest a realistic portrayal. Yet the draftsman has also incorporated a considerable amount of abstraction. A white belly with a few hatching marks rises off the body. The frog's head is pulled back, showing a smooth throat blending into the chest and abdomen. Hardly a hint of a face remains, except for the small bulge that we recognize as an eye in shadow. We could see a live animal emerge, however, in the digits of the limb, caught in the still image, spread wide as if clutching at empty air—a vestige of the naturalistic animal's liveness.

What might be called a pendant to this image is a second image of a frog, this one with its face turned entirely away (figure 2.5). The apparatus dominates the image in a more complete way than in the previous figure, and it seems in some details that parts of the body are entirely exchanged for technological devices. Where the head might be is a tube, which the text explains has been inserted in the larynx to facilitate artificial respiration so that the researcher can keep the lung inflated during the micro-

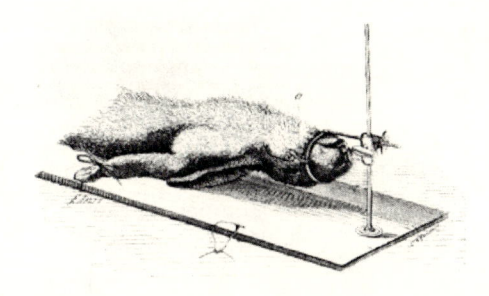

2.6 Apparatus for immobilizing rabbits. Louis-Antoine Ranvier, *Traité technique d'histologie* (Paris: F. Savy, 1875).

scopic examination. A plate holding a portion of the creature's lung partly obscures the wound out of which the organ has been drawn. The torso remains a blank white sphere (no hatching at all) and the two legs splay off symmetrically creating an angular diamond shape that contrasts with the round shapes of the viewing apparatus and the frog's own belly. Again, it is the silhouette of a raised paw that carries the weight of animation, with digits that seem to grasp anxiously at the white space between image and text.

As noted, while in France Eliot's Mr. Lydgate worked with rabbits as well as frogs, and so it is not surprising that mammals, too, make an appearance in Ranvier's text. In fact, he offers the image of a rabbit, albeit only the upper part with chest and head, clamped to an ingenious contraption, secured with hooks and bows of string or wire and an iron bit around the muzzle (figure 2.6). This animal has been made ready to participate in experimentation and pedagogical demonstration, certainly, but the illustration also offers a strange engagement between human experimenter and experimental animal. The draftsman and printer have shown great attention to detail here: shadowing has been depicted, the wood grain has been etched with detailed lines, and the individual hairs of the fur stand up off the chest. In this instance, the student is offered more than a hint at this animal's face—a nose; a mouth; a pinned-back ear; and, not least of all, a singular staring eye emerging out of dark shadow like an obscure black hole. It is an eye, however, that shows the barest, if any, sign of life. The fur rising from the chest, a field of wiry lines, appears more animated. With wire surrounding its snout and the face proper nearly lost to dark shadow, we can recognize the animal's succumbing to the facelessness of the tortured.

Still, this rabbit's face allows us to consider the possibility of a face-to-face encounter between student and animal in the actual classroom, and how it might be addressed pedagogically, perhaps in a text. Animals brought into the physiology laboratory, or as Claude Bernard called it, the "ghastly kitchen," that is, the space of their almost sure demise, all came with faces.[50] In human terms, we might return to Emmanuel Levinas, who finds the ethics of life and death crystallized in the face, for "the face is what one cannot kill, or, at least it is that whose meaning consists in saying: 'thou shalt not kill.'"[51] Is it that these face-to-face encounters engender a certain relationship, a sympathy, even empathy? If so, we can wonder how the rabbit's eye could possibly hinder the student getting psychologically past the violent acts committed for the physiology or histology laboratory lesson.

The student himself must be disciplined in order to engage repeatedly in such acts of discipline. After all, the classroom is a place of discipline. Bruno Latour commented specifically on the nineteenth-century physiology laboratory as a cultural training ground, as a space for disciplining a variety of bodies and tools, claiming that any consideration of this should "bear on strategic ways of disciplining-animals-to-disciplining-colleagues-to-disciplining students-to disciplining-instruments." Here, indeed, he insists, is how the "strategic path of the laboratory will appear."[52] Then, what of the face?

Conclusion: The Face

Deleuze and Guattari noted that the face is produced "only when the head ceases to be a part of the body; and only when the body, head included, has been decoded and has to be overcoded by something we shall call the Face." Still, crowded on that (sur)face are body parts—mouth, eyes, nose—the sense organs of cognition, and now reflectors of thoughts and feelings, and emotions, and as we have seen, organs for dissection and study in the medical school classroom. This thing we call a face Deleuze has also described, then, as an "organ-carrying plate of nerves." With these parts, these organs and nerves, feelings, thoughts, and emotions ("quivering lips, there a brilliance of look") are charted, always changing and making up an "intensive series of movements tending towards a critical instant," which culminates in, for example, joy, pain, desire, surprise, fear, paroxysm.[53] In its expressive role, Deleuze and Guattari assert, the face takes on three roles: individuating, socializing, and communicating. The face is the prime identifier of its owner in a way no other actual body part is. It is that aspect of the body that has initial social contact with another (meeting face-to-face). Through its automatic acts of expression (raised eyebrow, flared nostril) the face is also the main conduit of visual and aural communication, that is, affective communication with other subjects. Thus, the face both transcends the body and—being made up of sense organs—is locked into the body's very material relationships to interior emotion and physical sensation, and it is also the body's cognitive interface with the exterior world. The question is whether the nonhuman "face" does the same.

Analogically, though, it is possible to take what Deleuze and Guattari claim are the two basic components of the face, the white wall of signification and the black hole of subjectification, and imagine their actions across the faces of all animals. With this white wall/black hole system, the corporeal head, human or animal, is de-territorialized from the body and put to work circulating in alternate hierarchies, strata of the sign (signification) and of the world of the conscious (subjectification): it "surpasses the body." Imagine the black hole of consciousness, the black hole of the rabbit's eye, a face of a prisoner or one who is tortured, a face intended to be hidden or lost or denied, "facing" its torturer, its "despot." But the despot must be conditioned not to see this other face: it is to be effaced.

With this, I would like to end by turning to two faces together, that of the experimenter and that of the experimental animal. To do so, I will return to Michael Lynch's discussion of the "naturalistic animal," hovering, as it does in his interpretation, over the bits of tissue and organs, the analyzed animals, of laboratory work. Lynch himself is explicitly aware of the presentation of an amorality accompanied by a necessary desensitization of the scientists in their laboratory work on animals. He juxtaposes this to commonsense relationships between humans and animals (e.g., owners and their dogs, birds, and cats), which recognize an emotional life of nonhuman beings, not least of all in pets and higher mammals. (Note: we need only look to Darwin's *The Expression of the Emotions in Man and Animals* to see that Victorian society certainly recognized emotional life in animals, at least their pets.)[54]

To make his point, Lynch evokes a scene from Frederick Wiseman's remarkable documentary *Primate*, a film exploring the world of scientific research on animals. We have already encountered observations of headless frogs, and now there is the headless mammal. In his film Wiseman asks viewers to contemplate the decapitation of a rhesus monkey in preparation for dissecting its brain. As Lynch describes it, "the camera lingers on the human-like features of the head as it rests on the bench while a scientist casually discusses the procedure."[55] In turn, he then observes that "the 'dead-pan' facial expressions, mechanical gestures, and 'flat' tonal qualities of the scientist's voice projects an indifference to the gory scene that the camera makes explicit." Our scientist has learned his lesson about coping with life, death, pain, and the experimental "Other" animal. In the end, these two faces match each other in their lack of emotion (one dead, one numbed to violence); they are no longer individualized—experimenter and experimental animal—but their faces seem related in that both their gazes are unseeing eyes, black holes, held there, projected there on the white screen.

Notes

1. Christopher Lawrence, "Incommunicable Knowledge: Science, Technology and Clinical Art in Britain, 1850–1914," *Journal of Contemporary History* 20 (1985): 503–20.
2. Ibid., 503.
3. William D. French, *Antivivisection and Medical Science in Victorian Society* (Princeton, NJ: Princeton University Press, 1975), 296.
4. George Eliot, *Middlemarch* (Oxford: Oxford University Press, 1997), 138–39.
5. See Stephen Jacyna, "A Host of Experienced Microscopists: The Establishment of Histology in Nineteenth-Century Edinburgh," *Bulletin for the History of Medicine* 75 (2001): 226–27.
6. Lorraine Daston and Peter Galison, *Objectivity* (Cambridge, MA: MIT Press, 2007), 326.
7. Joseph Lister, "Observations on the Muscular Tissues of the Skin," *Quarterly Journal of Microscopical Science* 1 (October 1853): 262–68.
8. Patricia Helen Bracegirdle, "The Establishment of Histology in the Curriculum of the London Medical Schools, 1826–1886" (PhD dissertation, University College, London, 1996), 302–74.
9. Ibid., 355 n245.
10. Gilles Deleuze, *Cinema 1: The Movement Image*, trans. Hugh Tomlinson and Barbara Habberjam (Minneapolis: University of Minnesota Press, 1986), 87–88. Also see: Gilles Deleuze and Félix Guattari, "Year Zero: Faciality," in *A Thousand Plateaus: Capitalism and Schizophrenia*, trans. Brian Massumi (Minneapolis: University of Minnesota Press, 1987), 167–91.
11. Deleuze and Guattari, *A Thousand Plateaus*, 115–16.
12. Emmanuel Levinas, "The Paradox of Morality," in *Animal Philosophy: Essential Readings in Continental Thought*, ed. Matthew Calarco and Peter Atterton (New York: Continuum, 2004), 49.
13. Tom Regan, *The Case for Animal Rights* (Berkeley: University of California Press, [1983] 2004), 21–25.
14. Jacques Derrida, *The Animal That Therefore I Am* (New York: Fordham University Press, 2008), 7.

15. Matthew Calarco, *Zoographies: The Question of the Animal from Heidegger to Derrida* (New York: Columbia University Press, 2008), 63.

16. Lionel Beale, *The Microscope and Its Application to Clinical Medicine* (London: Samuel Highley, 1854), 2.

17. Lionel Beale, *The Medical Student, a Student in Science: The Introductory Lecture Delivered at the Opening of the Twenty-fourth Session of the Medical Department of King's College, London, October 1, 1855* (London: J Churchill, 1955).

18. Lionel Beale, "Advertisement," *Archives of Medicine* (London: John Churchill, 1857), 3.

19. Lionel Beale, *How to Work with the Microscope* (London: Harrison, 1880), 367. Beale also informs his readers that "swinging very rapidly through the air also destroys life very suddenly, without causing the contraction of the muscles, which seriously interferes with the preparation of successful injections" (122).

20. Ibid., 367–68. Beale uses the term "bioplast" to refer to the biological cell.

21. Beale, *Archives of Medicine*, 2.

22. Beale, *How to Work with the Microscope*, n.p. Students were strongly encouraged, as occurred in most classes, to make drawings themselves after what they viewed in the microscope.

23. Michael Lynch, "Sacrifice and the Transformation of the Animal Body into a Scientific Object: Laboratory Culture and Ritual Practice in the Neurosciences," *Social Studies of Science* 18 (1988), 267.

24. Claude Bernard, *An Introduction to the Study of Experimental Medicine*, trans. Henry Copley Green (New York: Collier Books, 1961), 111.

25. Stewart Richards. "Aneasthetics, Ethics, and Aesthetics: Vivisection in the Late Nineteenth-Century British Laboratory," in *The Laboratory Revolution in Medicine*, ed. Andrew Cunningham and Perry Williams (Cambridge: Cambridge University Press, 1992), 142–69. Also see, Nicholas Jardine, "The Laboratory Revolution in Medicine as Rhetorical and Aesthetic Accomplishment," in Cunningham and Williams, *The Laboratory Revolution in Medicine*, 307.

26. Richard D. French, *Antivivisection and Medical Science in Victorian Society* (Princeton, NJ: Princeton University Press, 1975), 48.

27. Paul White, "The Experimental Animal in Victorian Britain," in *Thinking with Animals: New Perspectives in Anthropomorphism*, ed. Lorraine Daston and Gregg Mitman (New York: Columbia University Press, 2005), 75.

28. Lewis Carroll, "Some Popular Fallacies about Vivisection," *Fortnightly Review* 23 (1875): 854. Quoted in French, *Antivivisection*, 303.

29. Susan Buck-Morss, "Aesthetics and Anaesthetics: Walter Benjamin's Artwork Essay Reconsidered," *October* 62 (Fall 1992): 9–10.

30. Ibid., 27–28.

31. The following description in full is found in William Rutherford, *Outlines of Practical Histology* (London: J. and A. Churchill, 1876), 4–6.

32. *The Monthly Microscopical Journal* 14 (1875): 287–88.

33. Rutherford, *Outlines of Practical Histology*, 112, 90.

34. Barthes, "The Blason," in *S/Z* (New York: Hill and Wang, 1974), 120.

35. Jonathan Sawyer, in his study of the Renaissance "emblazoned" body, considers how such attempts at stark anatomical realism make this literary endeavor a "riot of aesthetic and scientific exploration." He suggests that the blazon and anatomy text "both sought to gaze upon the body which they dismembered, piece by piece." See Jonathan Sawyer, *The Body Emblazoned: Dissection and the Human Body in Renaissance Culture* (London: Routledge, 1995), 195–98.

36. French, *Antivivisection*, 187–91. In these early years of the guidelines, many experimental practitioners were denied licenses, including Rutherford who was turned down in 1877 and in 1880.

37. Beale, *How to Work with the Microscope*, 373.

38. Richards, "Anesthetics, Ethics," 153; Richards, "Drawing the Life-Blood of Physiology: Vivisection and the Physiologists' Dilemma, 1870–1890," *Annals of Science* 43 (1986): 27–56; Richards, "'Conan Doyle's 'Challenger' Unchampioned: William Rutherford, FRS (1839–99), and the Origins of Practical Physiology in Britain," *Notes and Records of the Royal Society* 40 (1986): 193–214; French, *Antivivisection*, 187.

39. Eliot, *Middlemarch*, 141.

40. T. H. Huxley, "On Elementary Instruction in Physiology," *Science and Culture, and Other Essays* (New York, 1890), 100–101.

41. French, *Antivivisection*.

42. Rutherford, *Outlines of Practical Histology*, 104.

43. Thomas Huxley, "Has a Frog Got a Soul; and of What Nature Is That Soul, Supposing It to Exist?" *Metaphysical Society Papers*, 2 vols., Bodleian Library, Oxford, 2657c.1.

44. George Lewes, *Sea-Side Studies at Ilfracombe, Tenby, The Scilly Isles, & Jersey* (Edinburgh: William Blackwood and Sons, 1858), 329–34.

45. Lewes, *Sea-Side Studies*, 335. "Even among humans, however, the degree of susceptibility to pain differs. For example, "it is much less in savages than in highly-civilized men, as it seems also to be less in wild animals than in the domesticated . . . less in men leading an active out-of-door life than in those leading a sedentary intellectual life; less in women than in men; less in persons of lymphatic than in person of nervous temperaments." Martin Pernick has discussed the mid-nineteenth century belief that the ability to experience pain was related to intelligence, memory and rationality, and so "like the lower animals, the very young [infants] lacked the mental capacity to suffer." Pernick, *A Calculus of Suffering" Pain, Professionalism and Anesthesia in America* (New York: Columbia University Press, 1985), 150. Also see Lucy Bending, *The Representation of Bodily Pain in Late Nineteenth-Century English Culture* (New York: Oxford University Press, 2000), 123.

46. Charles Darwin, *The Expression of the Emotions in Man and Animals*, 3rd ed. (Oxford: Oxford University Press, 1998), 42.

47. Paul White notes this in White, "The Experimental Animal." Also, Jed Mayer makes the point that British antivivisectionists would reprint images of vivisected animals found in Continental publications because they were more

graphic than representations found in British texts. Jed Mayer, "The Expression of the Emotions in Man and Laboratory Animals," *Victorian Studies* 50, no. 3 (Spring 2008): 414.

48. Gerald F. Yeo, *A Manual of Physiology for the Use of Junior Students of Medicine* (London: J. and A. Churchill, 1887), 260–61.

49. Ibid., 639.

50. Claude Bernard, *An Introduction to the Study of Experimental Medicine*, trans. Henry Copley Greene (New York: Henry Schuman, 1949), 15. "If a comparison were required to express my idea of the science of life, I should say that it is a superb and dazzlingly lighted hall which may be reached only by passing through a long and ghastly kitchen."

51. Emmanuel Levinas, *Ethics and Infinity: Conversations with Philippe Nemo*, trans. Richard Cohen (Pittsburgh, PA: Duquesne University Press 1985), 87.

52. Bruno Latour, "The Costly Ghastly Kitchen," in Cunningham and Williams, *The Laboratory Revolution in Medicine*, 301–2.

53. Deleuze, *Cinema 1*, 87–88. Also see: Deleuze and Guattari, "Year Zero."

54. In particular, Jed Mayer picks up on this. Darwin's volume, Mayer notes, was published in 1872 at a moment of high Victorian protest against the idea that his evolutionary theory proposed humans as mere "animal." His suggestion is that Darwin intended *The Expression of the Emotions in Man and Animals* to emphasize the primacy of our (human) emotional connections to animals as a way of persuading readers of our evolutionary connections." Mayer, "The Expression of the Emotions," 402 n49.

55. Lynch, "Sacrifice and the Transformation," 267–68.

chapter three

Photography and Medical Observation

Scott Curtis

In his 1865 landmark book on experiment in the medical sciences, French physiologist Claude Bernard writes,

> Observers, we said, purely and simply note the phenomena before their eyes. They must be anxious only to forearm themselves against errors of observation which might make them incompletely see or poorly define a phenomenon. To this end they use every instrument, which may help make their observations more complete. Observers, then, must be photographers of phenomena; their observations must accurately represent nature. We must observe without any preconceived idea; the observer's mind must be passive, that is, must hold its peace; it listens to nature and writes at nature's dictation.[1]

By explicitly comparing observation and photography, Bernard not only champions photography as an exemplary scientific instrument, he also seems to make claims about what observation should be. Above all, according to this passage, it should be passive. Observation should not intervene, like experiment; instead, it should merely record, describe, even transcribe "nature's dictation." It should watch or listen intently, silently, and, it is implied, at one's leisure. Instruments such as photography are used to insure this state of pure receptivity. They help the scientists arm themselves against the error of the "preconceived idea," which might color the true picture of nature with the taint of human bias. Of all the instruments one might use, Bernard holds up photography as the one that best represents the ideals of scientific observation.

Closer examination of this connection between photography and observation, however, reveals that it is much more complex, especially with regard to "passivity." Even Bernard insists that there is no such thing as purely passive observation; his own characterization, he admits, is a heuristic conceit that breaks down in practice: "At first sight . . . this distinction between the experimenter's activity and the observer's passivity seems plain and easy to establish. But as soon as we come down to experimental practice we find that, in many instances, the separation is very hard to make, and that it sometimes even involves obscurity."[2] The dichotomy between "active" experimentation and "passive" observation seems to be a by-product of the nineteenth-century concern with objectivity and subjectivity and the subsequent attempt to distinguish collective scientific practice from individual theorizing.[3] Before the nineteenth century, at least since ancient Greece, observation was considered a very active process.[4] Since Bernard's time, however, philosophers of science have often demoted observation to a secondary, passive role to experiment.

Yet the history of medical observation resists this distinction between activity and passivity, even as Bernard himself struggled with it. Perhaps the most thorough discussion of medical observation is Michel Foucault's archeology of "the medical gaze." Foucault emphasizes medical perception's relationship to analysis as a process of simultaneously recognizing, separating, naming, and acting upon some disease element. For Foucault, observation is inseparable from the analytic function of diagnosis.[5] Michael Hau, on the other hand, describes medical observation as an active, Gestalt-like process of holistic apprehension; he argues that many German physicians at the turn of the century objected to the analytic, overly scientific approach to observation that had become fashionable in the medical community.[6] Even many contemporary philosophers, historians, and sociologists of science who have discussed observation, such as Norwood Russell Hanson, Dudley Shapere, Ian Hacking, Patrick Heelan, Trevor Pinch, Roberto Torretti, and David Gooding, have concluded that it is an incredibly complex, theory-laden procedure that is, above all, highly mediated by instrumentation and interpretation.[7]

Perhaps part of the lasting conceptual appeal of "passive" observation is due to Bernard's own metaphorical comparison (which was not uncommon) between ideal human perception in a scientific context and photography as a device that merely records what is in front of it. Indeed, most histories of objectivity and of the role of photography in science focus on the camera's perceived ability to document phenomena "objectively."[8] That is, they focus on the camera's privileged relationship to the world it documents—its ability to "automatically" (hence "passively") capture and fix phenomena at a particular moment in time and space, seemingly independently of human agency.[9] This aspect of photography has received the most attention and has certainly helped to shape conceptions of what scientific observation is and should be (and, in turn, has been shaped by these very ideals).

But this is not the only salient feature of photography, nor does this particular ability exhaust its utility in science. Instead, I propose we examine another pertinent relationship, namely between observer and image. Thinking about this relationship—as opposed to the relationship between camera and object—highlights certain features of observation that might otherwise go unremarked. Nineteenth-century discussions of medical observation, for example, insisted on careful, methodical observation of patients. These discussions therefore emphasized two important elements of observation. First, all observation had an important *temporal* dimension. An observation can be made quickly, to be sure, but over and over physicians warned their students against the dangers of hasty observation. Instead, observation was to be practiced carefully, with an eye to detail. Indeed, detail itself precludes hasty observation; if one is to attend to and assimilate the details, it takes time to do so. Of course, this implied that one had the luxury of time, that the researcher could proceed at an unhurried, even leisurely, pace, a possibility that separated the photograph from, say, live examination in the moment or even from the motion picture, the rush of which insistently pushes the viewer along. This luxury of time also separated the cultured, leisured class from the other classes. There was, paradoxically, an element of leisure in the labor of observation.

Second, observation seems to be a process of *correlation*, by which I mean establishing an often mutual or reciprocal relationship between objects or events, such as correlating lesions in a cadaver to disease elements in a living patient. Observation is an active process of comparison between the phenomenon before one's eyes and the knowledge one brings to it. Torretti calls it "the principle of the conceptual grasp": the observer grasps the object as a particular instance of some universal.[10] This involves a constant process of comparison between a variety of different elements past and present (Bernard called this "experimental reasoning"). Discovery depends on the ability to see new patterns from familiar data; Aristotle knew that distinguishing universals from particulars depended on habits of perception, which develop into memories and then into experience with patterns.[11] Medical observation in particular depended heavily on seeing symptoms or patients as part of a series and comparing different elements of that series.

My emphasis on pattern recognition and the duration of the observational act is meant also to tease out the aesthetic qualities of scientific observation. By this I mean not only that scientific observation, with its contemplative and admiring stance toward the beauty of nature, is a practice very much akin to art appreciation. I also mean that there is a marked cultural investment in the expertise that comes with a trained, scientific eye. Skillful, accurate observation is a mark of learning; physicians take pride in their ability to see patterns and details that are not available to laymen. For most physicians during the nineteenth century, when the legitimacy of the medical profession was only emerging, their emotional investment in this mode of viewing was palpable. As a form of cultural capital, scientific observation functions very much like the cultivation of distinctive aesthetic taste.

Duration and correlation were not the only characteristics of medical observation, of course, but I focus on these aspects because they corresponded to certain formal features of photography. Like Bernard, I will compare photography and observation, but in a descriptive, rather than normative sense. I am interested in the reciprocal relationship between the two, especially how the photograph facilitated, encouraged, and amplified certain patterns of observation that were emerging in nineteenth-century medicine. Certain observational strategies seemed to find purchase in aspects of the photographic image. Or, to put it another way, the training in observational methods that physicians underwent—whether "analytic" or "holistic"—found in photography an amiable partner. If these methods—careful attention, accurate description, and correlation across cases—were already in place before or without photography, with the rise of photography they could be applied to a "working object" in ways similar to how they were applied to natural phenomena. This could also be said of other representational technologies; the moulage, or wax model, for example, was very helpful for studying dermatological cases.[12] But photography had advantages that the moulage did not, most notably the relative ease with which one could create a series of images, what we could call its "repeatability." But it also had features in common with the moulage, especially the rich texture and detail of the image. No single feature makes photography unique. But I will argue that the combination of repeatability and detail made it an especially agreeable, even privileged mode of representation for nineteenth-century medicine.

So this essay has two goals: to offer a brief survey of the types, uses, and venues of medical photography in the nineteenth century, and to suggest some connections between photography and emerging practices of observation in medicine during this time. Specifically, I hope to show that the temporal and correlative aspects of medical observation correspond to specific features of photography, namely the *detail* of the photographic image, and its *repeatability*. To do this, I will chart in a preliminary way the advantages and applications of photography in medicine and place them in the context of discussions about the methods and goals of medical observation in the nineteenth century. What was medical photography during the nineteenth century? How was it used? What hopes did its proponents have for it, and how did these hopes express the needs of the discipline? How did the use of photography conform to ideals of observation and vice versa? First, however, it would be useful to describe the types, venues, and uses of nineteenth-century medical photography.

Varieties of Medical Photography

Over the course of the nineteenth century, photography became an increasingly important medium for medical illustration. By 1894 it was even possible to find an article complaining of "The Craze for Photography in Medical Illustration," which serves as a grumpy indication of the discipline's growing investment in the technology.[13] Even though it still competed with drawings and engravings, photography eventually became the default mode of representation. Among the thousands of medical photographs created in Europe and the United States in the nineteenth century, we can discern a variety of categories, which I will distill into three broad genres: photos of visible public spaces, photographs of hidden private spaces, and portraits.[14] Depictions of public, medical spaces such as hospitals, battlefield hospitals, sanatoria, and asylums, were common. Photographs of hospitals were designed to document the setting, but also to assure a skeptical public of the facility's charity and cleanliness. Likewise, photographs that focus on public health by depicting the state of public houses or streets tried to emphasize the need for or success of systematic sanitation measures. Photographs of educational settings such as lecture halls and dissection rooms were also common, but those of operating rooms and actual surgeries were rather rare until the 1890s, perhaps because surgery is a messy business. But as conventions developed for the depiction of the surgical space as a private, immaculate, and technologically modern arena, such photographs became more commonplace.[15]

Photographs of public and semipublic places emphasized the visible world, but a significant portion of photography for medical purposes also documented the hidden space of the human body. We can divide this genre by technology: microscopic, endoscopic, and radiological. Among the earliest and most prominent examples of microphotography are the illustrations by Alfred Donné (1801–1878), a French physician and professor, and his assistant Léon Foucault (1819–1868). Donné presented his microphotographic representations of various bodily secretions to the French Academy of Sciences in 1840 and published his atlas in 1844.[16] Later in the century, German physician Robert Koch (1843–1910) established the criteria for the legitimate photographic representation of microscopic specimens. After Koch's illustrations of the 1870s and 1880s, microphotography became a much more accepted practice in histology and pathology.[17] Endoscopic photography involves using a tubular instrument to visualize the interior of a hollow organ, such as the bladder. Probably the most well-known name in the early history of this technique is Max Nitze (1848–1906), who is credited with developing the first modern practical cytoscope, which was able to magnify and to view the bladder's interior. Between 1891 and 1894, he created a means of photographing the views from his cytoscope and published his results in his urological atlas.[18] Finally, after Wilhelm Roentgen (1845–1923) discovered X-rays in 1895, there were immediate attempts to apply this technology in medicine and to capture the image for medical research and diagnosis.[19] We may also include chronophotography and cinematography in this genre, since they were often used to explore natural phenomena hidden to the naked eye. French physiologist Etienne-Jules Marey (1830–1904) is the iconic figure here; in the 1880s and 1890s, his chronophotographs of human and animal movement transformed ephemeral movement into scientifically acceptable visual evidence, providing the basis upon which motion studies could be counted as a legible and legitimate area of inquiry.[20]

Perhaps the largest category of photographs during this period is the medical portrait, including a surprisingly common type, the portrait of the physician. A photographic portrait taken in the doctor's office or library conferred a sense of dignity and modernity at a time when the general public might have been still quite skeptical of the medical profession. These portraits, as *cartes-des-visite*, were often given to potential clients, sent to colleagues, and exchanged at meetings. Additionally, one finds a large number of commemorative photographs showing groups of physicians at professional meetings, which together certainly served as visual documentation of medicine's growing legitimacy.[21] Indeed, we could even say that photography and the physician reinforced each other's position: the presumed objectivity of the photograph reinforced the desired objectivity

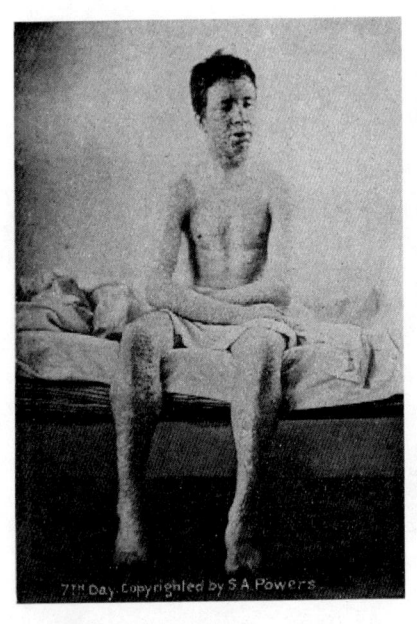

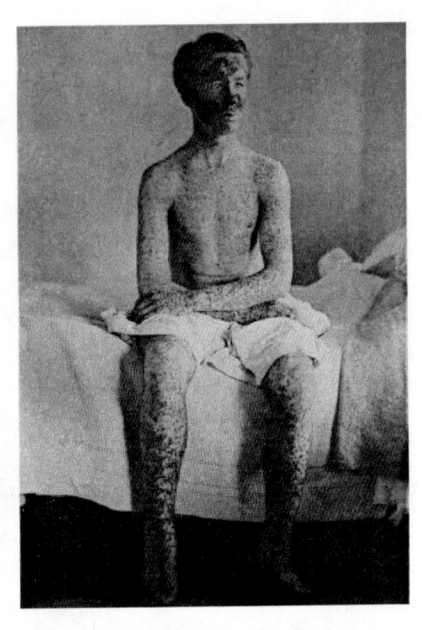

3.1 and 3.2 A typical nineteenth-century clinical portrait, here depicting the progression of smallpox eruptions over the course of days. (The patient survived.) From Samuel A. Powers, *Variola: A Series of Twenty-One Heliotype Plates Illustrating the Progressive Stages of the Eruption* (Boston: Samuel A. Powers, 1882).

of the physician, while the physician's authority underwrote the evidentiary status of the documentary photo.[22]

But when we think of the medical photograph as portrait, it may be the clinical portrait that most likely comes to mind (figures 3.1 and 3.2). Photographs of afflicted individuals were used by a variety of specialties across a range of settings.[23] Early examples of clinical portraits are Hugh W. Diamond's (1809–1886) photographs of inmates at the Surrey County Asylum in the 1850s, used to demonstrate "the physiognomy of insanity."[24] Similarly, Max Leidesdorf (1818–1889) in Germany assembled an atlas of "psychiatric illnesses" in the 1860s,[25] while Jean-Martin Charcot (1825–1893) and Albert Londe (1858–1917) took their famous portraits of patients displaying symptoms of hysteria at Paris's La Salpêtrière hospital beginning in the late 1860s.[26] A. de Montméja (1841–?), a Parisian physician and photographer who worked at the Hôpital Saint-Louis, completed the images for his colleague Alfred Hardy's (1811–1893) photographic atlas of skin diseases in 1868.[27] And so on, throughout the century.[28]

Clinical photographs tend to focus on extreme or abnormal cases, a tendency that corresponds to a general difference between scientific and medical thinking. As Ludwig Fleck has argued, science looks for the typical, normal phenomenon, while medicine attends to precisely the atypical, abnormal, morbid phenomena. This is so because medicine is faced with a huge range of individuality with no clear boundaries between health and illness, so it is only with the morbid case that the physician can clearly see the difference along the continuum. So this is often where medicine starts in defining the normal from the pathological.[29] The photographs we see in medicine generally match this concern, while also mediating between the physician and the patient. Photography thus also participates in a larger trend in Western medicine that segregates illness from the everyday and reinforces the uneven balance of power between patient and physician. We can see this in the uses to which photography was put and the venues in which it appeared.

Uses and Venues

The images under discussion here—such as operating room photos, endoscopic explorations, clinical portraits—were largely confined to expert communities. They were used in medical instruction or circulated in professional gatherings or journals. If photographs were used in public education, they were of a more innocuous type, stripped of their potentially sensationalist subject matter. In fact, before 1885 photographs were used sparsely in public education or media. After the development of instantaneous photography, dry-plate emulsions, and half-tone printing in the last fifteen years of the century, photographs were much more common in print media and public lectures. Medical images were much more common, too, due to the public excitement over Robert Koch's isolation of the tuberculosis bacillus (1882), Louis Pasteur's rabies vaccine (1885), the discovery of X-rays (1895), and other late-century advances in medical research.[30] The growing acceptance of the germ theory of disease meant that photomicroscopic images were more likely to illustrate public lectures on medical topics, while the increasing use of antiseptic techniques provided a new reason to picture and promote hospital facilities—not to mention the thousands of new X-ray images that found their way into every nook and cranny of the public sphere at the turn of the century. But clinical photographs were still mostly for expert eyes only. Indeed, the medical community explicitly frowned upon the public circulation of these images; photographs of the afflicted were painfully similar to purposefully sensationalist postcards of circus freaks, and when they circulated beyond the medical community, they were often used in the same way. So most public images of medicine from 1885 onward generally consisted of physician portraits, medical facilities and procedures, and microscopic (or X-ray) images, all of which were suitable for health education campaigns.

But the clinical portrait photograph was an important facet of medical training, precisely because its status as a document made it a convenient and viable substitute for live demonstrations of patients. While the live demonstration was a major advance in medical education—a huge step from the centuries-long, scholastic tradition of learning medicine only from ancient texts—it had its own set of challenges. According to an American student taking classes at the Allgemeine Krankenhaus in Vienna in 1865, each professor was provided with a lecture room near his ward: "At the time of lecture this room is filled in with 'specimens' in the shape of men and women who are transferred from the other wards for the occasion. These patients are looked upon and spoken of as 'material' for the medical instruction and as such are submitted to examination by the students without much reference to any feelings which they as men and women may have on the subject."[31] Patients did not submit gladly, apparently. In another letter, the student draws a sketch of his routine at the hospital, which includes "scolding and pitching into the patients for coming late (wh. they always do in Vienna)."[32] While photography would never replace patients in bedside instruction, of course, physicians were pleased to substitute photographs and slides for patients in lecture, if not to alleviate the obvious ethical concerns, then at least to present all the students with a larger, projected view.[33] Many physicians, such as prominent German surgeon and professor Theodor Billroth (1829–1894), maintained their own photographic collection for pedagogical purposes. Billroth published only one book of clinical photography, but it is known he had an even larger private collection, which he used for his medical school lectures.[34]

Furthermore, Billroth and others active in the scholarly community used photographs to demonstrate a diagnosis and persuade others of the chosen therapy. For example, after the 1880s, the discussions of individual cases filling the proceedings of the Berlin Medical Society were often accompanied by photographic images that circulated among the participants.[35] These photos were frequently the basis for demonstration and debate. From the 1860s onward, there arose a number of periodicals and publications designed to present these findings in photographic form. The College of Physicians of Philadelphia, for example, established the *Pho-*

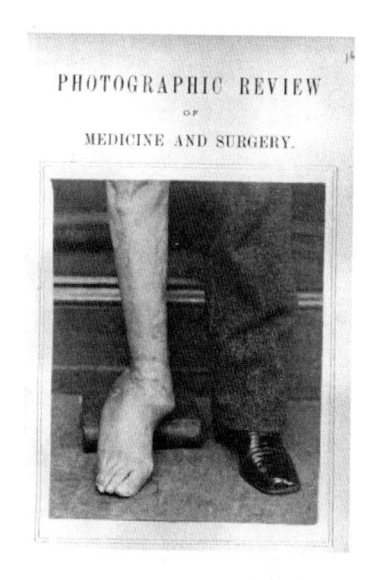

3.3 A frontispiece from the *Photographic Review of Medicine and Surgery*, published by J.B. Lippincott & Co. in Philadelphia, 1870–1872. Though short-lived, images from this journal are still used to illustrate certain cases or diseases.

tographic Review of Medicine and Surgery in 1870, which would present a photograph and a clinical explanation of the case in each issue (figure 3.3). Montméja started the *Revue Medico-Photographique* along the same lines in the 1880s, and Ludwig Jankau founded the *Internationale Medizinisch-Photographische Monatsschrift* in 1894. The kinds of photographs we find in these pages include case documentation, diagnostic aids, and testimonials of intervention outcomes, all testifying to what André Gunthert has called photography's "heuristic function."[36] We should also remark on the rise of technical books on the subject, such as Albert Londe's *La photographie médicale*,[37] and, of course, the photographic atlas.[38]

These photographs of medical cases usually became part of a disciplinary archive of images that could be tapped by students and practitioners. Hospitals such as the Saint-Louis in Paris, the Bellevue in New York, and the Charité in Berlin established photographic departments for just this purpose. A report from Bellevue in 1869 indicates that a photographic archive and department could be a magnet for the discipline: "Members of the medical profession begin to visit the Department periodically, for the purpose of obtaining such photographs as pertain to each one's more especial class of investigation. Many interesting cases of skin disease, fractures, and results of important surgical operations have been fully illustrated by series of photographs, which give opportunity for comparison and study not offered by any other means."[39] Unlike previous media, the photograph becomes, in all these various applications, a substitute for the patient—the human body becomes, through photography, a viable "working object" for medicine. As Bruno Latour notes, "Scientists start seeing something once they stop looking at nature and look exclusively and obsessively at prints and flat inscriptions."[40] Photography, in this respect, is an ideal inscription and archive. Indeed, of all the possible hopes the medical community had for photography, the dream of a universal and portable archive of cases is the most persistent.

Correlation, Series, Repeatability

Having outlined the types and uses of medical photography, we can now consider the relationship between medical observation and the photographic image. What did nineteenth-century physicians talk about when they talked about observation? First, they insisted on the difference between careful and careless observation, which lies in the ability to apply diligently a prescribed method. Textbooks at the time sought to outline this method for students and junior practitioners. British physiologist Thomas Laycock (1812–1876) wrote one such text, in which he made it clear: "The foundation of medical experience is observation of disease, and the requisites to successful observation are minuteness and accuracy. The clinical student must therefore make up his mind to be sedulously minute and carefully accurate in investigating the cases under his notice."[41] German pathologist Rudolph Virchow (1821–1902) described in detail his method of performing autopsies. Autopsies conducted in a haphazard way promoted interpretive error, he argued, so he "drew particular attention to the necessity of insisting—in autopsies for medico-legal purposes, as in everything else now—upon completeness of examination and exactness of method, both in the investigation and in note-taking, so that it might be decided subsequently, though not in anticipation, whether there was any significance or importance in what was observed, or whether it was accidental and unessential."[42] Virchow was particularly careful to describe exactly what and in what order should be observed in an autopsy. Laycock, Virchow, and others emphasize the importance of *method* in observation, and their equal emphasis on accuracy, detail, and completeness already echoes the rhetoric of the discourse on medical and scientific photography.

These authors therefore recognized, as did Bernard, that observation is never merely looking—it is also an intellectual process of comparison. The physician compares the diseased organ, for instance, with other organs around it, but also with his or her previous experience of that organ.

Virchow gives an example: "The freer the incision—always supposing that it is an even one—the larger will be the field of view, the more numerous will be the points of comparison between normal and abnormal parts, and the more exactly shall we be able to estimate the extent of the pathological territories."[43] Here Virchow's equation of the "field of view" with the "points of comparison" emphasizes the importance of correlation for observation. Likewise, Laycock argued that his method, essentially, "is based upon simple observation of the phenomena, and comparison of them with one another, and with the knowledge which the practitioner has acquired of similar phenomena, either by instruction or experience."[44] In practice, then, there is hardly room for "simple" observation, given the constant activity of isolation and recognition of elements in the field of view. Observation is always comparison and correlation. Because of this constant mental process, observation goes hand in hand with what Bernard called "experimental reasoning," but which was also often called "medical logic"—the principles of reasoning behind observational and experimental methods in medicine.[45]

Fleck and Foucault both argue that medical logic of the nineteenth century owed much to the concept of the series, especially the rise of statistical methods that used the series in computation. According to Foucault, this combination of medicine and statistics "opened up to investigation a domain in which each fact, observed, isolated, then compared with a set of facts, could take its place in a whole series of events whose convergence or divergence were in principle measurable."[46] Each case history became one in a series; facts, such as symptoms or signs, became significant only insofar that they were repeated. As one nineteenth-century physician proclaimed, "With each new case, one might think that we were presented with new facts; but they are merely different combinations, different subtleties: in the pathological state, there is never more than a small number of principal facts; all the others result from their combination and from their different degrees of intensity."[47] The finite number of combinations meant that each presumed "new fact" could be placed in a series of similar "facts." Through the repetition of these facts and their variation—and with its numerous case histories, the clinic allowed an almost endless repetition of symptoms and facts—researchers began to see the pattern that became the essence of the disease.[48]

Like the clinic, both the medical archive and the medical atlas demonstrate the practical connection between observation and correlation. Each showed that medical observation entails the isolation, orderly arrangement, and careful comparison of different examples of the object under examination.[49] Photography amplified and, to a certain extent, transformed the character of each of these institutions, in that the photograph became an easy substitute for the patient or the medical specimen. But more importantly, photography promoted a *proliferation of cases* that simultaneously affirmed, codified, and extended series logic. That is, this feature of photography corresponded to the logic of medical perception. As Fleck notes, "It is only numerous, very numerous, observations that eliminate the individual character of the morbid element."[50]

Here I want to emphasize the *repeatability* of the photographic image. By this I do not mean photography's reproducibility—the ability to make a number of prints of the same image—but its ability, especially after the development of instantaneous photography in the 1880s, to take a number of different shots of the same object, or to take a number of roughly the same views of different objects. This ability gave photography a tremendous advantage over hand-drawn illustration: it gave the physician a series for comparison. But it also corresponded to—even instantiated—popular principles of medical observation. When Laycock discusses the application of statistical methods to medical research, he insists that "it is most essential that the observations, facts, or events, be as nearly alike as practicable" and that "the number of observations must be considerable."[51] Photography's ability to isolate, frame, and repeat similar cases was a powerful aid in the standardization and multiplication of observational views. We should also note that arrangement of photographs in a series allowed not only their sequential organization, but also their *simultaneous* presentation. Georges Didi-Huberman has argued that Charcot's arrangement of his patients into living tableaux functioned like tables of data by organizing their signs into simultaneous events.[52] Photography allowed this same organization, and much more easily. In series photography, the sequence was important because it suggested a causal order or chronology, but the simultaneous display of images was arguably equally essential to the process of comparison and correlation. Series photography, as a research tool that could allow both sequential analysis and simultaneous display, succinctly articulated the ideals of medical observation and logic.

Detail, Presence, Contemplation

The photographic archive presented physicians with unlimited opportunities to observe and compare. It is true that looking at a photograph was different from bedside observation of a patient, which emphasized the temporal development of signs and symptoms. But sometimes doctors called upon photographs to fulfill this function as well, as in the work of American physician Samuel Powers, who used photos to track the development of smallpox in a patient over several days (figure 3.2).[53] Nevertheless, to the extent that *repetition* is a fundamental feature or application of the medium, it mimicked the clinic's emphasis on series.[54] The ease with which photography generated series of images of cases corresponded to the dream of a limitless well of evidence. This dream, however, presupposed the evidentiary status of the individual image. It presumed that each image was a window; seeing through enough of them could give the observer a vision of the whole field. It assumed, in other words, that each image was a fragment or representative of the whole archive. What aspect of the photographic image allowed this assumption? I would argue that it was not simply the automatic, mechanical character of the camera that underwrote its evidentiary status, but the photographic image's abundance of *detail*. Furthermore, photographic detail emulated and promoted an already established habit of observation—the complete description.

This is not to ignore the problems that photographic detail presented to nineteenth-century practitioners. Unlike human illustrators, who could select the salient elements of the specimen to include in the drawing and leave out the extraneous and potentially confusing details, the undiscerning eye of the camera included everything. It was often hard to tell exactly what was at issue in a photograph, what the illustration was actually meant to illustrate, because the camera could not emphasize, say, texture or shape. An editorial from 1886 voiced a common complaint: "photographs of diseased viscera often fail to instruct the observer, but rather remind him of Ovid's description of chaos."[55] For these researchers, the photograph had to learn how to "point" before it could be used as an instructional tool.[56] Richard Kretz (1865–1920), a professor of pathological anatomy at the University of Vienna, agreed that because it could be difficult to find the essential point in photography's "wealth of detail," "photography could not be used as a categorical replacement for drawing." Yet Kretz goes on to say that photography "in many cases is able to perform similarly and to render services that are virtually irreplaceable in scientific cases, as when the photograph of flagellum-bearing bacteria proves [their existence] in the most striking manner, because the photographic plate first presents to the expert [*bewaffneten*] eye, in the most detailed and unbiased condition, scarcely detectable objects with unquestionable clarity."[57] Despite the pedagogic problems that an abundance of detail could cause, for most physicians the "unquestionable clarity" of the photographic image was, along with the potential for objectivity ("unbiased") and discovery ("scarcely detectable objects"), its most highly prized attribute. Indeed, the clarity and texture of the image, its ability to represent the structure and randomness of the natural world, were arguably the most important features for its applications in medicine.

Discussions of photography's advantages to the discipline invariably emphasized its ability to *represent*—that is, substitute for—the object or patient. This is what physicians meant when they applauded photography as "*Naturgetreu*," or "faithful to nature." Of course, not every aspect of the photograph is perfectly faithful to nature, but in these discussions, "*Naturgetreu*" referred most often to the level of detail that allowed the photographic image to reproduce patterns of texture and variation. Kretz writes, "Photography is perfectly faithful to nature, that is, the images reproduce . . . all forms and proportions, the distribution of light and shadow in a completely correct manner."[58] Or Ludwig Jankau declares, "In medicine today, especially in practical medicine, the first requirement is that preparations and such, especially the conditions of the illness, are reproduced exactly as they are. And for reproductions faithful to nature, what can offer us more guarantee than photography?"[59] This "fidelity" referred not to color or depth or emphasis or any of the aspects of nature and observation that many complained photography could not represent well. Instead, it referred to the same qualities that brought the scientific curiosity to bear on nature in the first place: the abundant variations on patterns of similarity and difference found in the forms and random textures of natural phenomena. Because photography could replicate these forms and textures with such detail, it could act as a substitute or a representation of the patient.

This is also why photography was so often equated with immediate and efficient description. "A glance often teaches more than pages of description," intones Jankau,[60] while Kretz writes,

> Just as a sketch complements—even replaces—a long-winded topographical description, in cases that are difficult to describe, a good photographic image to which a brief commentary is added is able to say more than cumbersome descriptions. Especially in cases that are subject to a forensic examination or in accident photos in which the task is an assessment of the degree of disability, to the examining physician, an appropriately recorded photograph becomes the actual circumstances.[61]

Kretz may be suggesting that "a picture is worth a thousand words," as it were, but there is something more to be said here. Kretz describes an observer who directed his or her gaze to the photograph as a substitute for nature itself. The photograph was both an apparently agent-less record of nature, effectively used as a substitute for it, and an enunciation that functioned as a description of nature. The photograph was at once an image and a description—that which is visible is *already* wholly expressed. The abundance of detail in the photographic image worked in two, contradictory ways. It was at once a tangle of data that required prolonged (and perhaps frustrating) study, and a complete rendering that—if the image was scientifically legible and the eye expert—could be grasped immediately. Lorraine Daston has called this aspect of scientific observation "all-at-once-ness." Discussing Descartes's reversion to the language of vision in his quest for ontological bedrock, she notes that "Despite all the well-known illusions, the imprimatur of the real, the true, and the certain is the immediate, implicit all-at-once-ness of perception, especially vision."[62] This was especially true of the photograph and the observer's relationship to it. The density of the photographic image not only brought it closer to the ideal description—an immediate, acceleration of knowledge without residue or ambiguity—it also gave the image a *presence* unlike any previous medium.

Because it could present the object of study in this way, the researcher could use the photographic image as a source of information and discovery. The photograph, more than hand-drawn illustration, acted as a landscape over which the expert eye could roam, and within which discoveries could be made. The photograph was, to be certain, no substitute for observation—the still, flat image could not replicate the act of observation with its intricate process of recognition and comparison. But it did fix a view, which allowed a leisurely, contemplative approach that was fertile ground for detection and correlation. And contrary to the popular conception of the scientist, the aesthetic, contemplative approach was absolutely central to the practice of scientific and medical observation. As Billroth rhapsodized, "Solitary, meditative observation is the first step in the poetry of research, in the formation of scientific phantasies, the reality of which we then test with the tools of logic, mathematics, physics and chemistry. Our tests will be the more successful the better we have learned to handle these tools. The diseased organism, the patient, must be observed in just this way, thoughtfully, and in a state of mental solitude and meditation."[63] Contemplation was an integral part of "the poetry of research."

The aesthetics of scientific observation began with the contemplative gaze and the search for pattern. Because of its density and texture, the photograph could trigger this meditative gaze, and because of its repeatability, it encouraged the search for pattern. One British physician suggested, "A good radiograph in some respects may be said to resemble a painting by Turner. Without intuition or previous study the one is almost as incomprehensible as the other, but as we gaze the wealth of detail rises before our vision until finally we are able to interpret the meaning of streaks and shadows that to the untrained eye are meaningless."[64] Here the medical gaze takes on a quasi-aesthetic character. What connected the radiograph and the Turner was the observer's search for pattern. For the critic, the pattern was given, whole, in the painting. For the physician or scientist, the pattern in nature must be discerned from a series of examples. Like art, the photograph presented a new point of view to challenge one's own. Like nature, the photograph presented not an organized field of view or worldview, but the appearance of raw data from which the observer discerned a larger pattern correlated from a series of examples. The line between the aesthetic and the scientific was perhaps not as stark as we might think.[65] Indeed, in the following passage, Billroth describes the importance of imagination in medical training:

> Easily to reproduce sense impressions, to visualize a highly complicated sequence of events, to be able without difficulty to imagine all the different possible results of normal and abnormal processes—in short, to see a thousand things at once and to sense their interrelations—these are the essential qualifications for scientific research. The ready capacity to reproduce sense impressions imaginatively, the sharpening of the powers of observation, the orderly arrangement of mental images in vivid and living pictures, facility in the so-called inductive method of reasoning—these are the qualities that the teacher of science and of medicine must develop in his students; and to this end a certain amount of preparation is indispensable.[66]

Unwittingly, perhaps, Billroth sketches here precisely why photography would be so valuable for scientific and medical research. Not because it documented nature, not because it was objective—although we would be foolish to deny the importance of these properties—but because it aided observation and imagination. It "reproduces sense impressions," "visualizes a highly complicated sequence of events," "sharpens the powers of observation," and allows the researcher "to see a thousand things at once and to sense their interrelations." Photography, more easily and more persuasively than any hand-drawn illustration, functioned like the physician's own powers of observation and imagination.

Conclusion

So I am suggesting that certain features of photography—especially those that obtain during the act of viewing the image—corresponded to certain aspects of practical medical observation. Examining the interaction between image and observer highlights characteristics different from those that show up when we focus on the relationship between object and camera. Shifting our attention in this way brings us to a broader understanding of the acceptance of photography as an instrument in medicine and the sciences. That is, photography was not just a handy tool readily available and hence quickly adopted. Nor was its mechanical, automatic character the only reason for its rise in popularity. It also presented researchers with a set of features that spoke to established and emerging principles and habits of observation. It was not just how it was *made* that gave the photograph its scientific legitimacy, but how it was *looked at*.

Notes

1. Claude Bernard, *An Introduction to the Study of Experimental Medicine*, trans. Henry Copley Greene (New York: Macmillan, 1927), 21–22.

2. Bernard, *Experimental Medicine*, 6. The bulk of Bernard's opening chapter is in fact devoted to troubling the too-easy distinction between activity and passivity in experiment and observation. For more on Bernard's thoughts on the relationship between experiment and observation, see Sebastian Normandin, "Claude Bernard and *An Introduction to the Study of Experimental Medicine*: 'Physical Vitalism,' Dialectic, and Epistemology," *Journal of the History of Medicine and Allied Sciences* 62, no. 4 (October 2007): 495–528.

3. The most complete discussion of the relationship between objectivity and subjectivity in science is Lorraine Daston and Peter Galison, *Objectivity* (New York: Zone Books, 2007). On the general distrust of the senses and subjectivity, see Jonathan Crary, *Techniques of the Observer* (Cambridge, MA: MIT Press, 1990).

4. Gianna Pomata "A Word of the Empirics: The Ancient Concept of Observation and Its Recovery in Early Modern Medicine," paper presented to the Science in Human Culture Program, Northwestern University, April 2009. For the eighteenth-century view of observation, see Daston and Galison, *Objectivity*. Perhaps part of what is at issue is the definition of "activity" in relation to observation; scientists up to the nineteenth century readily conceded that observation was an active, *cognitive* process, but for Bernard and others, it was helpful to distinguish between activities that *physically* intervened and those that did not.

5. Michel Foucault, *The Birth of the Clinic: An Archaeology of Medical Perception*, trans. A. M. Sheridan Smith (New York: Vintage Books, 1973), esp. 107–22.

6. Michael Hau, "The Holistic Gaze in German Medicine, 1890–1930," *Bulletin of the History of Medicine* 74, no. 3 (Fall 2000): 495–524.

7. Norwood Russell Hanson, *Patterns of Discovery* (Cambridge: Cambridge University Press, 1969); Dudley Shapere, "The Concept of Observation in Philosophy and Science," *Philosophy of Science* 49 (1982): 485–525; Ian Hacking, *Representing and Intervening* (Cambridge: Cambridge University Press, 1983); Patrick A. Heelan, *Space-Perception and the Philosophy of Science* (Berkeley: University of California Press, 1983); Trevor Pinch, "Towards an Analysis of Scientific Observation: The Externality and Evidential Significance of Observational Reports in Physics," *Social Studies of Science* 15, no.1 (February 1985): 3–36; Roberto Torretti, "Observation," *The British Journal for the Philosophy of Science* 37, no. 1 (March 1986): 1–23; David Gooding, *Experiment and the Making of Meaning: Human Agency in Scientific Observation and Experiment* (Dordrecht: Kluwer Academic Publishers, 1990). All of these authors were struggling with the once-vexing question of the role of theory in scientific observation.

8. See Daston and Galison, *Objectivity*. On photography and science, see Jon Darius, *Beyond Vision* (Oxford: Oxford University Press, 1984); M. Susan Barger and William B. White, *The Daguerreotype: Nineteenth-Century Technology and Modern Science* (Washington, DC: Smithsonian Institution Press, 1991); Ann Thomas, ed., *Beauty of Another Order: Photography in Science* (New Haven, CT: Yale University Press, 1997); Jennifer Tucker, *Nature Exposed: Photography as Eyewitness in Victorian Science* (Baltimore: Johns Hopkins University Press, 2005); Corey Keller, ed., *Brought to Light: Photography and the Invisible, 1840–1900* (New Haven, CT: Yale University Press, 2008). Kelly Wilder, *Photography and Science* (London: Reaktion, 2009) discusses photography and scientific observation extensively.

9. This relationship between the photograph and the world it depicts has often been called "indexical," after C. S. Peirce's theory of signs, but I will try to avoid this contested and often confusing designation, which has generated reams of discussion. Among the latest contributions, see especially the issue on "Indexicality: Trace and Sign," *differences* 18, no. 1 (2007), edited by Mary Ann Doane. For helpful discussions about its limited usefulness, see Martin Lefebvre, "The Art of Pointing: On Peirce, Indexicality, and Photographic Images," in *Photography Theory*, ed. James Elkins (New York: Routledge, 2007), 220–44; Robin Kelsey, "Indexomania," *Art Journal* 66, no. 3 (Fall 2007): 119–22; and Tom Gunning, "What's the Point of an Index? or, Faking Photographs," in *Still Moving: Between Cinema and Photography*, ed. Karen Beckman and Jean Ma (Durham, NC: Duke University Press, 2008).

10. Torretti, "Observation," 8.

11. Aristotle, *Posterior Analytics*, 2.19.100a4–8, quoted in Lorraine Daston, "On Scientific Observation," *Isis* 99, no. 1 (2008): 100.

12. On wax models, see Thomas Schnalke, *Diseases in Wax: The History of the*

Medical Moulage, translated by Kathy Spatschek (Chicago: Quintessence Pub. Co., 1995), and Schnalke, "Dissected Limbs and the Integral Body: On Anatomical Wax Models and Medical Moulages," *Interdisciplinary Science Reviews* 29, no. 3 (2004): 312–22.

13. William Keiller, MD, "The Craze for Photography in Medical Illustration," *New York Medical Journal* (23 June 1894): 788–89. In the same issue there are a variety of articles that use photography for medical illustration, giving some substance to this complaint.

14. For overviews of the origins of medical photography, see Alison Gernsheim, "Medical Photography in the Nineteenth Century," *Medical and Biological Illustration* (London) 11, no. 2 (April 1961): 85–92; Renata Taurek, *Die Bedeutung der Photographie für die medizinische Abbildung im 19. Jahrhundert* (Köln: F. Hansen, 1980); Andreas-Holger Maehle, "Wie die Photographie zu einer Methode der Medizin wurde: Aus der Geschichte der Medizin-Photographie im 19. Jahrhundert," *Fortschritte der Medizin* 104, no. 15 (1986): 63–65; Daniel M. Fox and Christopher Lawrence, *Photographing Medicine: Images and Power in Britain and America since 1840* (New York: Greenwood Press, 1988); Andreas-Holger Maehle, "The Search for Objective Communication: Medical Photography in the Nineteenth Century," in *Non-Verbal Communication in Science Prior to 1900*, ed. Renato G. Mazzolini (Firenze: Leo S. Olschki, 1993), 563–86; Gunnar Schmidt, *Anamorphotische Körper: Medizinische Bilder vom Menschen im 19. Jahrhundert* (Köln: Böhlau, 2001). See also the collection of essays devoted to medicine and photography in a special issue of *History of Photography* 23, no. 3 (Autumn 1999), ed. Rachel A. Dermer. On the impact of photography on medical practice, see Stanley Joel Reiser, *Medicine and the Reign of Technology* (Cambridge: Cambridge University Press, 1978).

15. Fox and Lawrence, *Photographing Medicine*, 49–54.

16. Alfred Donné, *Cours de microscopie complémentaire des études médicales. Anatomie microscopique et physiologique des fluides de l'économie* (Paris: J.-B. Baillière, 1844), 36. See also G. Richet, "Daguerre, Donné et Foucault, trois francs-tireurs créent la microphotography," *Médecine/Sciences* (Paris) 13, no. 1 (January 1997): 45–48; and William Tobin, "Alfred Donné and Léon Foucault: The First Applications of Electricity and Photography to Medical Illustration," *Journal of Visual Communication in Medicine* 29, no. 1 (March 2006): 6–13. Donné hired an engraver to trace illustrations from the original daguerreotypes, but he also included plates in limited editions, giving the reader a choice between the two modes of illustration.

17. See, for example, Robert Koch, "Verfahrungen zur Untersuchung, zum Conserviren und Photographiren der Bacterien," *Beiträge zur Biologie der Pflanzen* 2, no. 3 (1877): 399–434; and Koch, *Zur Untersuchung von pathogenen Organismen* (Berlin: Norddeutschen Buchdruckerei und Verlagsanstalt, 1881–84). On Koch and microphotography, see Thomas Schlich, "'Wichtiger als der Gegenstand selbst': Die Bedeutung des fotografischen Bildes in der Begründung der bakteriologischen Krankheitsauffassung durch Robert Koch," in *Neue Wege*

in der Seuchengeschichte, ed. Martin Dinges and Thomas Schlich (Stuttgart: Franz Steiner Verlag, 1995), 143–74; Thomas Schlich, "Repräsentationen von Krankheitserregern. Wie Robert Koch Bakterien als Krankheitsursache dargestellt hat," in *Räume des Wissenschaftliches Repräsentation, Codierung, Spur*, ed. Hans-Jorg Rheinberger, Michael Hagner, and Bettina Wahrig-Schmidt (Berlin: Akademie Verlag, 1997), 165–90; and Jennifer Tucker, *Nature Exposed: Photography as Eyewitness in Victorian Science* (Baltimore: Johns Hopkins University Press, 2005), 159–93. For early twentieth-century discussions of the importance of Koch's work for the establishment of microphotography as a scientific tool, see Richard Neuhauss, *Lehrbuch der Mikrophotographie* (Braunschweig: Harald Bruhn, 1898), 232–40; or Kurt Laubenheimer, *Lehrbuch der Mikrophotographie* (Berlin: Urban & Schwarzenberg, 1920), 1–3.

18. See Max Nitze, *Kystophotographischer Atlas* (Wiesbaden: J.F. Bergmann, 1894). See also Wolfgang G. Mouton, Justin R. Bessell, and Guy J. Maddern, "Looking Back to the Advent of Modern Endoscopy: 150th Birthday of Maximilian Nitze," *World Journal of Surgery* 22, no. 12 (December 1998): 1256–58; M. Reuter, "The Historical Development of Endophotography," *World Journal of Urology* 18, no. 4 (September 2000): 299–302; and Harry W. Herr, "Max Nitze, the Cystoscope and Urology," *Journal of Urology* 176, no. 4 (October 2006): 1313–16.

19. The literature on X-ray imaging is vast, but for our purposes here, the most important overviews are Bernike Pasveer, "Knowledge of Shadows: The Introduction of X-ray Images in Medicine," *Sociology of Health and Illness* 11, no. 4 (December 1989): 360–81; Bernike Pasveer, "Shadows of Knowledge: Making a Representing Practice in Medicine: X-ray Pictures and Pulmonary Tuberculosis, 1895–1930," (PhD thesis, University of Amsterdam, 1992); Lisa Cartwright, *Screening the Body: Tracing Medicine's Visual Culture* (Minneapolis: University of Minnesota Press, 1995); Monika Dommann, *Durchsicht, Einsicht, Vorsicht: Eine Geschichte der Röntgenstrahlen, 1896–1963* (Zürich: Chronos, 2003); and Andrew Warwick, "X-rays as Evidence in German Orthopedic Surgery, 1895–1900," *Isis* 96, no. 1 (March 2005): 1–24.

20. On Marey, see especially Marta Braun, *Picturing Time: The Work of Etienne-Jules Marey (1830–1904)* (Chicago: University of Chicago Press, 1992); and François Dagognet, *Etienne-Jules Marey: A Passion for the Trace*, trans. Robert Galeta with Jeanine Herman (New York: Zone Books, 1992). On moving images in science and medicine, in addition to Cartwright's *Screening the Body*, see Martin Weiser, *Medizinische Kinematographie* (Dresden: Theodor Steinkopff, 1919); F. Paul Liesegang, *Wissenschaftliche Kinematographie* (Düsseldorf: Liesegang, 1920); Anthony Michaelis, *Research Films in Biology, Anthropology, Psychology, and Medicine* (New York: Academic Press, 1955); and Virgilio Tosi, *Cinema before Cinema: The Origins of Scientific Cinematography*, trans. Sergio Angelini (London: British Universities Film & Video Council, 2005).

21. Fox and Lawrence, *Photographing Medicine*, 21–27.

22. Nora L. Jones, "The Mütter Museum: The Body as Spectacle, Specimen, and Art" (PhD dissertation, Temple University, 2002), 153.

23. Comparing these two types of portraits—which, we should be careful to note, was not a common practice—we see that one of the by-products of the genre is an implicit contrast between the healthy or "normal" physician and the "pathological" patient, a distinction that was probably helpful, if not explicit, during medicine's budding professionalization. As concerns about privacy and sensationalism rose, conventions for photographing patients gradually moved away from those of portraiture; these changes also coincided with the growing legitimacy of the medical profession. For more on the conventions of portraiture in clinical medicine, see Chris Amirault, "Posing the Subject of Early Medical Photography," *Discourse* 16, no. 2 (Winter 1993–94): 51–76.

24. Hugh W. Diamond, "On the Application of Photography to the Physiognomic and Mental Phenomena of Insanity," in *The Face of Madness: Hugh W. Diamond and the Origin of Psychiatric Photography*, ed. Sander L. Gilman (New York: Brunner/Mazel, 1976). See also Sander L. Gilman's introduction to the same volume, "Hugh W. Diamond and Psychiatric Photography"; and John Cule, "The Enigma of Facial Expression: Medical Interest in Metoposcopy," *Journal of the History of Medicine and Allied Sciences* 48, no. 3 (July 1993): 302–19.

25. Maximilian Leidesdorf, *Lehrbuch der psychischen Krankheiten* (Erlangen: Enke, 1865).

26. J.-M. Charcot, Paul Marie Louis Pierre Richer, Georges Albert Edouard Brutus Gilles de la Tourette, and Albert Londe, *Nouvelle iconographie de la Salpêtrière, clinique des maladies du système nerveux* (Paris: Lecrosnier et Babé, 1888–1892). See also Albert Londe, *La photographie instantanée: théorie et pratique* (Paris: Gauthier-Villars, 1886); Albert Londe, *La photographie médicale; application aux sciences médicales et physiologiques* (Paris: Gauthier-Villars, 1893); Ulrich Baer, "Photography and Hysteria: Toward a Poetics of the Flash," *Yale Journal of Criticism* 7, no. 1 (1994): 41–76; and Georges Didi-Huberman, *Invention of Hysteria: Charcot and the Photographic Iconography of the Salpêtrièlre*, trans. Alisa Hartz (Cambridge, MA: MIT Press, 2003).

27. Alfred Hardy and A. de Montméja, *Clinique photographique de l'hôpital Saint-Louis* (Paris: Librairie Chamerot et Lauwereyns, 1868). See also Norbert Kuner and W. Hartschuh, "Möglichkeiten und Grenzen früher Fotografie in der Dermatologie: Die *Clinique photographique de l'hôpital Saint-Louis* von 1868," *Der Hautarzt* 54, no. 8 (August 2003): 760–64. For more on Montméja and nineteenth-century medical photography in France, see Monique Sicard, Robert Pujade, and Daniel Wallach, *À corps et à raison: Photographies médicales, 1840–1920* (Paris: editions Marval, 1995).

28. In the United States, for example, George A. Otis (1830–1881), assistant surgeon of the United States Army, was instrumental in collecting wartime medical photographs into a monumental, multivolume edition of *The Medical and Surgical History of the War of the Rebellion (1861–65)*, which was published in the 1870s and 1880s. See Joseph K. Barnes, Joseph Janvier Woodward, Charles Smart, George A. Otis, and D. L. Huntington, *The Medical and Surgical History of the War of the Rebellion (1861–65)*, 3 vols. (Washington: Government Printing Office,

1870–1888). Most of the illustrations are engravings from photographs, but there are some splendid examples of medical photography. For more on American medical photography, see the work of Stanley B. Burns, especially A *Morning's Work: Medical Photographs from the Burns Archive and Collection, 1843–1939* (Santa Fe, NM: Twin Palms Publishers, 1998). Germany's Heinrich Curschmann (1846–1910) directed the Leipzig Clinic, where he enthusiastically developed darkroom facilities in order to document case histories. His *Klinische Abbildungen* from 1894 is a masterpiece of the genre. See Heinrich Curschmann, *Klinische Abbildungen: Sammlung von Darstellungen der Veränderung der Äusseren Körperform bei inneren Krankheiten* (Berlin: Julius Springer, 1894). See also Heinrich Curschmann, "Beiträge aus der photographischen Sammlung der medizinischen Klinik," *Internationale medizinisch-photographische Monatsschrift* 1 (1894): 225–28; and Edgar Goldschmid, *Entwicklung und Bibliographie der Pathologisch-anatomischen Abbildungen* (Leipzig: K. W. Hiersemann, 1925).

29. Ludwik Fleck, "Some Specific Features of the Medical Way of Thinking," in *Cognition and Fact: Materials on Ludwik Fleck*, ed. Robert S. Cohen and Thomas Schnelle (Dordrecht: Kluwer Academic Publishers, 1986), 39–46. See also Georges Canguilhem, *The Normal and the Pathological*, trans. Carolyn R. Fawcett in collaboration with Robert S. Cohen (New York: Zone Books, 1991).

30. Bert Hansen, *Picturing Medical Progress from Pasteur to Polio: A History of Mass Media Images and Popular Attitudes in America* (New Brunswick, NJ: Rutgers University Press, 2009).

31. Clare [Blake] to Dear Pater, Vienna, 9 November [1865], Clarence John Blake Papers, Francis A. Countway Library of Medicine, Harvard University, quoted in John Harley Warner, *Against the Spirit of System: The French Impulse in Nineteenth-Century American Medicine* (Princeton, NJ: Princeton University Press, 1998), 304.

32. Clare to Sister Agnes, Vienna, 29 March 1869, Clarence John Blake Papers, Francis A. Countway Library of Medicine, Harvard University, quoted in Warner, *Against the Spirit of System*, 311.

33. For discussion of the projection of images in medical education, see Sigmund Theodor Stein, *Das Licht im Dienste wissenschaftlicher Forschung: Handbuch der Anwendung des Lichtes und der Photographie in der Natur- und Heilkunde* (Leipzig: Spamer, 1877); and Sigmund Theodor Stein, *Die optische Projektionskunst im Dienste der exakten Wissenschaften: ein Lehr- und Hilfsbuch zur Unterstützung des naturwissenschaftlichen Unterrichts* (Halle: W. Knapp, 1887). See also Henning Schmidgen, "Pictures, Preparations, and Living Processes: The Production of Immediate Visual Perception (Anschauung) in Late-Nineteenth-Century Physiology," *Journal of the History of Biology* 37, no. 3 (October 2004): 477–513.

34. Theodor Billroth, *Chirurgische Klinik in Zürich: Stereoscopische Photographien von Kranken* (Zurich, 1865). Billroth was born and educated in Germany, but after posts in Berlin and Zurich, he accepted a position as professor of surgery and director of the surgical clinic at the University of Vienna, for which he is best

known. On Billroth's enthusiasm for new media technologies, see Ernst Kern, *Theodor Billroth, 1829–1894: Biographie anhand von Selbstzeugnissen* (München: Urban & Schwarzenberg, 1994), 73–75. For a general overview of Billroth's legacy, see Erna Lesky, *The Vienna Medical School of the Nineteenth Century*, trans. L. Williams and I. S. Levij (Baltimore: Johns Hopkins University Press, 1976); Robb H. Rutledge, "Theodor Billroth: A Century Later," *Surgery* 118, no. 1 (July 1995): 36–43; and Tatjana Buklijas, "Surgery and National Identity in Late Nineteenth-Century Vienna," *Studies in History and Philosophy of Biological and Biomedical Sciences* 38, no. 4 (December 2007): 756–74.

35. An example, picked more or less at random, is Adolf Magnus-Levy, "Ueber Organ-Therapie beim endemischen Kretinismus," *Verhandlungen der Berliner medicinischen Gesellschaft* 34, Part II (1903): 350–57. See especially the discussion of this presentation on 22 July 1903 in Part I, pp. 246–49. No photos are published with the paper, but they discuss the photographs that were passed around among the audience. Many such uses of photographs can be found in the *Verhandlungen* and similar proceedings.

36. André Gunthert, "La rétine du savant: La fonction heuristique de la photographie," *études photographiques* 7 (May 2000): 29–48.

37. Albert Londe, *La photographie médicale: Application aux sciences médicales et physiologiques* (Paris: Gauthier-Villars, 1893). See also Peter Geimer, "Picturing the Black Box: On Blanks in Nineteenth Century Paintings and Photographs," *Science in Context* 17, no. 4 (2004): 467–501.

38. On the scientific atlas, see Daston and Galison, *Objectivity*.

39. J. Frey, "Report of the Photographic Department of Bellevue Hospital for the Year 1869," in *Tenth Annual Report of the Commissioners of Public Charities and Correction of the City of New York for the Year 1869* (Albany, NY: Charles van Benthuysen and Sons, 1870), 85. Mark Rowley has transcribed the reports of the Bellevue Photographic Department; they are available at his "Cabinet of Art and Medicine" website, www.artandmedicine.com/ogm/Reports.html.

40. Bruno Latour, "Drawing Things Together," in *Representation in Scientific Practice*, ed. Michael Lynch and Steve Woolgar (Cambridge, MA: MIT Press, 1988), 39.

41. Thomas Laycock, *Lectures on the Principles and Methods of Medical Observation and Research* (Philadelphia: Blanchard and Lea, 1857), 64.

42. Rudolph Virchow, *Post-Mortem Examinations*, trans. T. P. Smith, 3rd ed. (Philadelphia: P. Blakiston, Son & Co., 1895), 12.

43. Ibid., 40.

44. Laycock, *Lectures*, 94.

45. On medical logic, see Friedrich Oesterlen, *Medical Logic*, trans. G. Whitley (London: Sydenham Society, 1855); Frederick P. Gay, "Medical Logic," *Bulletin of the History of Medicine* 7 (1939): 6–27; Lester S. King, "Medical Logic," *Journal of the History of Medicine and Allied Sciences* 33, no. 3 (July 1978): 377–85; and Lester King, *Medical Thinking: A Historical Preface* (Princeton, NJ: Princeton University Press, 1982).

46. Foucault, *Birth of the Clinic*, 97.

47. P. J. G. Cabanis, *Du degré de certitude de la médecine*, 3rd ed. (Paris: Caille et Ravier, 1819), 125, quoted in Foucault, *Birth of the Clinic*, 99.

48. For more on the series as an important component of medical thinking, see Bernike Pasveer, "Representing or Mediating: A History and Philosophy of X-ray Images in Medicine," in *Visual Cultures of Science: Rethinking Representational Practices in Knowledge Building and Science Communication*, ed. Luc Pauwels (Hanover, NH: University Press of New England, 2006), 41–62.

49. Of course, I do not want to argue that this method is unique to medicine, since it is an age-old practice in science as well. But my argument assumes the successful integration of scientific principles into medical observation, even if this integration was at times contested. For more on this debate, see John Harley Warner, "Ideals of Science and Their Discontents in Late Nineteenth-Century American Medicine," *Isis* 82, no. 3 (September 1991): 454–78.

50. Fleck, "Some Specific Features of the Medical Way of Thinking," 40.

51. Laycock, *Lectures*, 169.

52. Didi-Huberman, *Invention of Hysteria*, 24–25.

53. Samuel A. Powers, *Variola: A Series of Twenty-One Heliotype Plates Illustrating the Progressive Stages of the Eruption* (Boston: Samuel A. Powers, 1882).

54. Photography also fits into the emphasis on repetition, demonstration, and visual stimuli common to nineteenth-century German pedagogy, especially the tradition of *Anschauungsunterrricht* ("visually based method of instruction"). This tradition was especially important to nineteenth-century medical education's investment in demonstration, as opposed to mere book learning. I cannot develop this connection here, but for an example, see S. Stricker, "Ueber den Anschauungs-Unterricht in den medicinischen Schulen," *Medizinische Jahrbücher* 82 (1886): 120–52, esp. 135.

55. "Photography in Pathology," *The British Medical Journal* (23 January 1886): 162–63.

56. On "visual pointing" in medical photography, see Martin Kemp, "'A Perfect and Faithful Record': Mind and Body in Medical Photography before 1900," in *Beauty of Another Order: Photography in Science*, ed. Ann Thomas (New Haven, CT: Yale University Press, in association with the National Gallery of Canada, Ottawa, 1997), 122. The abundance of detail was only one problem, however. Researchers quickly recognized that the photograph did not always depict faithfully and that what they saw was not always what they got in the image. For example, until the introduction of orthochromatic and panchromatic photographic emulsions in 1884 and 1905 respectively, the reddish-yellow color of anatomical specimens came out too dark in photographs unless special precautions were taken. See Gernsheim, "Medical Photography in the Nineteenth Century," 87. Kelly Wilder, in *Photography and Science*, rightfully emphasizes that standardization was definitely not photography's strong suit in the nineteenth century. For more positions against photography, see Daston and Galison, *Objectivity*, 161–79.

57. Richard Kretz, "Die Anwendung der Photographie in der Medicin," *Wiener*

klinische Wochenschrift 7, no. 44 (1 November 1894): 832. All translations are my own unless otherwise indicated.

58. Ibid., 832.

59. Ludwig Jankau, "Die Photographie im Dienste der Medizin," *Internationale medizinisch-photographische Monatsschrift* 1, no. 1 (January 1894): 2. See also Jankau, *Die Photographie in der praktischen Medizin* (München: Seitz & Schauer, 1894).

60. Jankau, "Die Photographie im Dienste der Medizin," 3.

61. Kretz, "Die Anwendung der Photographie in der Medicin," 833.

62. Lorraine Daston, "On Scientific Observation," *Isis* 99, no. 1 (2008): 110.

63. Theodor Billroth, *The Medical Sciences in the German Universities: A Study in the History of Civilization*, trans. William H. Welch (New York: Macmillan, 1924), 52–53. Originally published in 1876.

64. Dr. J. F. Halls Dally, "On the Use of the Roentgen Rays in the Diagnosis of Pulmonary Disease," *The Lancet* (27 June 1903): 1806.

65. Lorraine Daston, "Fear and Loathing of the Imagination in Science," *Daedalus* 127, no. 1 (Winter 1998): 73–95.

66. Theodor Billroth, *The Medical Sciences in the German Universities*, 102.

chapter four

Cinematography without Film
Architectures and Technologies of Visual Instruction in Biology around 1900

Henning Schmidgen

Johann Nepomuk Czermak was obviously satisfied, probably even proud: "In order to give an idea of the viewability of this surprising demonstration, I only mention that the diameter of the silhouette of the still-beating heart appearing on the wall was about two meters." On December 21, 1872, the physiologist had inaugurated his "Private Laboratory at the University of Leipzig." The climax of the ceremony was the extremely enlarged projection of an isolated, still contracting frog heart in the darkened lecture hall (figure 4.1). As Czermak underlined in his opening speech, more than 400 spectators were able, by means of this "surprising demonstration" to observe minute details in the contraction of the frog heart "that escape or are hardly visible to the unaided eye."[1]

The "viewability" (*Schaubarkeit*) Czermak referred to in this context, was part of his comprehensive conception of visual instruction in experimental physiology. At the beginning of his speech, he had emphasized the crucial role of *Anschauung* in physiological teaching. For the teaching physiologist, the decisive challenge was, so he explained, "to create and acquire the means for physiological demonstrations in a hitherto unachieved completeness and extension, so that—once this would have been accomplished—the attempt would become possible to treat physiology for the first time by means of authentic and generally conceivable representations based on immediate visual perception" (*unmittelbarer Anschauung*).[2]

It was personally gratifying to him, then, to list the special features of the laboratory building that he had erected using his own financial means,

4.1 Czermak's frog heart projection in the Leipzig Spectatorium (1872).
Reproduced from: Anonymous, "Czermak's physiologisches Privatlaboratorium und Amphitheater in Leipzig," *Illustrirte Zeitung* 1556 (1873): 305.

in particular the architectural design and technological equipment of the lecture hall. For Czermak, this hall was not a mere auditorium, but at the same time and above all a *spectatorium*, a "viewing hall."

In this spectatorium the seat rows were arranged in the form of a horseshoe around the lecturer so that the distance between the "experimenter's arena" and the audience was reduced to a minimum. By arranging the ascending seat rows according to strict geometrical principles, spectators could focus their gaze on this arena without disturbing or obstructing each other. In the "Optical Room," situated at the top and behind the last row of the seats were installed two limestone light projectors. With these, images were projected on a circular screen behind and well above the lecturer. Czermak used these projections toward the end of his speech to demonstrate various "images, preparations and living processes" in the hall, now darkened: photographs of embryos, polished sections of bones, a colored photograph of a dog's knee and a cross section of human skin tissue. While the subjects of these images still belonged to the research areas of developmental biology, histology, and anatomy, he moved into the realm of experimental physiology when presenting his "artificial circulation model." In this demonstration, an excised, still-contracting heart cut out of a frog served as the "natural motor" driving a salt solution through rubber and glass tubes.

After Czermak's assistants had prepared the model, they installed it in front of one of the projectors situated at the back of the room. In the darkened theater, the silhouette of the still-contracting heart became visible on the circular screen. In his concluding remarks, Czermak emphasized that this demonstration was not just pedagogical but also of scientific value. Because of its considerable enlargement it became evident "that the irregular tetanic and peristaltic contractions during the gradual necrosis of the heart due to increasing heat displayed a variety that up until now one had hardly a sufficient notion of."[3] By visual means the Leipzig physiologist thus played in two ways the role of "foreman of creation."[4] Czermak had isolated a central organ of all animal life and kept it alive outside of the body, exhibiting its functioning in front of a large audience. At the same time, he literally brought to light how death took possession of this organ.

The present paper investigates the architectures and technologies used for projecting living frog hearts in the physiological lecture halls of the late nineteenth and early twentieth centuries. From the 1860s, the frog heart became an increasingly important actor in physiological and pharmacological research. In this period, physiologists such as Elias von Cyon, Henry Bowditch, Luigi Lucian, Hugo Kronecker, Friedrich Martius, and Oscar Langendorff developed a growing interest in the innervation of this organ, its irritability and rhythmicity, and the problems of nutrition and fatigue.[5] In 1870, Oswald Schmiedeberg, one of the future "founding fathers" of experimental pharmacology in the German-speaking world, published his path-breaking "Investigations into Some Effects of Poison in the Frog Heart" in the *Arbeiten aus der Physiologischen Anstalt zu Leipzig*, edited by Karl Ludwig. Between 1880 and 1900, German physiology journals continued to publish numerous pharmacological studies exploring the effects of ether, strychnine, muscarine, ethyl alcohol and other substances on the functions of the frog heart.[6]

The growing interest in physiology and pharmacology went hand in hand with the development of new techniques for visualizing and measuring the activities of the frog heart, for example devices for registering changes in volume of the isolated frog heart, instruments for pumping various kinds of fluids through it, frog heart manometers, and apparatuses for the graphical recording of heart movements. However, the majority of these instruments were designed for laboratory use, not for demonstrations in the lecture hall. It was again Czermak who introduced such demonstration devices into the teaching of physiology. When giving popular lectures on the physiology of the heart in the Jena *Rosensaal* in 1868, he demonstrated the rhythmic movements of this organ by means of a "heart mirror," a cardioscope specifically designed for this purpose. A still-contracting heart, removed from a frog's chest, was placed on a small stand. Small pieces of cork, connected with angled steel sticks, were placed on the two heart chambers. Attached to the end of each stick was a light mirror plate that moved backward and forward while the corresponding heart chamber was contracting. By means of a device "similar to a magic lantern," a ray of light was directed at both mirror plates. The moving plates reflected the light onto a screen. In the darkened lecture hall the audience could thus see an enlarged image of the heart's movements.[7]

Against this background, the present paper examines the possibilities and problems of biological viewing created by in vivo projections of physiological and pharmacological manipulations of the isolated frog heart between 1910 and 1925. In particular, the paper draws on the case of the German physiologist and pharmacologist Carl Jacobj (1857–1944) who made extensive use of "episcopic" frog heart projections in his lectures and seminars at the pharmacological institute of Tübingen University.[8] Powerful light sources, especially the carbon arc lamp, permitted Jacobj to create projections not only by passing light through an object (transparent, or *diascopic*, projection), but also onto an object, that is, by having the object directly reflecting the light ray (opaque, or *episcopic*, projection). Starting in 1908, he used episcopic projection to show, in a manner as direct as possible, simple pharmacological experiments in living organisms and organs to a growing number of students.[9]

Jacobj was one of the pioneers of pharmacological science in Germany. In the 1880s, he had studied medicine in Göttingen, Leipzig, Tübingen, and Strassburg, with Karl Ludwig and Oswald Schmiedeberg among others. Under the direction of Schmiedeberg, he received his medical degree in Strassburg in 1887, for which he completed the study "On the excretion of iron from the animal body after subcutaneous and intravenous injection." In the following decade, Jacobj became Schmiedeberg's assistant in the newly established Institute for Pharmacology in Strassburg. After a brief interlude at the National Institute of Health in Berlin, Jacobj held the position of assistant professor of pharmacology at the University of Göttingen from 1897 to 1907. Shortly after, he was promoted to full professor and given the directorship of the new pharmacological institute in Tübingen, where he remained until his retirement in 1927.

As Sabine Waldmann-Brun has recently shown, Jacobj's reputation as an excellent teacher played a decisive role in his promotion from Göttingen to Tübingen.[10] It is hardly surprising, then, that in planning and organizing his new institute, Jacobj placed special emphasis on the design of the lecture hall and its technical equipment. In particular, he aimed at creating a "lecture projection facility" (*Vorlesungsprojektionseinrichtung*) that served the special needs of pharmacological teaching. The architecture and technology of this facility was meant to allow for a kind of "*Anschauungsunterricht*" that would no longer rely on static illustrations (e.g., the presentation of charts and slides), but rather be structured as a seamless integration of dynamic projections (e.g., the living frog heart).

The distinctive feature of Jacobj's projection images consisted in being a kind of cinematography without film. It is true that, from early on, pioneers of the graphic method attempted to set their chronophotograhic images into motion. In the early 1880s, Eadweard Muybridge created a projection device called the "zoopraxiscope" in order to project such images from rotating glass disks to give the impression of motion. In contrast, Etienne-Jules Marey was not interested in projection per se, but rather in applying this technique for analyzing motion by "slowing down some movements and speeding up others."[11] After the advent of cinematography around 1900, scientists such as Charles-Emile François-Franck, Lucienne Chevroton, and Charles Richet used chronophotography and cinematography for capturing the pulsing of the heart in various kinds of organisms.[12] But it was not before the 1930s that similar films—for example, by René Lutembacher and Emil von Skramlik—were made for and used in the academic teaching of biology.[13] What, around 1915, appeared on the screen of Jacobj's lecture hall in Tübingen, was all together different. Jacobj presented colored movement images. Their repeated projection did not rely on chronophotographic or cinematographical shots that had been made beforehand, but on the renewed presence of a living being that was "directly" projected.

In order to better grasp this kind of image production, this chapter highlights the architectures and technologies where it took place. As Norton Wise has recently observed, "historians of science traditionally have devoted relatively little of their attention to the means of producing images."[14] Against this dominant stream, the following study offers a detailed reconstruction of the heterogeneous assemblages that were devised and applied for creating movement images of the living in the visual instruction of biology. As will be shown in detail, the Tübingen lecture projection facility was based on the "Universal Projection Apparatus" by Leitz. Around the turn of the century similar projection devices were offered by various instrument makers in Europe and the United States, for example, the "Epidiascope" by Zeiss in Jena and the "Universal Projectoscope" by Stoelting in Chicago. When used for the projection of simple animal experiments, these projection devices were transformed into rather complex assemblages of organic and mechanical parts, electrical current and light rays, lenses, carbon rods and mirrors, frogs, wooden boards, and water jars. In order to describe their material complexity, Jacobj in his writings used the term *Anordnung*, meaning "arrangement," whereas other historical actors went so far as to speak about their projection devices as "complex organisms." Tied together by means of a cast iron frame, the combined elements of these image-producing assemblages reached their maximum effect within architectural conditions that allowed for projections "from behind" (a technique often used in magic lantern shows, in particular phantasmagorias, and the early cinema). Jacobj was convinced that only this kind of screen practice guaranteed the synchronicity of word and image, concept and object that according to him was the decisive feature of all visual instruction.[15]

As will be demonstrated, the Tübingen projection technology and architecture was a hybridization of two distinct traditions in the material culture of scientific visualization: on the one hand, the separation between the lecture hall and an adjacent preparation room that emerged in experimental physiology in the 1870s and, on the other hand, the use of episcopic projection for creating images of macroscopic organ preparations in experimental pathology in the 1880s. Since the late 1870s, physiological institutes such as the one in Budapest were provided with preparation rooms behind the lecture halls that were used for arranging demonstration experiments that were brought into the lecture hall on rolling tables. However, institutes of pathology, for example, in Vienna and Berlin, were pioneering episcopic projections of preparations and experiments. Leading figures in these contexts were Salomon Stricker (1834–1898) and Carl Kaiserling (1869–1942).[16] Jacobj's lecture projection facility brought together these architectures of physiology and apparatuses of pathology, insofar as, at the Tübingen institute, it was possible to project simple experiments by means of episcopic projection out of the preparation room into a lecture hall where the corresponding images became visible on a screen.

The reconstruction of this material culture of image production shows that, in contrast to Marey's famous techniques for graphic recording, episcopic projections of the moving frog heart did not simply serve to "picture time."[17] Rather, these projections made time "palpable" in various ways. In fact, Jacobj based his visual practice on a physiological theory of the primacy of the visual in both the emergence and distribution of knowledge. As a result, a rationalized mode of teaching replaced the idea of *Anschauung* that in many nineteenth-century physiologists was still provided with romantic overtones. In Czermak's visual practices it was crucial to demonstrate that instruments of experimental physiology could create an image of life that was just as *anschaulich* as the image provided, say, by natural history or embryology. Jacobj's central problem, however, was how an increasing number of students would be able to absorb the rapidly growing body of knowledge produced in and by the experimental life sciences in a limited period of time. In other words, Jacobj's presentation of movement images of the living was not an end in itself. It served a specifically modern economy of time and attention.[18]

Visual Instruction and the Management of Time

Anschauung was one of the key concepts of philosophical idealism and romantic biology in late eighteenth- and early nineteenth-century Germany. In the writings of Immanuel Kant, Johann Wolfgang von Goethe, and Alexander von Humboldt, its meaning oscillated between "visual perception" and "intuition." Kant famously stated: "Thoughts without contents are empty, intuitions [*Anschauungen*] without concepts are blind."[19] In Goethe and Humboldt, sensory *Anschauung* designated an aesthetic principle guaranteeing the unity of the dispersed, manifold, and heterogeneous facts with which natural research was confronted. Goethe and Humboldt aimed at a comprehensive "view" (*Ansicht*) of nature as embodied in perceptible totalities or concrete "archetypes" that were apt to presume empirical knowledge and at the same time stimulate and direct further research. However, in the writings of the Swiss pedagogical reformer Johann Heinrich Pestalozzi, *Anschauung* was chiefly used to imply visual perception. Following Jean-Jacques Rousseau, Pestalozzi stressed the importance of vision, criticizing an exclusive textual mediation of knowledge through books and lectures.[20]

Starting in the late 1830s, physiologists highlighted the role of visual perception in acquiring experimental knowledge. As William Coleman and Richard Kremer have shown, Purkinje and other Prussian pioneers of physiological science followed Pestalozzi's arguments against an education completely relying on written texts and spoken words.[21] Even before the foundation of the first large-scale institutes for physiology in the late 1860s, physiologists such as Czermak and Jan Evangelista Purkinje (one of Czermak's academic teachers) used phenakistoscopes, mirrors, magic lanterns, and other optical media in order to create a new image of life in front of academic and general audiences.[22] The striking feature of this image of life was to be a movement image that itself was in movement— not a fixation of organic kinematics by means of relatively stable lines and points (as in the graphic method), but a dynamic representation of the equally dynamic movements of the living body. By means of animated drawings, and above all through the projection of prepared organs and organisms, the basic facts of physiology as a new kind of *anatomia animata* were made "anschaulich," or visually evident, and this contributed significantly to gaining recognition and respect for the emerging new science.

It is hardly surprising, then, that in the 1870s, when laboratories of physiology were created in growing numbers, the architecture of lecture halls and the available technologies for visual instruction received heightened attention. By means of wall charts and devices for hanging large drawings, through movable screens and powerful projectors, the use of flags, rolling tables, and entire series of specifically designed demonstration devices— e.g., Emil du Bois-Reymond's "contraction telegraph"—physiologists created in their teaching facilities a thoroughly temporalized visual representation of the functions of life with an almost immersive quality. In 1879, young Stanley Hall, who at that time was studying physiology in Berlin and other German cities, found it quite obvious that the innovative projections and demonstrations had transformed the physiological lecture hall into "a sort of theater."[23] And for many physiologists this turn away from mere lectures to carefully prepared spectacles was a logical step in the development of their discipline. With respect to the history of other sciences, du Bois-Reymond had declared at the opening of his Berlin physiological institute in 1877: "The physiological lecture hall must become a show stage [*Schaubühne*] for natural phenomena, just as the physical and chemical lecture hall, and from then on physiologists needed a teaching and research laboratory fitted to their specific needs."[24]

No doubt that Jacobj was familiar with these developments. As he explained in 1919, in his article "Visual Instruction and Projection" (*Anschauungsunterricht und Projektion*), the primary role of vision in the process of creating and distributing objective knowledge dictated that the teaching methods at German high schools and universities increasingly relied on visual means, for example, by inserting figures into text-books or using wall charts and demonstrations in the classroom. Jacobj continued that this kind of *Anschauungsunterricht* facilitated the acquisition of new knowledge, while at the same time—and this was decisive for him—"accelerated" it. Under the precondition that word and image were presented simultaneously, instruction could restrict itself to mere explanation of the image in question: "The symbolically descriptive word images [*Wortbilder*] of concepts can be replaced by the simultaneously created visual image [*Anschauungsbild*] that represents the factual object of observation immediately and in all its details, so that it [i.e., the visual image] is imprinted in a faster, stronger, and more sustainable way on the conscious mind and, as a consequence, on memory."[25]

Through the discussion of "immediate visual perception" of an object during class lectures, students came to better understand the object under view. More importantly, immediate visual perception saved time since it was sufficient to direct the students' attention to the details of the image shown by the instructor. As a result, Jacobj no longer understood "*Anschauungsunterricht*" within the framework of some romantic aesthetic. To him it had become a question of efficient time management.

For Purkinje, and even Czermak, "immediacy" had referred to a spatial closeness of the spectator to an object, an experience that afforded a special visual clarity or, as Czermak put it, "intense viewability." In Jacobj, the same term, *unmittelbar*, designated a temporal relation, a momentariness, or synchronicity of that which was viewed and that which was heard. Precisely along these lines Jacobj underscored that "by means of simultaneous co-action of the two sense impressions of hearing and viewing" instructors were able to achieve "a swifter and easier absorption of the material in shorter time."[26] According to him, the "immediate, simultaneous presentation" of word and image, concept and object was the only possible way for transferring the "vast amount of knowledge required by the profession and by life in the short time period of academic training," a strategy that obviously reflected Jacobj's confrontation with an ever growing number of students—between 45 and 75 in the years until 1918, and between 100 and 215 in the years from 1919 until 1923.[27]

Jacobj also offered a detailed criticism of the traditional methods of demonstration and projection. The basic pattern of this criticism was not unexpected. The spatial and temporal separation between general lectures and specific demonstrations, the distribution of images, books, preparations, and instruments around the lecture hall as well as the use of wall charts or hanging drawings, schemes, tables, and formulae—all this did not correspond to the "principle of unity of word and visual image" that, to Jacobj, formed the definition of *Anschauungsunterricht*. Thus he argued, when circulating books or instruments, the respective object moved through the seat rows, from eye to eye, whereas a wall chart remained permanently visible to all, even if the lecture had long come to its end. Jacobj even criticized the attempt to integrate standard projections into lectures. In this case, the usual "intermediary pauses" (*Zwischenpausen*) resulted in the decoupling of concept and image, and this was to be avoided: "Above all, the darkening of the auditorium when presenting images was an utterly annoying disturbance."[28]

This led Jacobj to another point. In addition to asynchronicities, visual instruction could be severely impaired by various means, all which might disturb and distract the student's attention. He argued that the prompt hanging and taking away of charts did not offer any solution to the problems described, since the corresponding handling of the charts by the lecturer or his assistant also distracted students. In projections, the noises of the darkening mechanisms, the operation of curtains and shades, the "disappearance" of students and professors in the dark as well as the noise of the working projector in the lecture hall and unavoidable calls such as "More focus, please! Too much! Wrong! Stop!"[29] and eventually the "blinding" of eyes when returning to daylight were immensely distracting, disturbing, and irritating. To Jacobj it was evident that the highly questionable result of all visual teaching proceeding along these lines did nothing but waste time. In contrast, the visual instruction method he had developed in Tübingen since 1908 appeared to be much more promising. It was based on the "principle of placing the apparatus in an adjacent room while using a transparent screen."[30] In the Tübingen Institute, this principle translated itself into an architectural structure where a preparation room situated behind the lecturer and his experimenting table allowed for projections from this darkened room into the lecture hall. In other words, the temporal unification of the spoken word and the projected image was produced by means of a separation in space (figure 4.2).

Next to the experimenting table was a large window connecting the lecture hall and the preparation room. The interior of the latter was painted black to avoid reflection. The window was in fact a screen of 2 meters square, consisting of fine linen covered with paraffin. This interface between front and back stage was of prime importance in Jacobj's lecture projection facility: it separated the image production by the assistant on

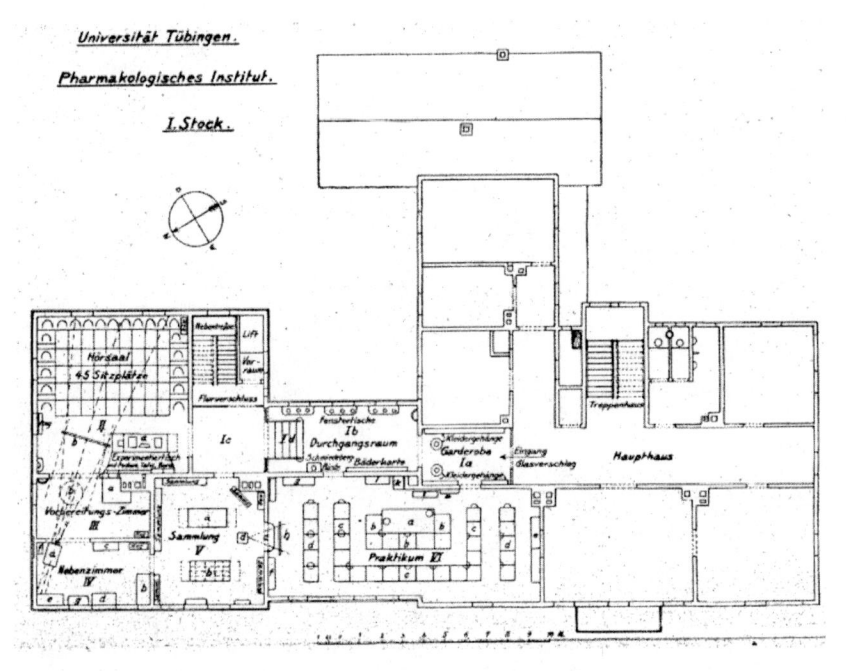

4.2 Plan of the Tübingen Institute for Pharmacology, first floor, with lecture hall and projection facility, ca. 1928 (left). Reproduced from: Carl Jacobj, *Das pharmakologische Institut zu Tübingen und seine Einrichtungen für Unterricht und Forschung*, Tübingen: Buchdruckerei der Tübinger Studentenhilfe, 1927, table I.

the back stage from the image consumption by the students sitting close to the front stage, while at the same time connecting them to each other, whereas Jacobj, standing next to the screen, could control the circulation of the images. No wonder, then, that, already in a 1910 article, he had reported extensively that only "after an extended series of preliminary trials" (*Vorversuchen*)[31] had it been possible to achieve the required semi-permeability of the linen screen by means of a special procedure.

According to Jacobj, the projection on the backside of this screen made it possible to avoid the darkening of the lecture hall. At the same time, a moveable blind allowed for showing the images prepared by the assistant at exactly the moment they were needed by the instructor. Jacobj explained that the projection from the front presented clear disadvantages, "since it means drawing curtains, arranging the projector, and having the speaker's personal contact with the students disappear, on account of their losing sight of each other."[32] At Tübingen, all necessary preparations for the projection, including the changing and focusing of images, were done by an assistant placed in the preparation room. The assistant would follow the lecture and had to know when to react to what the instructor was saying. Jacobj summarized the advantages of his system in the following way: "There is no disturbance attendant upon the projecting itself, no interruption of the speaker's flow of thought, all the students can see the image at the same time, at exactly the proper moment, and when this image has served its purpose, it can be made to disappear."[33]

Preparation Rooms and Episcopes

In his article, Jacobj suggested that it was the pharmacological institute in Strassburg that had first used a comparable lecture projection facility. For ten years Jacobj had worked as the lecture assistant to Schmiedeberg, and his long-term experience behind the scenes obviously influenced how he arranged devices and designed the lecture hall at Tübingen. In fact, the Strassburg building of the Institute for Pharmacology, constructed between 1883 and 1887 according to plans by architect Otto Warth, had a preparation room directly behind its auditorium.

As a contemporary description of this institute reads, this preparation room served to store the "animals meant for demonstration" for as long as they would be needed for a lecture. The room was also used as a storage site for other demonstration objects, such as drugs, chemicals, medicinal products, figures, and so on.[34] Jacobj did not refer to the fact that similar spatial structures had already appeared in other physiological institutes, which had explicitly embedded in their design ambitious projects for visual instruction.

This was certainly the case for the Budapest Institute for Physiology. Built between 1873 and 1876, the physiologist Andreas Eugen Jendrassik had planned the facility's layout with architect Antal Szkalnitzky. As other physiologists before him, Jendrassik was eager to offer "immediate visual perceptions [*unmittelbare Anschauungen*] of the object" of physiological science.[35] Similar to du Bois-Reymond, Jendrassik assumed that physiology, just like chemistry and physics, was capable of "making sensual" (*versinnlichen*) oral presentations by means of experimental demonstrations, thus turning the latter into the "experiential basis" (*Erfahrungsgrundlage*) for all theoretical considerations.[36] Implicitly alluding to Czermak, Jendrassik added that, when planning the new institute, his aim had been to integrate not just an auditorium, but specifically a "spectatorium" into the building. Jendrassik also touched upon issues of time management when he explained: "As soon as physiology aims at communicating experience-based knowledge to larger circles, it has to strive for appropriate means to guarantee that large groups will be given simultaneous insight into the articulated complex of organic processes."[37]

The "appropriate means" to which Jendrassik alluded were, first, the rather generous layout of the auditorium, giving space to 200 students; second, the calculated arrangement of the seat rows with respect to the large experimenting table; third, the illumination of the lecture hall by means of gas lamps that could be switched on by means of electricity; and fourth, the presence of blackboards and other equipment for presenting large drawings as well as projection devices and automatic shutters. In particular, it was the creation of a "preparation room for experiments" adjacent to the back of the lecture hall that should serve the purposes of his *Anschauungsunterricht*. The preparation room was connected to the auditorium by a door so that "tables with the animals prepared for experiment or other apparatus . . . can be transferred to the lecture hall."[38] This was facilitated by means of a system of tracks fixed to the floor. This system allowed for pushing the rolling tables back and forth. A revolving disk allowed them to move left and right in front of the seat rows so that the largest possible number of students was able to see the experiment. The surfaces of some of the these tables could even be inclined so that it became possible, as Jendrassik explained, "to turn the vivisection board with the animal fixed onto it toward the spectators as any given case requires it."[39] The connecting door between lecture hall and preparation room was covered with various boards that could be moved horizontally, among them large opal glass plates that could be used for "light projections from the preparation room situated behind."[40]

Similar separations between front and back stage can be found in other physiological institutes of the period, for example the one in Würzburg (founded in 1892) and Turin (1894). These spatial arrangements not only served the preparation of demonstration experiments. In Würzburg, for example, a sciopticon, an early form of the slide projector, was used to project images on the back side of a frame of 1.3 square meters, covered with a specially prepared canvas. One of the main advantages of this arrangement was the vibration-free position of the projector placed on a table made of stone in the back room. The projector could be handled without disturbing the students.[41]

Apparently, the projections realized in the physiological institutes at Budapest, Würzburg, and Turin, were mainly diascopic. In contrast, epi-

4.3 Stricker's use of episcopic projection technology in the lecture hall, ca. 1890. Reproduced from: Anonymous, "Das Episkop," *Ueber Land und Meer: Deutsche Illustrirte Zeitung*, 64, no. 26 (1890): 1028.

scopic projections were initially used at institutes for pathology. However, in these institutes the separation between lecture room and preparation room often did not exist. Instead, all devices required for episcopic projection were placed in the auditorium itself. The pioneer of this technique was the former Brücke student Salomon Stricker. Starting in 1868, Stricker directed the Institute for General and Experimental Pathology at the University of Vienna. His starting point was the projection of microscopic preparations. Since the early 1880s, he projected such preparations by means of a Dubosq lantern and a vertically mounted microscope. By means of an enhanced "projection microscope," manufactured according to Stricker's needs by the well-known optical workshop of Simon Plössl in Vienna, he presented, in the 1880s and 1890s, improved projections of microscopic preparations during his lectures and at scientific meetings and congresses, such as the annual meeting of German naturalists and physicians in Berlin in 1886 (figure 4.3).

In the same period, Stricker started to use projection and other technologies to show animal experiments in the lecture hall. In experiments on respiration, he worked with enlarged shadows and flags that were fixed to the chest wall of the experimental organism, while the kymographic registration of the animal's circulation took place on a glass plate that, while the experiment was taking place, moved horizontally through a sciopticon. In this way, "even students in the most remote seats [were able] to perceive all details of the pulse curve with highest precision."[42]

In addition, Stricker applied episcopic projection for demonstrating in vivo experiments in the lecture hall. The episcopic projection device that he made use of was developed by his assistant Max Reiner. Similar to Jacobj in the field of pharmacology, Stricker used this device mainly for showing heart movements. Looking back on the history of physiology since Harvey, he underscored the importance of this kind of demonstration: "This form of observation [i.e., the observation of the isolated heart] might be one of the simplest, but it has also allowed the highest achievements."[43] With respect to teaching, he also highlighted that it was "of highest importance to show the pulsating heart to the future physician."[44]

By means of episcopic projection, Stricker demonstrated in his lectures, among other physiological phenomena, the influence of the heart nerves for inhibition and acceleration and the behavior of the heart in suffocation, and this "in the most visible and instructive [*anschaulichen*] way," as his assistant Gärtner noted.[45] Stricker himself chose a different register when describing the effects of his heart projections: "The pulsing heart appears on the white wall as an enlarged, relief-like [*plastisch*], and, if I may dare to say, it is a living image that, in all its details, is perfectly visible to hundreds in the audience."[46] Stricker alluded here to the genre of "tableaux vivants," highly popular in the nineteenth century, suggesting the power of the new, lifelike images of physiological processes and phenomena.

Shortly after, episcopic projections were used in the teaching of pathology at the Charité in Berlin. As in Stricker's institute, the starting point was the facilitation of diascopic projections of microscopic preparations. In order to bring the macroscopic preparations of Virchow's substantial collection under the eyes of large groups of students, the curator of the pathological museum, Carl Kaiserling, devised a "universal projection apparatus." Kaiserling closely cooperated with the optical manufacturer Ernst Leitz in Wetzlar. The first description of the new projection device was published by Kaiserling in 1906. It was meant for both episcopic and diascopic projections. Within a large frame of cast iron, its main components were a collapsible carbon arc lamp with perpendicular carbon rods, an optical bench to which various kinds of attachments could be fixed (for example, a slide changer), and a small kind of table that served to place objects under the beam of light. It was the surface of this table that Kaiserling used to produce projections of macroscopic preparations during the lectures of the institute's director, Johannes Orth. One method consisted in taking the preparations out of their jars and reducing their glossiness "by dabbing away the fluid," the other in directing the light beam through the jar and the fluid surrounding the preparation in question.[47]

Paradoxically, Kaiserling described his projection device as a "complex organism that one has to know precisely."[48] He underscored that its appropriate use during lectures required "quite a bit of good will, lots of patience, and long practice."[49] At the same time, he stressed the necessity of using this device if one really wanted to offer effective training in pathology. In his eyes, academic training in pathology was challenged most by the "lack of time and large number of listeners."[50] As we have already seen, Jacobj in Tübingen confronted similar circumstances. As in Kaiserling, episcopic projection was an efficient technology in order to teach in rapid ways a growing number of students "how to see" as a scientist and as a physician.

The "Pandidascope" and Its Critics

Technically speaking, the lecture projection facility that Jacobj used at his pharmacological institute from 1908 was based on Kaiserling's Universal Projection Apparatus. According to his wishes, Leitz had furnished Jacobj with a modified projection apparatus "with specific arrangements for all projections that are important in the teaching of pharmacology, e.g. the projection of the living frog heart and muscle."[51] These modifications mostly concerned the size of the table inside of the projection apparatus. Over the years, Jacobj made other minor modifications, in cooperation with the "well-known Tübingen workshop for precision mechanics, E. Albrecht."[52] In order to highlight his contributions to the Kaiserling apparatus, Jacobj spoke of the Tübingen "Pandidascope" (figure 4.4).

The striking feature of this apparatus was the "extraordinarily powerful lamp," which resulted in an "excellent light power of the projection images."[53] As in Kaiserling, the lamp consisted of perpendicular carbon rods. With 30 amperes current intensity and a voltage of 60, it produced a brightness of 10,500 candles. The lamp and its lenses were movable on an optical bench and permitted a stream of light to be directed upon almost any kind of object without any change in the direction of the optical axis. One could attach optical devices such as microscopes or slide changers.

110 · *The Educated Eye*

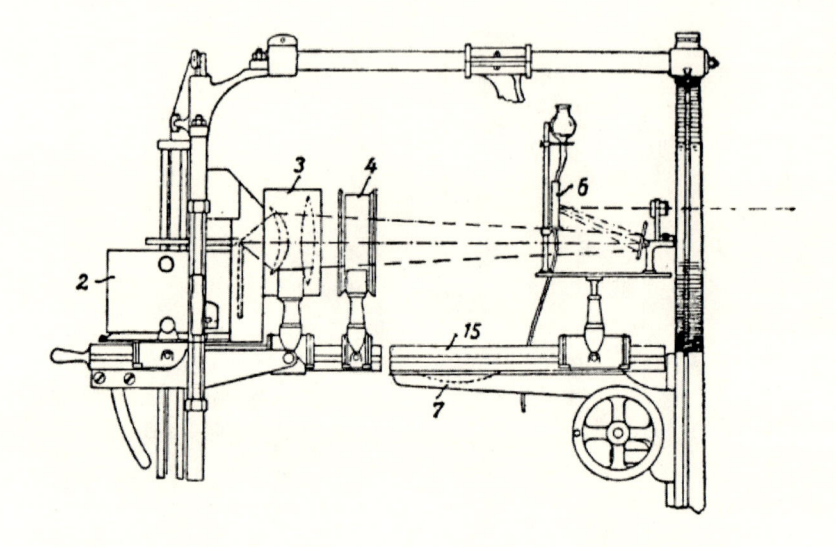

4.4 Diagram showing the use of Jacobj's projection apparatus for frog heart projections. Reproduced from: Carl Jacobj, "Visual Instruction and the Projection Method," *Methods and Problems of Medical Education* 6 (1927): 264.

As Jacobj explained, this arrangement allowed for different kinds of projections; for example, under direct light the projection of vertical slides of spectra and of microscopic objects was possible, while under reflected light one could project large horizontal slides of opaque objects. Similar to the design of Jendrassik's lecture hall in Budapest, the Tübingen projection apparatus was mounted on fixed tracks so that the distance to the screen and hence the size of the projected image could easily be varied.

For projecting the functioning of a frog heart, the following arrangement was used. At 20 centimeters behind lenses 3 and 4 Jacobj placed a water container provided with a glass front and set into a horizontal plate raised on two small pillars: "A frog whose chest wall has been opened so as to lay bare the heart is spread upon a small piece of board and the whole is immersed in the water, with the board so attached to two flat springs screwed into the sides of the container that the heart is directly in front of the opening of a tube that comes up at an angle from below and that terminates near the middle of the container and close to the glass front. The tube carries a flow of neutral salt solution to the heart from a second container above the first one to prevent the heart from being injured by the strong heat generated by the light."[54]

The results of this arrangement were remarkable. Evocative of Czermak's much earlier statements, Jacobj explained that this arrangement "makes it possible to project the tiny frog heart, hardly one centimeter square, with such brilliancy that even magnified over 400 times its life processes can be directly seen with the naked eye by sixty to eighty spectators."[55] Here again, the experimenting life scientist in the lecture hall appeared as a "foreman of creation." He not only controlled a central function of organic life, at the same time he made it visible to a large number of spectators.

The principles and techniques of Jacobj's visual instruction method were met with considerable international interest. In 1927, his article "Anschauungsunterricht und Projektion" was published in English translation in *Methods and Problems of Medical Education*, a journal edited and published by the Rockefeller Foundation between 1924 and 1932. In 1928, the same journal published a description of the Tübingen pathological institute by Jacobj, based on his extended German publication from the year before.[56] On the pages of the Rockefeller Foundation journal, the young Morton S. Biskind (1906–1981), at this point Fellow in Pharmacology at the School of Medicine at Western Reserve University, responded to Jacobj. In his article, "The Demonstration of Biological Experiments by Optical Projection Methods," Biskind underscored the importance of visual instruction in the teaching of physiology and pharmacology.[57]

At the same time, Biskind criticized Jacobj's method in several respects and suggested an alternative method for projecting frog hearts in the lecture room. On the one hand, he objected that the Tübingen pharmacologist did "not enter sufficiently into the details of the projection of demonstration experiments."[58] On the other hand, he raised some general doubts as to the use of the episcopic projection apparatuses that were available on the market. In Biskind's eyes, these apparatuses presented "serious disadvantages." They were too large and cumbersome, their interior gave only limited space for objects and experiments and, above all, they were expensive.

Biskind did not offer any general discussion of the theoretical or scientific foundations of *Anschauungsunterricht*, as Jacobj had done. In his introduction, he offered a sketchy history of the magic lantern and its use for scientific purposes: from Baptista Porta's camera obscura to Athanasius Kircher's magic lantern and Salomon Stricker's episcopic projection method. The body of the paper, however, focused on the technical issues involved in visual instruction of future physiologists and pharmacologists. The overall tone of his article was pragmatic. Whereas Jacobj had discussed the scientific foundations of his visual instruction method at length and used ample space for the criticism of traditional projection methods, Biskind restricted himself to suggestions and advice for other practitioners.

In particular, the projection methods that he suggested were meant to correspond to the following criteria: adaptability to the working conditions of a variety of experiments, simplicity and ease of adjustment, and use of the less expensive apparatus whenever possible. As a result, Biskind did not refer to a specific projection apparatus by a specific manufacturer and did not speak about tailored modifications of such apparatuses. Rather he presented various projector elements or modules made by various manufacturers and firms that could be used to assemble projection devices in flexible ways. The components he discussed in detail were "lighting," "lenses," and "screens."

Similar to Jacobj, Biskind argued in favor of translucent projection from behind the screen (the pharmacological lecture and demonstration room at Western Reserve was provided with appropriate architecture) but recommended tracing cloth, not linen, as the projection surface. Drawing on concrete examples, Biskind then described the combination of all required modules for projecting the frog heart, intestinal preparations and uterine strips, intact and excised rabbit intestines, as well as physical models. Biskind was obviously eager to offer descriptions for practical use. Thus he described the preparation of the required technological components as well as the required organisms in detail, while dealing at the same time with the "adjustment" of technological and biological components to one another for the eventual projection. He listed firms and prices of various devices, gave practical hints ("It is advisable to wear dark glasses during this adjustment and during all manipulation of the frog while projecting"),[59] and concluded his article with a checklist of optical equipment. For projecting the frog heart, he suggested the following procedure: The frog is placed horizontally on a board where it may be subjected to any desired manipulation during the actual projection. . . . The short focus lens (6.5- to 8.5-inch focal length) is used with a reversing mirror. The projection distance for effective magnification should be 7 to 9 feet. For illumination, a 15-ampere current carbon-arc is employed. . . . The back condensing lens is removed, leaving only the front lens of 7.5-inch focal length. This permits having the light source at a considerable distance from the object (i.e., the frog heart), which allows sufficient room for the insertion of a cooling tank in the path of the beam.[60]

As Biskind had explained earlier in his article, the cooling device should be a water chamber of 15 centimeters in thickness, preferably "a rectangular museum jar."[61] If one used smaller jars, for example, 5 centimeters, the water would heat and begin to boil within an hour. The problem was not only that the heat would eventually make the frog useless for projection but that an optical problem arose as well: "The bubbles in the hot liquid interfere seriously with the transmission of light."[62] Biskind then presented another method for cooling the frog during the projection process. This method relied on removing the excess heat as it is generated in the object. As was the case in Jacobj's arrangement, this heat removal was achieved by placing the entire frog inside a water chamber in which a continuous stream of cold water was circulating. Biskind described this method in considerable detail, emphasizing that it would make it possible to keep a frog's heart beating normally for over six hours under continuous intense illumination.[63] In contrast, the time span for viewing under the normal cooling method was approximately only one hour.

Jacobj, in his publication on the same topic, did not address the time span during which the functioning frog heart could be projected. Biskind's rather extended discussion of this point suggests that this was one of the details he had missed in the article of his senior colleague from Tübingen. Jacobj had restricted himself to discussing issues of time management exclusively with respect to the relation between lecturer and student, focusing on the synchronicity of the lecturer's word and the projected image. Biskind had a different focus. To him, the extension of projection time was an achievement in itself. The comparison with earlier procedures of projection seems indeed striking. In Czermak's projection of the isolated frog heart, projection time was a matter of minutes in which the audience could witness the transition from lively contraction to gradual necrosis. In Biskind's projection of the heart laid bare it was a question of hours, almost days. What became possible was an extended manipulation of this organ—even though lectures would hardly take six hours.

Conclusion

The imagery of experimental physiology has often been identified with the "graphical method." As impressive as the epistemic productivity of this method certainly was, and as attractive as its aesthetics aspects remain until today, the up-to-now emphasis by historians on traces, curves, and graphs of bodily functions have blocked from view a highly diversified production of physiological images that were produced, from the late nineteenth century to the early twentieth century, in order to bring the functioning of organs and organisms to "immediate visual perception," or *unmittelbare Anschauung*. Since the 1870s, phenakistoscopes, magic lanterns, and rolling tables transformed the physiological lecture hall into a kind of theater where the contractions and pulsations of organic life were shown by means of various kinds of movement images—animated drawings, shadow projections, "contraction telegraphs"—to growing audiences. At the turn of the twentieth century, improved lamp technology made possible episcopic projections of in vivo experiments. Leitz's Universal Projection Apparatus and similar devices were used for this purpose in various contexts—in physiology, pathology, and pharmacology.

The recurring motif of these projections was the isolated heart of the frog—a "simple" demonstration, as one of the historical actors admitted, but still one that was deemed to be crucial in the training of future scientists and physicians. The persistence of this motif over a period of more than sixty years (from Czermak's cardioscope to Biskind's modular projection method) reflects the central importance of heart, circulation, and respiration in all forms of animal and human life. It is also due to the relative ease with which the effect of physiological and pharmacological interventions could be demonstrated in this organ. However, the respective publications by Czermak, Jacobj, and Biskind speak also about the gaze as a means of power to control the organisms. Episcopic projections in the lecture hall made it possible to share this gaze with a large group of spectators. Jacobj presented this multiplication of gazes as a process that was rooted in physiology. To him, it was a physiological fact that vision is the key to objective knowledge. In a sense, then, it was not just the organism of the frog that was subjected to vision. His students, as sensate organisms, were part and parcel of this spectacle.

The graphical method has often been described as a means for "picturing time." Something similar can obviously be said with respect to the architectures and technologies of visual instruction in biology around 1900 that were presented and discussed in this chapter. The projection of the contracting frog heart embodies as well as illustrates the pulse of time. The change between systole and diastole that this organ displays marks duration for the eye just as the ticking of a clock marks it for the ear. In contrast to the products of the graphical method, however, the projected images did not leave any durable trace. This is because here time is not only pictured, but manifests itself and is *felt* by the historical actors in various forms and formats: as the time that the gaze is offered to see the organ in its functioning during a lecture, as the time that is needed for the preparation of the organism, the projection device, and their mutual adjustment, but equally as the time that went into the production of the screen tissue, as the time that was required for the technical development of universal projection apparatuses and for the building of institutes, and eventually as the time that is available for the emergence and the transfer of scientific knowledge.

What the heterogeneous projection assemblages devised in the experimental life sciences of the early twentieth century thus make clear is that time becomes "palpable" in complex ways. As a consequence, the history of these assemblages inscribes itself into the history of modernity. Since the emergence and evolution of these projection assemblages is largely independent of the rising technology of cinematography, they are not simply part of the prehistory of the cinema. What the projection technologies of Jacobj, Biskind, and other scholars show is a kind of deconstructive history of cinematography, a history that disassembles the black box of the "basic cinematographic apparatus" in order to redistribute its elements without making any claim to be complete: the movement image, the projector, the screen, the seat rows, and so on. In the place of a filmstrip that can be shown again and again, without any substantial alteration, the projection assemblages presented in this chapter integrate a living object. This object has to be exchanged and replaced by a new, similar living object, if the projection is meant to start again. The result can be understood as a cinema of life in which the repetition of the show remains tied to the differences of organic individuals.

Notes

This chapter was finalized in the context of the interdisciplinary research project "The Experimentalization of Life: Configurations between Science, Art, and Technology (1830–1930), Max Planck Institute for the History of Science, Berlin (Department III: Hans-Jörg Rheinberger). Numerous historical sources quoted in this chapter are online, available in the Virtual Laboratory of this project (http://vlp.mpiwg-berlin.mpg.de/index_html). Thanks to Oliver Gaycken, Robyn Smith and the editors of this volume for their stimulating comments and suggestions.

1. Johann Nepomuk Czermak, "Ueber das physiologische Privat-Laboratorium an der Universität Leipzig," in *Gesammelte Schriften in zwei Bänden* (Leipzig: Verlag von Wilhelm Engelmann, 1879), 2:144.
2. Czermak, "Ueber das physiologische," 119.
3. Ibid., 144.

4. Claude Bernard, *An Introduction to the Study of Experimental Medicine*, trans. A. M. Henry Copley Greene (New York: Dover Publications, 1957), 18.

5. See, for example, Elias von Cyon, "Über den Einfluß der Temperaturveränderungen auf Zahl, Dauer und Stärke der Herzschläge," *Arbeiten aus der Physiologischen Anstalt zu Leipzig* (1866): 77–127; Henry P. Bowditch, "Über die Eigenthümlichkeiten der Reizbarkeit, welche die Muskelfasern des Herzens zeigen," *Arbeiten aus der Physiologischen Anstalt zu Leipzig* (1871): 139–76; Luigi Luciani, "Eine periodische Funktion des isolirten Froschherzens," *Arbeiten aus der Physiologischen Anstalt zu Leipzig* (1872): 113–96; J. Steiner, "Zur Innervation des Froschherzens," *Archiv für Anatomie, Physiologie und wissenschaftliche Medicin* (1874): 474–90; Hugo Kronecker, "Ueber die Speisung des Froschherzens," *Archiv für Physiologie* (1878): 321–22; Friedrich Martius, "Die Erschöpfung und Ernährung des Froschherzens," *Archiv für Physiologie* (1882): 548–62; Oscar Langendorff, "Studien über Rhythmik und Automatie des Froschherzens," *Archiv für Physiologie* (1884), Suppl.: 1–133.

6. See Oswald Schmiedeberg, "Untersuchungen über einige Giftwirkungen am Froschherzen," *Arbeiten aus der Physiologischen Anstalt zu Leipzig* (1870): 41–52; Hugo Kronecker, "Ueber die Wirkung des Aethers auf das Froschherz," *Archiv für Physiologie* (1881): 354–57; Moritz Löwit, "Beiträge zur Kenntniss der Innervation des Herzens, Dritte Mittheilung: VI. Die Deutung einiger Giftwirkungen am Froschherzen," *Archiv für die gesammte Physiologie des Menschen und der Thiere* 28 (1882): 312–42; Richard Rhodius and Walther Straub, "Studien über die Muskarinwirkung am Froschherzen bei erhaltenem Kreislauf, bes. über die Natur des Tetanus des Herzens im Muskarinzustand und die der negativ inotropen Wirkung auf die Herzmuskelzuckung," *Archiv für die gesammte Physiologie des Menschen und der Thiere* 110 (1905): 492–512; Hermann Dold, "Über die Wirkung des Äthylalkohols und verwandter Alkohole auf das Froschherz," *Archiv für die gesammte Physiologie des Menschen und der Thiere* 112 (1906): 600–622.

7. Johann Nepomuk Czermak, "Populäre pysiologische Vorträge, gehalten im akademischen Rosensaale zu Jena in den Jahren 1867–1868–1869," in *Gesammelte Schriften in zwei Bänden* (Leipzig: Wilhelm Engelmann, 1879), 2:2–15. On the history of other techniques to visually represent the function of the heart, see Robert G. Frank, "The Telltale Heart: Physiological Instruments, Graphic Methods, and Clinical Hopes 1854–1914," in *The Investigative Enterprise: Experimental Physiology in Nineteenth-Century Medicine*, ed. William Coleman and Frederic L. Holmes (Berkeley: University of California Press, 1988), 211–90.

8. On Jacobj, see Paul Pulewka, "Jacobj, Carl," in *Neue Deutsche Biographie*, ed. Historische Kommission bei der Bayerischen Akademie der Wissenschaften, Hufeland—Kaffsack, (Berlin: Duncker & Humblot, 1974), 10:239–40; Sabine Waldmann-Brun, *Carl Jacobj: Leben und Werk* (Inauguraldissertation zur Erlangung des Doktorgrads der Medizin, Medizinische Fakultät der Eberhard-Karls-Universität, Tübingen, 2008).

9. On the history of projection techniques and technologies, see Heinrich Dilly, "Lichtbildprojektion—Prothese der Kunstbetrachtung," in *Kunstwissen-

schaft und Kunstvermittlung, ed. Irene Below (Gießen: Anabas, 1975), 153–72; Thomas L. Hankins and Robert J. Silverman, "The Magic Lantern and the Art of Scientific Demonstration," in *Instruments and the Imagination* (Princeton, NJ: Princeton University Press, 1995), 37–71; Jens Ruchatz, *Licht und Wahrheit: Eine Mediumgeschichte der fotografischen Projektion* (München: Fink, 2003). For historical projection manuals that address the issue of physiological demonstrations, see Sigmund Theodor Stein, *Die optische Projektionskunst im Dienste der exakten Wissenschaften: Ein Lehr- und Hilfsbuch zur Unterstützung des naturwissenschaftlichen Unterrichts* (Halle a.S.: Knapp, 1887), 113–22; Lewis Wright, *Optical Projection: A Treatise of the Use of the Lantern in Exhibition and Scientific Demonstration*, 4th ed., London: Longman, 1906, 230–41. See also Oskar Zoth, *Die Projectionseinrichtung und besondere Versuchsanordnungen für physikalische, chemische, mikroskopische und physiologische Demonstrationen am Grazer physiologischen Institute: Als Leitfaden bei Anlagen und Versuchen* (Wien: Hatleben, 1895). For historical literature on episcopic projections, see, for example, Richard Neuhauss, *Lehrbuch der Projektion*, Zweite umgearbeitete Auflage (Halle a.S.: Knapp, 1908, 95–103); Simon Henry Gage and Henry Phelps Gage, *Optic Projection: Principles and Use of the Magic Lantern, Projection Microscope, Reflecting Lantern, Moving Picture Machine* (Ithaca, NY: Comstock Publishing Company, 1914), 166–99.

10. Waldmann-Brun, *Carl Jacobj*, 85–86.

11. For Muybridge, see Charles Musser, *The Emergence of Cinema: The American Screen to 1907*, New York: Scribner's Sons and Simon & Schuster, 1990 [History of American Cinema; 1], 48–54. For Marey, see Marta Braun, *Picturing Time: The Work of Etienne-Jules Marey (1830–1904)*, Chicago: University of Chicago Press, 1992, 174. In the early 1880s, Marey started to use a device called "polygraphe à projection" for projecting the process of inscribing physiological curves by means of a Dubosq lantern in the lecture hall. See Ch.-E. François Franck, "Procédés pour exécuter les figures destinées aux demonstrations à l'aide des projections," *Journal de Physique Théorique et Appliquée* 10, no. 1 (1881): 406–8. See also Zoth, *Die Projections-Einrichtung*, 50.

12. See Adolf Nichtenhauser, *A History of Motion Pictures in Medicine*. Unpublished manuscript, Bethesda, MD (ca. 1950), vol. II, 50–57.

13. See, for example, Emil von Skramlik, *Die Automatiezentren im Froschherzen* [Hochschulfilm-Nr. C 248], Jena 1938. On Lutembacher, see Thierry Lefebvre, "Les graphies du coeur: Les dispositifs oubliés de Dr Lutembacher," in *Images, science, movement: Autour de Marey* (Paris: L'Harmattan, 2003), 79–94.

14. Norton Wise, "Making Visible," *Isis* 97, no. 1 (2006): 75.

15. For the historiographical importance of the concept "screen practice," see Musser, "Toward a History of Screen Practice," in *The Emergence of Cinema*, 15–54.

16. On Stricker see Erna Lesky, *The Vienna Medical School of the Nineteenth Century*, trans. I. S. L. L. Williams (Baltimore: Johns Hopkins University Press, 1976), 499–506.

17. On Helmholtz, Marey, and the "graphic method," see Merriley Borell, "Extending the Senses: The Graphic Method," *Medical Heritage* 2, no. 2 (1986): 114–221; Anson Rabinbach, *The Human Motor: Energy, Fatigue, and the Origins of Modernity* (Scranton, PA: Basic Books, 1990); Braun, *Picturing Time*, 1992; François Dagognet, *Etienne-Jules Marey: A Passion for the Trace* (New York: Zone Books, 1992); Soraya de Chadarevian, "Graphical Method and Discipline: Self-Recording Instruments in Nineteenth-Century Physiology," *Studies in History and Philosophy of Science, Part A*, 24, no. 2 (1993): 267–91; Robert M. Brain and Norton Wise, "Muscles and Engines: Indicator Diagrams and Helmholtz's Graphical Methods," in *Universalgenie Helmholtz: Rückblick nach 100 Jahren*, ed. Lorenz Krüger (Berlin: Akademie-Verlag, 1994), 124–45; John W. Douard, "E.-J. Marey's Visual Rhetoric and the Graphic Decomposition of the Body," *Studies in History and Philosophy of Science, Part A* 26, no. 2 (1995): 175–204; Robert M. Brain, "The Graphic Method: Inscription, Visualization, and Measurement in Nineteenth-Century Science and Culture" (PhD dissertation, University of California, 1996); Laurent Mannoni, *Etienne-Jules Marey: La mémoire de l'œil* (Milano: Mazzotta, 1999); Robert Michael Brain, "The Pulse of Modernism: Experimental Physiology and Aesthetic Avant-Gardes circa 1900," *Studies in History and Philosophy of Science, Part A* 39, no. 3 (2008): 393–417; Henning Schmidgen, *Die Helmholtz-Kurven: Auf der Spur der verlorenen Zeit* (Berlin: Merve, 2009).

18. Jonathan Crary, *Suspensions of Perception: Attention, Spectacle, and Modern Culture* (Cambridge, MA: MIT Press, 1999). On the history of modernity as closely tied to the emergence of "palpable" time, see Mary Ann Doane, *The Emergence of Cinematic Time: Modernity, Contingency, The Archive* (Cambridge, MA: Harvard University Press, 2002), 4.

19. Immanuel Kant, *Critique of Pure Reason*, trans. P. Guyer & A. W. Wood (Cambridge: Cambridge University Press, 1998), 193–94.

20. On *Anschauung* in philosophical context, see Werner Flach, *Zur Prinzipienlehre der Anschauung* (Hamburg: Meiner, 1963); Manfred Frank, "Intellektuale Anschauung: Drei Stellungsnahmen zu einem Deutungsversuch von Selbstbewußtsein: Kant, Fichte, Hölderlin/Novalis," in *Die Aktualität der Frühromantik*, ed., Ernst Behler and Jochen Hörisch (Paderborn: Ferdinand Schöningh, 1987), 96–126; Friedrich Kaulbach, "Anschauung," in *Historisches Wörterbuch der Philosophie, Bd. 1: A–C*, ed. Joachim Ritter (Darmstadt: Wissenschaftliche Buchgesellschaft, 1971), 340–47. On *Anschauung* in the context of pedagogy, see Ines Roeder, *Das Problem der Anschauung in der Pädagogik Pestalozzis* (Weinheim: Beltz, 1970). On the importance of *Anschauung* in Romantic German biology, see Timothy Lenoir, *The Strategy of Life: Teleology and Mechanics in Nineteenth-Century German Biology* (Dordrecht: Reidel, 1982), 54–111; Lynn K. Nyhart, *Biology Takes Form: Animal Morphology and the German Universities, 1800–1900* (Chicago: University of Chicago Press, 1995), 35–64.

21. William Coleman, "Prussian Pedagogy: Purkyne at Breslau, 1823–1839," in *The Investigative Enterprise: Experimental Physiology in Nineteenth-Century Medicine*, ed. William Coleman and Frederic L. Holmes (Berkeley: University

of California Press, 1988), 15–64; Richard L. Kremer, "Building Institutes for Physiology in Prussia, 1836–1846: Contexts, Interests, Rhetoric," in *The Laboratory Revolution in Medicine*, ed. Andrew Cunningham and William Perry (Cambridge: Cambridge University Press, 1992), 72–109.

22. Henning Schmidgen, "Pictures, Preparations, and Living Processes: The Production of Immediate Visual Perception (Anschauung) in Late-Nineteenth-Century Physiology," *Journal of the History of Biology* 37 (2004): 477–513.

23. Stanley Hall, "The Graphic Method," *Nation* 745 (9 October 1879): 238.

24. Emil du Bois-Reymond, "Der physiologische Unterricht sonst und jetzt" [1877], in Estelle du Bois-Reymond (ed.), *Reden von Emil du Bois-Reymond in zwei Bänden* (Leipzig: Verlag von Veit & Comp, 1912), 1:637.

25. Carl Jacobj, "Anschauungsunterricht und Projektion," *Zeitschrift für wissenschaftliche Mikroskopie und mikroskopische Technik* 36, no. 4 (1919): 275.

26. Ibid., 277.

27. Carl Jacobj, *Das pharmakologische Institut zu Tübingen und seine Einrichtungen für Unterricht und Forschung* (Tübingen: Buchdruckerei der Tübinger Studentenhilfe, 1927), 7.

28. Jacobj, "Anschauungsunterricht und Projektion," 285.

29. Ibid., 295.

30. Ibid., 289.

31. Carl Jacoby [sic], "Vorführung der Einrichtung des pharmakologischen Instituts für Demonstrationen im Unterricht an der Hand der Besprechung einiger pharmakologischer Wirkungen," *Münchener Medizinische Wochenschrift* 15 (1910): 829.

32. Jacobj, "Anschauungsunterricht und Projektion," 287.

33. Ibid., 297.

34. Anonymous, "Das pharmakologische Institut," in *Festschrift für die 58. Versammlung Deutscher Naturforscher und Ärzte: Die Naturwissenschaftlichen und Medicinischen Institute der Universität und die Naturhistorischen Sammlungen der Stadt Strassburg* (Strassburg: Heitz, 1885), 121.

35. Andreas E. Jendrassik, *Das neue physiologische Institut an der Universität zu Budapest* (Budapest: Kön. Ung. Universitäts-Buchdruckerei, 1877), 14.

36. Jendrassik, *Das neue physiologische Institut*, 14.

37. Ibid.

38. Ibid., 21.

39. Ibid., 22

40. Ibid., 21.

41. R. von Horstig, "Die Universität und ihre Anstalten: Das physiologische Institut," in *Würzburg, insbesondere seine Einrichtungen für Gesundheitspflege und Unterricht: Fest-Schrift gewidmet der 18. Versammlung des Deutschen Vereins für öffentliche Gesundheitspflege*, ed., Karl B. Lehmann and Julius Röder (Würzburg 1892), 259. Angelo Mosso, *L'Institut Physiologique de l'Université de Turin: A l'occasion du XI Congres International de Médicine tenu à Rome en 1894* (Turin: Vincent Bona, 1894).

42. Gustav Gärtner, "Stricker's Unterrichtsmethode," in 30 *Jahre experimentelle Pathologie: Stricker Festschrift* (Leipzig: Deuticke, 1898), 59.

43. Salomon Stricker, *Skizzen aus der Lehranstalt für experimentelle Pathologie* (Wien: Hölder, 1892), 3.

44. Ibid.

45. Gärtner, "Stricker's Unterrichtsmethode," 61.

46. Stricker, *Skizzen aus der Lehranstalt*, 4.

47. Carl Kaiserling, "Ueber die Schwierigkeiten des demonstrativen Unterrichts und seine Hilfsmittel, insonderheit einen neuen Universal-Projectionsapparat," in *Arbeiten aus dem Pathologischen Institut zu Berlin: Zur Feier der Vollendung der Instituts-Neubauten*, ed., Johannes Orth (Berlin: Hirschwald, 1906), 118–119, 123.

48. Kaiserling, "Ueber die Schwierigkeiten," 123.

49. Ibid.

50. Ibid.

51. Jacobj, "Anschauungsunterricht und Projektion," 298.

52. Ibid.

53. Ibid., 299.

54. Ibid., 308.

55. Ibid., 309.

56. Carl Jacobj, "Visual Instruction and the Projection Method," *Methods and Problems of Medical Education* Sixth Series (1927): 257–264; Jacobj, "Institute of Pharmacology, University of Tübingen," *Methods and Problems of Medical Education* 10th Series (1928): 89–104.

57. Morton S. Biskind, "The Demonstration of Biological Experiments by Optical Projection Methods," *Methods and Problems of Medical Education*, 10th Series (1928): 309–21.

58. Biskind, "The Demonstration of Biological Experiments," 309.

59. Ibid., 315.

60. Ibid.

61. Ibid., 311.

62. Ibid.

63. Ibid.

chapter five

Cinema as Universal Language of Health Education

Translating Science in *Unhooking the Hookworm* (1920)

Kirsten Ostherr

> In my opinion nothing will surpass in effectiveness a "movie" in demonstrating public health conditions.
> —Letter from Arkansas State Board of Health to Rockefeller Foundation, January 5, 1918

> If you insist upon the extreme smallness of the plasmodium, the countryman will regard it as imaginary and of no importance.
> —Letter from H. R. Carter to Dr. Ferrell, IHB, January 9, 1922

In 1909, John D. Rockefeller dedicated a small fortune from his Standard Oil Company profits to the establishment of a new philanthropic entity, the Rockefeller Foundation, whose mandate was "to promote the well-being and to advance the civilization of the peoples of the United States and its territories and possessions and of foreign lands in the acquisition and dissemination of knowledge, in the prevention and relief of suffering, and in the promotion of any and all of the elements of human progress."[1] Within a few months, the Foundation established the Rockefeller Sanitary Commission for the Eradication of Hookworm Disease in the U.S. South, the first of many national and international projects in public health sponsored by the Rockefeller Foundation. As historians of the Rockefeller Foundation have observed, "Before the founding of the WHO in 1948, [the Rockefeller Foundation's International Health Division] was arguably the world's most important agency of public health work."[2] In an era when departments of public health only existed in the major urban centers of the United States and Western Europe, the Rockefeller Foundation was at the forefront of disease eradication and health promotion worldwide.

In 1920 the Foundation released a film to aid in these endeavors, titled *Unhooking the Hookworm*, produced in association with Coronet Films of Providence, Rhode Island.[3] Convinced by widespread enthusiasm for the motion picture as the ideal medium of visual education in this period, Rockefeller Foundation researchers also produced films on malaria in the 1920s and polio in the 1930s, followed by a series of films produced in association with Walt Disney in the early 1940s to teach literacy and basic health principles to viewers in Latin America.[4] Although the Rockefeller Foundation scientists were intimately involved in the production of *Unhooking the Hookworm*, by the 1940s they essentially had contracted out to Disney, participating minimally in the development of the screenplay, camerawork, and editing.[5] Why this shift in production philosophy? As articulated by Fred Gates, Rockefeller's principal philanthropic aide in this period, the members of the Health Division of the Rockefeller Foundation believed that "disease is the supreme ill of human life, and it is the main source of almost all other human ills—poverty, crime, ignorance, vice, inefficiency, hereditary traits, and many other evils."[6] But when it came time for the Rockefeller Foundation scientists to implement their biomedical approach to public health, they found that these supposed secondary effects of disease played a primary role in obstructing their ability to communicate with the target populations for the film. Although these researchers continued to believe in the power of moving images to persuade and instruct, they also had to acknowledge that the motion picture was not the transparently legible instrument of universal communication that they had hoped.

This essay will examine what happened between *Hookworm* and the Disney pictures when the Rockefeller Foundation was convinced to (1) contract out the production of its films, (2) replace the original mixture of live action, cinemicroscopy, and stop-motion animation with 100 percent cel animation, and (3) target illiterate audiences with simple messages that did not include any scientific information and only conveyed the most basic of ideas. As we will see, contemporary theories of visual pedagogy assumed that animation held an intrinsic appeal for "simple-minded" audiences, in part because animation minimizes the amount of visual information that the image conveys and is therefore deemed easier to comprehend, and consequently more "entertaining" as well. However, as a non-indexical mode of representation, animation introduces a problem for scientific representation.[7] Unlike documentary film, which is popularly conceived as possessing a privileged relation to reality, animation operates at a remove from the real, and thus lacks the immediacy so often associated with the unique powers that made motion pictures ideal pedagogical instruments. The obvious intervention of the animator's hand in the otherwise scientifically sound mechanical reproduction of images raised a tension between objectivity and entertainment that played a crucial role in shifting the Rockefeller Foundation researchers' approach to cinematic health education.[8] Significantly, a different sense of immediacy—that is, simple comprehensibility rather than unmediated indexicality—ultimately positioned cel animation as the ideal genre for health education.[9]

Production and Exhibition

The Rockefeller Foundation was made up of scientists, not filmmakers, so it should come as no surprise that this group approached the production of *Unhooking the Hookworm* as technically and methodically as they approached their laboratory experiments. Part of their preproduction research concerned the feasibility of using motion pictures as pedagogical tools; a sort of market research letter was sent out to all of the state boards of health in the United States, inquiring about the facilities they had available to them and their opinion of the use of motion pictures in health education. Many replies expressed great interest in showing health films while noting the state's dependence on commercial theaters to provide the technology, as in a 1917 letter from the Mississippi State Board of Health, which indicates that the Board "does not possess a moving picture machine but we have been able to get a great many moving picture shows to use films which we furnished upon a cooperative basis."[10] Although such cooperation did enable the publicity for some health initiatives to reach the public, the role of the private, for-profit sector in cinematic public health education would be a source of much debate and skepticism in this transitional period of film history. A 1918 letter from the Arkansas State Board of Health highlights the problem of audience segregation that could arise with the use of commercial theaters: "[N]either the State nor the counties have any facility for using moving picture films, thus we will be dependent on the moving picture houses, and these picture shows do not reach the class of people in this State that it is desired to reach with hookworm films."[11] The "class of people" referenced here was the rural, uneducated, working class of the southern United States; because this population included both black and white viewers, the issue of racial spectatorship would become an important consideration for the Rockefeller researchers, as I will discuss in more detail below.

Despite obstacles to public exhibition of the film, *Unhooking the Hookworm* was nonetheless seen by large audiences, in part because educational films were still shown alongside entertainment features in movie theaters at this point in film history.[12] (These genres would not be formally segregated until 1934.[13]) Moreover, nontheatrical exhibition also thrived in this era, as evidenced by a 1917 letter from the president of Educational Films Corporation of New York, who boasted, "The present active management of the company has had the satisfaction of bringing about something the film trade said two years ago was impossible, i.e., to produce and sell (not give away) pictures that would be at least 50 percent instructive, truly educational but which would be at the same time sufficiently entertaining for theatres, churches, schools, lyceums, etc., to pay real money for."[14] However, another letter from the same production company, dated only two months later, expressed a somewhat tempered enthusiasm as it informed the Rockefeller Foundation of the differences between commercial and nontheatrical exhibition, complaining, "I know too well the character of the average theater manager to expect for a moment that he will pay money for a picture of this kind. He knows it will not bring in an additional dollar, consequently he will not spend a dollar for it." The letter goes on to suggest that *Unhooking the Hookworm* would be more eagerly received by "Health boards, Town Committees or Civic bodies of any kind . . . as they are not money making institutions and it is something they need and should have for their community work."[15] As we will see, the question of what makes a health film "sufficiently entertaining" for moviegoers to spend time or money on it would vex the Rockefeller scientists throughout their involvement in motion picture production.

Signs pointing toward the future separation of educational and entertainment film exhibition recur throughout the *Hookworm* correspondence. Alternately blaming greedy theater managers and simple-minded audiences, many proponents of educational film considered their work to be in conflict with the facile pleasures offered by Hollywood, and yet, many eventually recognized (often with exasperation and dismay) that their films would never be as successful as Hollywood fare unless they acquiesced to the public demand for "entertainment." Nonetheless, the future of cinematic education seemed bright enough in 1921 for Harry

Levey, president of Educational Production for National Non-Theatrical Motion Pictures, Inc., to claim boldly in the *New York Times*, "Industrial and educational films are a by-product of the motion-picture industry, and in my opinion they are destined to become a more important part of the business than the production of pictures for entertainment. . . . Health boards throughout the country can put their lessons 'across' in pictures as they can in no other way."[16] The same company also promised to distribute *Unhooking the Hookworm* "over our free non-theatrical circuit."

Between commercial and nontheatrical screenings, the *Hookworm* film was clearly a successful health intervention, as demonstrated by a letter from a Kentucky state health officer, who sent the following confirmation of the film's effect in 1921: "You will be interested in knowing that as a result of showing [*Unhooking the Hookworm*] in the picture shows of Louisville more than 500 specimens were submitted to our own laboratory, and that the physicians of Louisville all report a large number of patients who have come to them for examination. Quite a considera[ble] number have been treated and are being relieved."[17] Such direct evidence of the impact of motion picture viewing is uncommon and not easily demonstrated; in this context it provides precisely the sort of quantifiable verification of the film's success that the administration-minded Rockefeller Foundation sought out.

However, the problem of appealing to the managers and audiences of moving picture shows continued to challenge the hookworm campaign. Despite the film's success in Louisville theaters, the Rockefeller Foundation was repeatedly admonished for its egg-headed misperception of the general public's cinematic preferences. This critique was presented succinctly by the American Social Hygiene Association in 1923: "Motion picture theatre audiences desire to be entertained and amused. It is generally held by motion picture theatre owners that their audiences resent propaganda or educational pictures and usually object to showing them."[18] The Rockefeller Foundation thus faced two key problems: how to solve the logistical challenge of exhibiting the film to poor, rural audiences, and how to get those audiences interested in an educational film, once the details of the screening were arranged. Indeed, while the proliferation of stool samples in Louisville seemed to offer unambiguous evidence of the audience's positive reception of the film, this response seems to be somewhat unique. The vast majority of commentators simultaneously praised the film's sophistication and technical accuracy and questioned the audience's tolerance for what might easily be regarded as condescending, not to mention just plain boring, lecturing on film. But *Unhooking the Hookworm* went to great lengths to avoid producing a preachy effect, mixing what were meant to be amusing and familiar scenes of country life with animated demonstrations of how the depicted activities pose a hidden threat of hookworm infection.

Nonetheless, comments from a Dr. J. A. LePrince indicate the Rockefeller Foundation's failure to comprehend the viewing tastes of the "general public." In a letter concerning the Foundation's next film project on malaria, LePrince questions the screenwriter's familiarity with his target audience, commenting that "the scenario would be good for an 'above the average' audience such as one consisting mostly of college graduates." LePrince goes on to demonstrate his own understanding of the potential obstacles to viewer engagement, noting, "If the film is to be shown at night to country folks who have been working hard all day in the field, it may possibly put some of them to sleep or get them thinking of other things that are more interesting." The letter concludes by observing somewhat ruefully, "we are not appealing to an audience of medical men. We are merely learning how to appeal to the public and there is much yet to be learned."[19] Despite his knowledge of the day-to-day demands on farm workers' attention spans, the doctor clearly shares the Rockefeller scientists' attitude of detached observation toward their audience, and it is precisely this attitude that ultimately leads the Foundation to adopt animation as its preferred medium of health education.

Further comments on the malaria film treatment by a Dr. Howard in a 1923 letter expand upon this notion of appealing to the average viewer: "It seems to me that the diagrams are too 'highbrow' and are likely to leave the average person somewhat confused. If it is desirable to repeat the explanation I would think it better psychology to do it by means of an animated cartoon showing the entire process. I believe that would stick in the mind much longer than the diagrams, and even than the photographs."[20] Here we begin to see how animation takes on a privileged — if problematic — status within theories of visual pedagogy in this period, as it becomes the default medium for communicating with the "average person." In addition to its presumed superiority as a medium for communicating with the mass audience, animation was also frequently employed in early health films as a means of representing organisms that are invisible to the naked eye. *Unhooking the Hookworm* was not alone in its commitment to accurate depictions of physiological phenomena, but like many films of its kind, *Hookworm* was incapable of strictly adhering to scientific realism in its representational techniques. The film's success as a health intervention depended on its ability to convince audiences of the authority and accuracy of its depiction of hookworm disease. To do so, Unhooking the Hookworm utilized documentary images whenever possible, but the filmmakers were forced to resort to animation for sequences involving internal bodily processes. Both because of its ostensible appeal for the simple mind of the "average person" and because of its lack of photographic indexicality, the animated image was treated as a debased form of representation. But as we will see, animation nonetheless proved indispensable to this and future Rockefeller health campaigns.

Despite the Foundation's seeming failure to reach the average moviegoer with its first film, Unhooking the Hookworm was still in constant demand sixteen years after its original release. A 1936 letter from the chief engineer and director of the Bureau of Sanitation in the state of Alabama praises the film's effectiveness, noting that twenty-one counties own copies of the film, and other counties are seeking funds to purchase motion picture equipment. Most significantly, the sanitarian urges the Rockefeller Foundation to remain actively engaged in motion picture production, pleading, "[O]ur educational programs will suffer considerably if the Rockefeller Foundation does not decide to revise these films.... With the increase in public health activities, there should be a continued demand for films on these subjects. It is believed that our County Health Units will continue the use of visual education provided suitable films are available, for if public health is to progress education must be kept up year after year."[21] Notwithstanding this well-reasoned appeal, the Rockefeller Foundation did not continuously update its health films, largely because this widespread enthusiasm was tempered by commentators who focused precisely on the limits of visual education's faith in the universal comprehensibility of film, even with the potential leveling effects of animation, when it came to racially and geographically diverse audiences. To explain the complicated reception of Unhooking the Hookworm both within the United States and abroad, it will be useful to review the film in more detail.

Unhooking the Hookworm

Unhooking the Hookworm begins with an intertitle that establishes the global scale of the problem under discussion: "In all these warm countries, dwells one of man's most dangerous enemies—the Hookworm." The film cuts to an image of a flat animated map of the globe with shading spreading over affected areas in the southern United States, South America, Africa, Italy, and South Asia. The film immediately shifts from the vastness of the global to the minute scale of the local by observing, "Here he is—and she also—beside a common pin." The next shot demonstrates the size of a hookworm beside a sewing pin, alongside a ruler that clearly marks the indexicality of the evidence. After this initial shift from animation to live action, the film changes register yet again, as the intertitle explains, "Under a microscope, they look like this," accompanied by cinemicroscopic footage of the worms squirming and writhing like snakes (a point underscored by a drawing of a snake at the bottom of the title card).

At this point, we shift back to a diagram of the parts of a male worm, with a pointer indicating the head, digestive organs, and nervous system. Superimposed above the male worm is an image of the female, with the ominous intertitle, "These coils are full of eggs—thousands are laid in a day." Here we cut to a graphic map of India: "Their victims are counted by the millions, from India," followed by documentary footage of a street scene. Despite the global reach of the film's opening sequence, no other national locale is identified in the rest of the film.

Unhooking the Hookworm uses an idealized scenario of childhood play as a framing device for the film. While this technique was meant to provide a universal point of identification, it was in fact the source of much criticism of the film, as we will see later on. After initially establishing the global scale of the hookworm problem, the film presents a general diagnosis: "An 'always tired' feeling is one of the signs of a mild case," followed by footage of a white boy of approximately ten years of age lying against a rock on the stoop of a small shack. The location is not identified, but given much of the feedback from viewers, is widely assumed to be the southern United States. The next shot shows a friend saying, "C'mon swimming," but the lethargic boy replies, "Naw. I'm too tired," at which point, his friend walks away. Here, the ravages of hookworm disease are depicted as interfering with the joys of a fun-filled, carefree youth.

After identifying the hookworm as a parasite and providing gruesome images of the worms feasting on unidentified bodily organs, we see microscopic images of hookworm eggs hatching, and the locations—such as pigpens and outhouses—where the baby hookworms might be found. This leads into a sequence on the insidious invisibility of the baby hookworms, whose seeming omnipresence poses a threat even in the most innocent of scenarios. As an intertitle notes, "Out in the fields, the baby

Cinema as Universal Language · 129

hookworms often crawl up blades of grass and enter drops of the morning dew." We see a close-up of dewy blades of grass, and the narration elicits further emotional identification from viewers: "It's hard to believe that the dew is alive with danger—but let us see." The hands of an off-screen scientist—presumably the narrator of the film—remove a drop of water from the grass to create a microscope slide. The next shot shifts location to a laboratory where the slide provides a new cinemicroscopic view of the worms. To underscore the malevolence of these tiny creatures, the next intertitle again compares them to snakes.

After providing ample scientific evidence of the danger lurking within the morning dew, we receive a more concrete demonstration of the process of contamination. A card stating, "Bare feet easily pick up baby hookworms from the dew," is followed by a medium shot of the same lethargic boy walking through a field of grass in bare feet. The camera cuts in to provide closer and closer views until the boy reaches down, pulls his toes apart, and begins to scratch. Now, the most concretely instructional sequence of the film begins as we move from live action to animation: "Here goes one of them, right into a pore of the skin." Through stop-motion animation, a clay model of a worm entering a skin pore is shown in graphic detail. As the infection progresses, "Irritation—the 'ground itch'—or 'dew itch'—then blisters and sores." We return to a live-action medium-long shot of the boy in the field, then cut in to a medium close-up of the boy scratching his foot, and finally an extreme close-up of his toes covered with sores.

The educational sequence continues with more assistance from special effects as we enter the interior of the human body. "From the blisters to the blood is a short journey, then—" a diagram of a human body with a pointer shows movement up from the foot through the veins and into the heart. We then see more cinemicroscopic views of worms as they traverse the body, moving from heart to lung to windpipe, and are then swallowed before they ultimately land in the intestines (figure 5.1). "In the bowels, they make themselves at home. Now, to 'work.' And while 'working,' they grow to full size. . . . A good grip with the mouth and they don't intend to let go." An enlarged diagram of a hookworm head is interpreted with the captions "poison fang" and "holding mechanism." Stop motion clay models create a repellent scene of numerous worms attached to model intestines. "It's not long before ulcers form—with frequent bleeding." As the clay worms detach, blood streams from the sores they leave behind.

In this brief, four-minute sequence, several different types of visual images are employed to create an educational and scientifically accurate

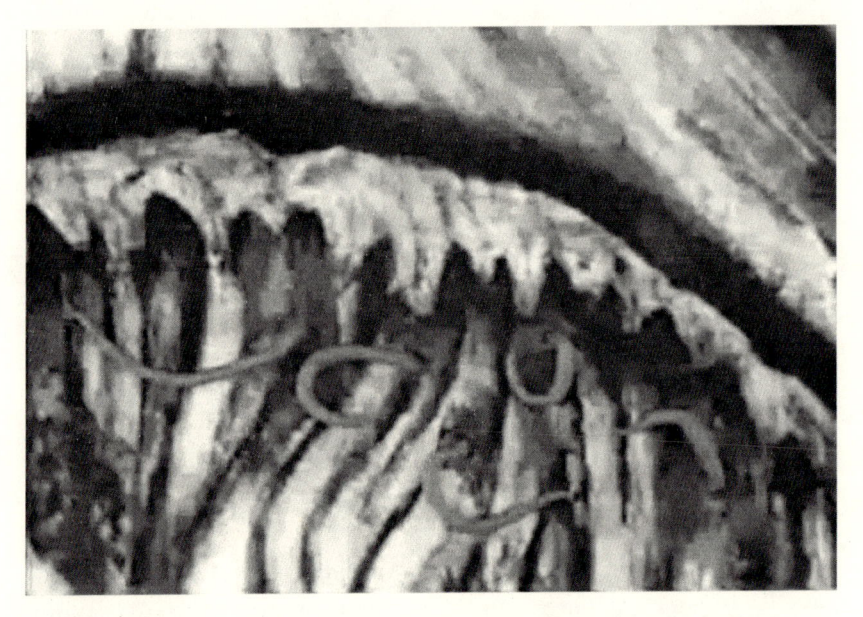

5.1 "Clever animation of hookworms attaching to the intestines." *Unhooking the Hookworm* (1920).

impression of the process of hookworm infection. While the heterogeneity of the footage might be seen as a creative effort at producing realistic special effects, some commentators found that the vividness of the portrayal actually interfered with their sense of the film's authenticity. A Rockefeller Foundation staffer observed, "A further criticism . . . relates to the use of 'faked' diagrammatic pictures, like that of the hookworm entering the skin, along with true microscopic photographs. It is interesting to hear the comments of educated people on such pictures. There would be no objection to the pictures if the legends stated or implied what they really are. Even our own staff are unable to answer questions as to the authenticity or origin of some of these clever photographs."[22]

Following the special-effects montage, we finally see the sequence that prompted the submission of hundreds of stool samples in Louisville, Kentucky. A man and his son enter a doctor's office. "Have you brought a specimen of the boy's stool?" The father pulls a jar out of his pocket and gives it to the physician, who leaves the examination room, enters an adjacent laboratory, and places a sample from the jar onto a slide. "The doctor finds hookworm eggs in the stool." There is then a close-up of microscopic images of the eggs along with the narration: "These eggs came from the worms causing the disease. A physician can give you medicine to kill them." We then see the doctor collect medicine from large pill bottles lining the shelves of the lab, confirm the son's diagnosis, and tell the father, "This medicine will cure him." Back at home, the father gives a pill to the boy. We see a close-up of the pill in his hand, followed by a medium shot of the boy swallowing the pill with water. Each step of the process is methodically explained through intertitles and demonstrated with live action, so that the film might serve a mimetic function for viewers who receive their own diagnosis and prescription for treatment. Thus, for the next step in the sequence, "In a few hours, Epsom salts to drive out the dead worms," we again see the boy drinking water from a glass. Finally, the results: "Here are some of the dead hookworms the medicine removed." A close-up of a pile of shriveled worms on a sheet of paper confirms the effectiveness of the regimen. "One treatment is frequently sufficient to effect a complete cure."

From here, the film makes a plea for prevention, not just cure, by demonstrating proper methods for constructing sanitary outhouses. *Unhooking the Hookworm* closes with a triumphant declaration: "Cured! 'Happy boyhood' is now a reality. And the home—made sanitary—will keep him so." Finally, the once lethargic boy runs off with his friend to go swimming. However, rural audiences often saw hookworm's interference with childhood play as insufficient motivation for seeking a cure. This response underestimated the larger threat posed by the parasite, as a Rockefeller staffer stationed in Dutch Guiana (now Suriname) observed in a 1922 letter: "the appeal of 'happy childhood' is not as great to the people under consideration as the improved health of the 'bread winner' would be."[23] Another Rockefeller worker in Ceylon (now Sri Lanka) did not object to the general principle of play, but noted that the present example was not culturally appropriate, and suggested "another game than 'going swimming,'" such as hackney racing or Mazoka.[24]

Evaluation and Techniques of Translation

Despite mixed results from their first foray into filmmaking, the Rockefeller Foundation scientists commenced production on a malaria film shortly after *Unhooking the Hookworm* was completed. Letters exchanged within the organization reveal the scientists' view of lessons learned from the reception of the *Hookworm* film, particularly regarding modes of address and interpretive variability among audiences. An interoffice memo from 1921 notes two critical points to observe in future film productions:

"(1) The audience to be reached should be kept constantly in mind. If this audience is the so-called 'general public,' certain principles of human nature must be recognized—i.e., the appeal of the immediate and the concrete. (2) Diagrams, maps, and information should be constantly interspersed with concrete pictures. Even animated diagrams and maps will not hold the attention of the average mind unless these devices interpret some vividly visualized situation."[25] These principles formed the basis for international adaptations of the *Hookworm* film in the mid-1920s. An additional suggestion that films should include a segment produced by members of the target population was particularly influential, but the demand for local specificity soon became overwhelming for an organization with the global reach of the Rockefeller Foundation.

For instance, *Unhooking the Hookworm* was adapted for use in Nicaragua by filming actual hookworm treatments that were given in already existing health centers.[26] In Java, the storyline of the sick white boy was replaced with a local tale of an ailing Javanese "everyman."[27] In the southern United States, the film was adapted for use with African American viewers. A letter exchanged between two Rockefeller scientists in 1936 provides a telling glimpse into prevailing views on the importance of racialized spectatorship in contemporary theories of visual pedagogy, and is thus worth quoting at length:

> The exceedingly diverse conditions under which a film may be employed makes it practically impossible to devise a film that would be equally effective before any audience anywhere in the world. I think this is very well exemplified by my experience in Mississippi. You may recall that we were furnished with several editions of the Foundation film in 16mm size. We soon gained the impression that the exhibition of this film to rural negro audiences made very little impression. The technical features were above their heads and the human interest continuity based on a scenario dealing with whites made little impression. No one realizes better than a southern negro the vast gulf that exists between the whites and the negroes, consequently the negro is not much impressed by scenes dealing with whites. On this assumption I duplicated the human interest scenario, with the cooperation of some negro actors, and cut out certain sections in one of the Foundation films and substituted these negro pictures. This modified film was enthusiastically received by negro audiences wherever it was shown and I am sure made the rural negroes appreciate malaria as a negro problem. The same remarks also apply to the exhibition of such a film in any foreign country.

I think it would be practicable for the Foundation to supply the technical portions of a film, plus the verbal sketch of a human interest scenario in a 16mm size and encourage staff members in different countries to supplement these by a human interest scenario locally acted plus local scenes of malaria control operations.[28]

The principle of racial identification articulated here attributes racial specificity to disease. Reviews repeatedly suggest that the film's pedagogical efficacy depends upon viewers' ability to see their own racial identity reflected back to them from the screen. This presumption is repeated in notes for a revised shooting script, directed toward a more "universal" audience through a racially diverse mise-en-scène that includes "representatives of several nationalities in native costume," and a "group of people, preferably of various shades and ages."[29] Regardless of the accuracy of this view of racialized spectatorship, the Rockefeller filmmakers persisted in their belief that cross-racial identification was impossible, and therefore, every racial group that might see the film should be included in it. However, it soon became apparent that this principle would lead to an unmanageable multiplication of film editions.

The tension between universality and particularity is also evident in the 1922 letter from Dutch Guiana quoted above, which offers suggestions for improving the *Hookworm* film by appealing to racialized spectatorship:

> Races other than the white race are easily distracted when a partially exposed white person is pictured. It invariably brings down a roar of laughter and continually distracts the audience from the point that is being made. I do not believe in elimination of the white patients, because then many white folks will continue to keep the impression that hookworm does not affect the white people. Introduction of pictures of various very sick patients from various countries would improve the film as nearly all races would then be represented. For instance; at the introduction of the film where the map shows the infected parts of the world; pictures of patients from India, China and elsewhere, could be easily introduced. It would not be necessary to bring in the name of the country as long as the people see their own race.[30]

In his discussion of Rockefeller hygiene films in colonial Java, Eric Stein argues that *Unhooking the Hookworm* often provoked unintended laughter among villagers, who likened the pale, gaunt, pot-bellied hookworm sufferers in the film to the *Gareng*, a comedic figure in their own popular folklore.[31]

One observer of a screening in the southern United States noted that, "A source of constant amusement to the rural people is the farmer who in the first place does not look sick, and in the second place does not dress like a farmer and wears a gold watch."[32] The prevalence of such unforeseen audience responses suggests that, while racial and cultural similarity between spectators and actors seemed vital to ensuring that a film's "message" was received, racial and cultural difference could be responsible not only for viewers' rejection of the message, but also for producing entirely new, even contradictory interpretations of the film. Media scholars now routinely acknowledge cinema's capacity for evoking a wide array of readings among viewers, but to the Rockefeller Foundation, the success of their visual pedagogy initiative depended on their ability to elicit uniform responses from diverse viewers of *Unhooking the Hookworm*. To this end, the scientists attempted to control for viewer variability by creating different versions of the film that were designed to anticipate and thereby foreclose any alternate interpretations.

As the Rockefeller Foundation studied the issues involved in preparing versions of the film for use around the world, feedback about racial and geographical determinants in audience response led to proposals for combined local and global film segments. An interoffice memo of 1925 described the strategy:

> I think that it would be well to divide the hookworm film into several more or less independent parts, each with its own heading, and to put the greater part of our biological and more strictly scientific descriptions into one part. The object would be to make it possible for people in foreign countries to take the scientific part bodily and insert it into their local films without any question of our responsibility for their film. In almost every country in which hookworm work is done it would be of advantage to have at least one section of the film prepared locally."[33]

This comment implicitly suggests that audience response is fragmented not only by race, class, education, and geography, but also by the division between science and popular entertainment. This final distinction ultimately concerns the core question of documentary realism that is seen as critical to the educational film's pedagogical effects.

The Rockefeller Foundation scientists had painstakingly employed every mode of visual representation available to present a depiction of hookworm that was both technically accurate and persuasively educational. However, based on many commentators' observations, the goal of education often conflicted with the principle of entertainment, which was nonetheless viewed as a prerequisite to pedagogical efficacy.[34] In other words, in order to be receptive to the film's educational message, audiences first had to be entertained by the film. But what made a film entertaining? According to the many health officers and film distributors who provided feedback to the Rockefeller Foundation, it was precisely the lack of technical details, coupled with the frequent insertion of lighthearted, slapstick humor, that made a film appealing to the masses of vulnerable viewers the Rockefeller Foundation hoped to reach.

And where did education fit in? The suggested revisions to *Unhooking the Hookworm* all point in the direction of eliminating precisely the information that would enable the film to serve as an effective pedagogical tool. While other scientists might be fascinated by the film's visual representation of a process that is usually invisible to the naked eye, audiences of the general public were apparently bored by this pedantic approach. As the proposed adaptations of *Unhooking the Hookworm* imply, the target demographic for this outreach effort was engaged by the narrative segments, not by the exposition of disease etiology. That is, they were drawn to the sequences that most resembled fictional storytelling—the sequences that Rockefeller staffers called the "human interest" part of the film. Not surprisingly, the hookworm sanitarians treated the "human" elements as regrettable—and disposable—necessities to attract viewers' attention for the core "scientific" part of the film. (Recall that, despite their fastidious attention to detail in the technical sequences of the film, the Rockefeller scientists were willing to allow local populations to make up their own narrative sequences and insert them whole into the film, regardless of continuity, as long as the scientific sequences remained.) Implicit in this attitude toward fiction is the assumption that pedagogy can only legitimately occur through nonfictional forms of representation.

Ironically, many motion picture reform efforts of the 1920s and 1930s operated on exactly the opposite premise; treatises such as the best-selling *Our Movie Made Children* claimed that fictional scenarios so powerfully capture the viewer's imagination that Hollywood films function as an unofficial and deeply immoral educational system for America's youth.[35] And yet, the Rockefeller filmmakers could not reconcile the fiction of the human interest segments with the nonfiction of the scientific contributions. By privileging the pedagogical power of nonfictional representation and insisting on the clear separation of the instructional sequences from the entertainment sequences, the Rockefeller Foundation undermined its own efforts at reaching the mass viewing audience. Frustrated by the seemingly impossible task of making an intelligent film appealing to the general public, the Foundation revised its objective; forsaking complex explanation (the life cycle of the hookworm) in favor of simple instruction ("wash your hands"), they called in Walt Disney.

Conclusion

While the view of film as a medium of worldwide pedagogical diplomacy was in keeping with prevalent ideas about visual education in the 1920s, the Rockefeller researchers found that the assumed universality of the film's scientific foundations actually required extensive translation to be intelligible to different audiences, both within and outside of the United States. Detailed discussion of the usefulness of particular depictions of local circumstances versus more "universal" images reveals that the goal of allowing a film to circulate globally requires the production of a generalizable language of cinema that refrains from particularity in favor of universalism in its representational forms. But this very shift signaled, for Rockefeller Foundation scientists, the elimination of scientific accuracy and pedagogical effectiveness.

Ruth Vasey has argued that during the Production Code era, Hollywood studios inadvertently created the category of the "other" by trying to avoid particular offenses to specific national and racial groups. Ironically, and as a result of quite different forces, the Rockefeller Foundation scientists ended up coming to the same conclusion. The researchers' methodical approach to assessing the effectiveness of the film screenings generated a great deal of feedback about viewer responses, and it became increasingly apparent that the only way for *Unhooking the Hookworm* to be an effective educational tool was to produce a virtually infinite number of versions of the film, each adapted to particular local circumstances.

Not surprisingly, this level of sensitivity demanded more involvement than the Rockefeller Foundation researchers could or willingly would sustain over time. Thus, it was precisely through the Rockefeller Foundation scientists' efforts to reach a global audience that the particularities of representation were ultimately rejected in favor of the broad generalizations of animation as employed in the Disney films. In the process, however, they also abandoned the project of scientific visual education, resorting instead to "benign children's entertainment."[36]

Notes

Letter from Arkansas State Board of Health to Rockefeller Foundation, January 5, 1918, Rockefeller Foundation Archives, RG 1, Series 100, Box 5, Folder 42, Films—Reports 1917–1927 (Part 3).

Letter from H. R. Carter to Dr. Ferrell, IHB, January 9, 1922, Rockefeller Foundation Archives, RG 1, Series: 100, Box 5, Folder 42, Films—Reports 1917–1927 (Part 3).

The author gratefully acknowledges the assistance of the Rockefeller Archive Center.

1. Original Deed of Trust, Rockefeller Archive Center, RG.2 B.24 f.241, 242. Cited in John Farley, *To Cast Out Disease: A History of the International Health Division of the Rockefeller Foundation (1913–1951)* (New York: Oxford University Press, 2004), 3.

2. Farley, *To Cast Out Disease*, 2.

3. Memo of 10-16-20: the film was eventually produced by Coronet Films of Providence, RI, a subsidiary of Educational Films Corporation. The Rockefeller Foundation Archives, RG 1, Series: 100, Box 5, Folder 40, Films—Reports 1917–1927. *Unhooking the Hookworm* may be viewed on the website of the Rockefeller Archive Center, at http://archive.rockefeller.edu/feature/hookworm.php

4. The films were: *Defensa Contra La Invasion (Defense Against Invasion)* (Walt Disney Productions, 1943); *Water—Friend or Enemy* (Walt Disney Productions, 1943); *The Winged Scourge* (Walt Disney Productions, 1943); *La Historia de José* (Walt Disney Productions, 1944). In the "Reading for the Americas" series: *José Come Bien* (Walt Disney Productions, 1944); *La Historia de Ramón* (Walt Disney Productions, 1944); *Ramón Esta Enfermo* (Walt Disney Productions, 1944); *El Cuidado del Niño* (Walt Disney Productions, 1945). In the "Health for the Americas/Salud para Las Americas" series: *El Cuerpo Humano* (Walt Disney Productions, 1944); *La Enfermedad Se Propaga (How Disease Spreads)* (Walt Disney Productions, 1944); *¿Qué es Enfermedad?* (Walt Disney Productions, 1945); *Es Facil Comer Bien* (Walt Disney Productions, 1945); *Insectos que Transmiten Enfermedades* (Walt Disney Productions, 1945); *La Limpieza Trae Buena Salud* (Walt Disney Productions, 1944); *Saneamiento Del Ambiente (Environmental Sanitation)* (Walt Disney Productions, 1945); *La Tuberculosis* (Walt Disney Productions, 1944); *Uncinariasis (Hookworm)* (Walt Disney Productions, 1945). For a discussion of some of these films, see Lisa Cartwright and Brian Goldfarb, "Cultural Contagion: On Disney's Health Education Films for Latin America," in *Disney Discourse: Producing the Magic Kingdom*, ed. Eric Smoodin (New York: Routledge, 1994).

5. For a discussion of Rockefeller Foundation filmmaking activities in the immediate pre- and postwar periods, see William J. Buxton, "Rockefeller Support for Projects on the Use of Motion Pictures for Educational and Public Purposes, 1935–

1954," Rockefeller Archive Center Research Reports Online, 2001, www.rockarch.org/publications/resrep/rronlinealpha.php

6. Gates, quoted in Fosdick, *History of the Rockefeller Foundation*, p. 23. Cited in Farley, *To Cast Out Disease*, 5.

7. For a discussion of indexicality in documentary and animation, and the problem it poses for scientific representation, see Kirsten Ostherr, *Cinematic Prophylaxis: Globalization and Contagion in the Discourse of World Health* (Durham, NC: Duke University Press, 2005), especially chap. 2, "'Noninfected but Infectible': Contagion and the Boundaries of the Visible." On the history of early animation, see Donald Crafton, *Before Mickey: The Animated Film 1898–1928* (Cambridge, MA: MIT Press, 1982).

8. On the presumed objectivity of mechanical reproduction of images in contrast to the subjectivity of the medical illustration, see Lorraine Daston and Peter Galison, "The Image of Objectivity," *Representations* 40 (1992): 81–123. See also Walter Benjamin, "The Work of Art in the Age of Mechanical Reproduction," in *Illuminations: Essays and Reflections* (New York: Harcourt, 1968).

9. The motion pictures produced by the World Health Organization (WHO) exemplify the preference for cel animation in cinematic health education; this author's archival research at the WHO has confirmed that the majority of films produced by the organization from the late 1940s through the late 1960s were created using this technique.

10. Rockefeller Foundation Archives, RG 1, Series: 100, Box 5, Folder 42, Films—Reports 1917–1927 (Part 3).

11. Letter dated January 5, 1918, Rockefeller Foundation Archives, RG 1, Series: 100, Box 5, Folder 42, Films—Reports 1917–1927 (Part 3).

12. A glimpse into the film's exhibition context is provided by an excerpt from an October 25, 1924, story in the *Anderson* (South Carolina) *Tribune*, titled, "Health Film Is Awarded Perfect Score by Critics." The article discusses numerous health films deserving of praise: "Animated cartoons of droll physical culturists, elaborate scenes of Greek dancers and pictures of the everyday modern man and woman impress the audience with the truth of the film's health teaching. Each of the exercises presented in as many part-reels is a lesson complete in itself told pleasantly, concisely, and with sufficient variety to make it effective health pedagogy. Top films are 'Keeping Fit' by Pathé and 'Well Born' by Carlyle Ellis for the Children's Bureau. Other high-scoring films: 'Jinks' by National Tuberculosis Association, 'The High Road,' by Carlyle Ellis for Young Women's Christian Association, 'Working for Dear Life' by Metropolitan Life Insurance Company, 'The End of the Road,' by American Social Hygiene Association, 'Unhooking the Hookworm' by the Rockefeller Foundation, 'Out of Everywhere' by the Worcester Film Corporation for the Connecticut State Dept. of Health and 'Preventing Diphtheria' by John Hancock Life Insurance Company. All of these films except 'Keeping Fit,' which is new, have been widely circulated among schools, churches, social institutions and clubs. . . . The rental cost of the films ranges between $1 and $5 a day plus transportation charges. The insurance companies lend them to reliable groups

of institutions free except for the transportation cost. As a result of this policy the Metropolitan film, 'Working for Dear Life' has been shown 2181 times this year to a reported attendance of over 850,000 persons." The Rockefeller Foundation Archives, RG 1, Series: 100, Box 5, Folder 42, Films—Reports 1917–1927 (Part 3).

13. On "aesthetic censorship" and the subsequent segregation of health films from Hollywood films, see Martin Pernick, *The Black Stork: Eugenics and the Death of "Defective" Babies in American Medicine and Motion Pictures Since 1915* (New York: Oxford University Press, 1999); Ostherr, *Cinematic Prophylaxis*, especially chap. 1, "Public Sphere as Petri Dish; or, 'Special Case Studies of Motion Picture Theaters which are Known or Suspected to be Foci of Moral Infection'"; and Susan Lederer, "Repellent Subjects: Hollywood Censorship and Surgical Images in the 1930s," *Literature and Medicine* 17, no. 1 (1998): 91–113.

14. August 28, 1917, letter from Mr. G. A. Skinner, President of Educational Films Corporation of New York, NY. The Rockefeller Foundation Archives, RG 1, Series: 100, Box 5, Folder 40, Films—Reports 1917–1927.

15. October 19, 1917, letter from Educational Films Corporation to RF/IHB. The Rockefeller Foundation Archives, RG 1, Series: 100, Box 5, Folder 40, Films—Reports 1917–1927.

16. *New York Times* Sunday May 8, 1921. Section 7 (Special Features), 2. "New Movie Expansion: Combination of Non-Theatrical Producers for Development of Educational-Industrial Films." (Extract from three-column (half-page) article.) The Rockefeller Foundation Archives, RG 1, Series: 100, Box 5, Folder 42, Films–Reports 1917–1927 (Part 3).

17. May 26, 1921, Memorandum concerning comments on the film *Unhooking the Hookworm*. Comments from varied sources, including: Dr. A. T. McCormack, State Health Officer, Louisville, KY. The Rockefeller Foundation Archives, RG 1, Series: 100, Box 5, Folder 40, Films—Reports 1917–1927.

18. 1923 letter to National Health Council, from Films Committee. Subject: Reply to IHB's letter concerning review of film "Unhooking the Hookworm." The Rockefeller Foundation Archives, RG 1, Series: 100, Box 5, Folder 41, Films—Reports 1917–1927 (Part 2). The American Social Hygiene Association was also involved in the production of educational films, including *The End of the Road* (1919); *The Naked Truth* (1927); *The Venereal Diseases* (1928); and *Health Is a Victory* (1942), among others.

19. The Rockefeller Foundation Archives, RG 1, Series: 100, Box 5, Folder 42, Films—Reports 1917–1927 (Part 3).

20. Letter from Dr. Howard dated November 15, 1923, Rockefeller Foundation Archives, RG 1, Series: 100, Box 5, Folder 42, Films—Reports 1917–1927 (Part 3).

21. Letter from G. H. Hazlehurst, chief engineer and director of the Bureau of Sanitation in the state of Alabama, circa December 21, 1936, Rockefeller Foundation Archives, RG 1, Series: 100, Box 5, Folders 39–44, Films 1922–1938, Folder 39.

22. W. A. Sawyer, Rockefeller Foundation Archives, RG 1, Series: 100, Box 5, Folder 40, Films—Reports 1917–1927.

23. December 2, 1922, letter from W. C. Hausheer, Ankylostomiasis Commis-

sion, Dutch Guiana to Dr. Howard, RF/IHB, Rockefeller Foundation Archives, RG 1, Series: 100, Box 5, Folder 41, Films—Reports 1917–1927 (Part 2).

24. W.P.J., Rockefeller Foundation Archives, RG 1, Series: 100, Box 5, Folders 39–44, Films 1922–1938, Folder 39.

25. Memorandum on proposed malaria film, signed GEV, circa November 25, 1921, Rockefeller Foundation Archives, RG 1, Series: 100, Box 5, Folder 42, Films—Reports 1917–1927 (Part 3).

26. A note attached to a document titled, "Brief Outline of Titles of Scenes Filmed in Nicaragua to Complete Film 'Unhooking the Hookworm' and Adapt it to Local Conditions" places a strong emphasis on the indexicality of the image by documenting exactly what was taking place when these scenes were filmed. For example, "Note: No special preparations were made for taking this picture. Everything used belongs to the Regular Equipment for giving treatments in a rural centre, the only change made being that of bringing the table and equipment outdoors instead of continuing treatments in the open corridor where they were under way when the photographer reached the scene. In other words the picture was not 'staged.' 120 treatments were given in this centre the day the picture was taken." Rockefeller Foundation Archives, RG 1, Series: 100, Box 5, Folder 42, Films—Reports 1917–1927 (Part 3).

27. See Eric A. Stein, "Colonial Theatres of Proof: Representation and Laughter in 1930s Rockefeller Foundation Hygiene Cinema in Java." *Health and History* 8, no. 2 (2006): 40.

28. Letter from Dr. Mark F. Boyd, RF/IHD in Florida, to Dr. Ferrell, RF in New York, April 13, 1936, Rockefeller Foundation Archives, RG 1, Series: 100, Box 5, Folders 39–44, Films 1922–1938, Folder 39.

29. Rockefeller Foundation Archives, RG 1, Series: 100, Box 5, Folders 39–44, Films 1922–1938, Folder 39.

30. Rockefeller Foundation Archives, RG 1, Series: 100, Box 5, Folder 41, Films—Reports 1917–1927 (Part 2).

31. Stein, "Colonial Theatres of Proof," 9.

32. Letter from Dr. Howard dated November 15, 1923, Rockefeller Foundation Archives, RG 1, Series: 100, Box 5, Folder 42, Films—Reports 1917–1927 (Part 3).

33. Interoffice memo of October 20, 1925, Rockefeller Foundation Archives, RG 1, Series: 100, Box 5, Folders 39–44, Films 1922–1938, Folder 39.

34. For further discussion of the tensions between entertainment and scientific accuracy in another genre of educational films, see Gregg Mitman, *Reel Nature: America's Romance with Wildlife on Film* (Cambridge, MA: Harvard University Press, 1999).

35. *Our Movie Made Children* popularized the findings of the Payne Fund Studies, a series of social scientific examinations conducted between 1930 and 1933 that studied the effects of movies on youth. Henry James Forman, *Our Movie Made Children* (New York: Macmillan, 1933).

36. See Lisa Cartwright and Brian Goldfarb, "Cultural Contagion."

chapter six

Screening Science

Pedagogy and Practice in William Dieterle's Film Biographies of Scientists

T. Hugh Crawford

> Observation and experiment are subject to a very popular myth. The knower is seen as a kind of conqueror, like Julius Caesar winning his battles according to the formula "I came, I saw, I conquered." A person wants to know something so he makes his observation or experiment and then he knows.
> —LUDWICK FLECK

In "Give Me a Laboratory and I will Raise the World," Bruno Latour asks, "If nothing scientific is happening in laboratories, why are there laboratories to begin with and why, strangely enough, is the society surrounding them paying for these places where nothing special is produced?"[1] Latour drew this rather startling question from his early anthropology of laboratory life, a study that helped launch a wave of science studies, which culminated in the hue and cry roused by Paul Gross and Norman Levitt's *Higher Superstition*. A National Association of Scholars meeting then had shrilly trumpeted, "Science Is Being Challenged,"[2] and the Sokal/*Social Text* "scandal" received unexpected national media attention. Latour was a central figure in these controversies which really were not about the validity of science per se, but rather had to do with what counts as science in contemporary culture.[3] The present essay inverts this debate. Instead of examining scientific practices—laboratory life or scientific writing—in order to demonstrate how they are or are not embedded in ideology and culture, I propose to discuss public entertainment as scientific practice. I want to see how the tools developed by these rogue sociologists and

141

historians work when applied to films that depict (and, as I will argue, are) science in action.

In the late 1930s and early '40s, Hollywood film biographies of scientists and inventors were enormously popular. MGM released *Young Tom Edison* and *Edison the Man* in 1940, and Mervyn LeRoy's *Madame Curie* premiered in 1944. Central to this subgenre are two films by William Dieterle: *The Story of Louis Pasteur* (1936) which received two academy awards and has been called the first docudrama, and *Dr. Ehrlich's Magic Bullet*, which was released in 1940. Produced at a crucial historical moment regarding the prestige of scientists in America, Dieterle's films provide a unique perspective on the production of scientific authority through popular entertainment. It is difficult to place Dieterle's science films in traditional genres. They are not action films; they partake of drawing room comedy and family melodrama, but those elements are peripheral to what is clearly their pedagogical function. Unlike related films of the period—*Young Tom Edison* and *Edison the Man*—which do not explain the actual principles underlying their subject's technological innovations, Dieterle devotes considerable time depicting scientists at work.[4] Dieterle's scientists are active agents making quick, incisive decisions in order to reveal scientific truth in spite of a dull, unimaginative, and often willfully ignorant populace. His message is that the practice of science requires strong moral conviction, extreme self-reliance, indomitable will, and the occasional support of colleagues, spouses, and families. Rather than simple "Pasteurs of microbes" (to use Latour's phrase), Dieterle's scientists are Rambos in the lab. In the depths of the Great Depression, this creation of hero-scientists and the depiction of science's seemingly inexorable progress marks an anxiety about parallel social progress, but perhaps what is more significant in these films, when viewed through the lens of recent sociology of science, is their (probably unconscious) commentary on the production of scientific knowledge.

Latour's work is particularly useful for this discussion, as he spends a great deal of time dealing with two particular elements of scientific practice that serve as the focus of Dieterle's films: theater and inscription. *Pasteur* is a careful representation of laboratory life and dramatizes Latour's concept of the "Theater of Proof," while *Ehrlich* carries these themes further and, at the same time, examines the role of inscription in science. However, if this analysis were to remain on this level—examining these two tropes in the films—it would simply be a discussion of the representation of scientific practice in a popular medium. The key insight Latour and his sometime colleague Steven Shapin provide to link the representation of science to actual scientific practice is the pedagogical function of the theater and the inscription. Both films show clear concern with scientific practice *and* pedagogy, and what they teach is that the terms are inextricably linked. Pedagogy is not merely what comes after science is completed; it is always already part of the project itself. Science and film require a heavy investment in teaching their respective audiences how to see, to read, and to warrant the accuracy of their perceptions.

Learning to See

> Objectivity in science cannot be proclaimed, it must be built.
> —LONDA SCHIEBINGER[5]

Pasteur has been called the first docudrama, a designation that aligns it with realism, objectivity, and pedagogy, and, at the same time, distances it from the major genres.[6] In a discussion of realism and documentary, Fredric Jameson argues,

> Photography as such—and even documentary film, which one might have expected to be central to these theories of realism, but whose claims are paradoxically resisted in the name of *fiction* film—both offer occasions for reification and, one would like to say, for ontic misreading, insofar as they are preeminently susceptible to reappropriation by an aesthetic of inert reflection, and tend to omit (or to allow us to ignore and forget) the dual structure of eventfulness constitutive of such photographic realism—the event on the side of the subject as well as that on the side of the object, the happening of the act of registration as well as the instant of history uniquely "registered" upon the bodies of the photographic "subjects."[7]

Even though Dieterle's films are carefully staged—indeed, they are about careful staging—Jameson's point can be usefully extended to the fictive realism of these films (their documentary quality) and to related theories of scientific realism. A reductive characterization of the latter is that its practitioners believe in a stable world "out there" waiting to be discovered by increasingly sophisticated scientific theories, procedures, and technologies. Science, then, is the patient, ongoing accumulation of increasingly accurate pictures of the world. What such theories fail to take into account (either by ignoring or minimizing its importance) is, to paraphrase Jameson, the dual structure of eventfulness constitutive of such scientific realism: the socially and technologically embedded circumstances of the production and consumption of scientific knowledge. Those issues have traditionally been the province of constructivists.[8]

Dieterle's films are aligned with what Jameson calls realism in fiction film. They do not foreground their production as do contemporary documentaries (or, in its own dizzying self-reflexivity, Vertov's *Man with a Camera*); however, they do foreground the *production* of scientific realism so that, even though they finally celebrate scientific "progress," their interest in the social element of scientific work and, perhaps a bit more unconsciously, in the technologies of presentation and objectification connects these films to the constructivist position. At times these films dismiss social pressures as in any way influencing scientific practice, but their subject matter is overwhelmed by such details. In an early refinement of the constructivist position, Latour and his collaborator Steve Woolgar emphasize that "[o]ur argument is not just that facts are socially constructed. We also wish to show that the process of construction involves the use of certain devices whereby all traces of production are made extremely difficult to detect."[9] This comment, when read in relation to Dieterle's films, helps reveal the paradox of documentary that underlies the very notion of both theater *and* practical objectivity.[10]

The production and exhibition of any significant detail demands a technology of observation (social and technical) and an audience, yet that detail at the same time must appear naked, untainted by those very technologies. As Latour and Woolgar conclude *Laboratory Life*:

> The result of the *construction* of a fact is that it appears unconstructed by anyone; the result of rhetorical *persuasion* in the agonistic field is that participants are convinced that they have not been convinced; the result of *materialization* is that people can swear that material considerations are only minor components of the "thought process"; the result of the investments of credibility, is that participants can claim that economics and beliefs are in no way related to the solidity of science; as to the *circumstances*, they simply vanish from accounts, being better left to political analysis than to an appreciation of the hard and solid world of facts.[11]

Perhaps it is not surprising that the dominant trope of Dieterle's films is the reproduction of the scientist's own public staging of experimental proof where *two* audiences come to know, believe, and warrant the veracity of Pasteur's claims. In such scenes, scientific proof is simply a matter of presenting the object so that all who come to see will know and understand it.

Discussing the production of objective knowledge in seventeenth-century British science, Shapin argues, "the simplest knowledge-producing scene one can imagine in an empiricist scheme would not, strictly speaking, be a social scene at all. It would consist of an individual, perceived as free and competent, confronting natural reality outside the social system."[12] The ideal of unmediated observation is the hallmark of Enlightenment science and is a rhetoric that has been exploited by scientists at least since Vesalius's public anatomies and Bacon's *Novum Organum*. However, as Shapin argues, this notion of free, objective observation is fraught with difficulties. Galileo learned much to his chagrin that those who peered through his telescope did not see what he was seeing; neither did the members of the French Academy of Medicine see the same microbes as Pasteur when peering through his microscope.

Shapin's point is that, ironically, scientific objectivity depends on teaching practitioners how to see, what to look for, and what conclusions to draw.[13] One form of technologically mediated observation Dieterle exploits (perhaps for the first time in popular film) is microcinematography.[14] The first encounter the film audience has with Pasteur is through his own eyes as he looks in a microscope. Although the audience does not yet know what it is seeing, Pasteur and Dieterle quickly provide an explanation — these are the microbes responsible for childbed fever. Significantly, in the film Pasteur himself parallels the objects he pursues. First the germs are on stage, then Pasteur. He pursues, locates, and pasteurizes microbes while, at the same time, he is pursued by the Academy of Physicians who clearly desire *his* elimination. It is also significant that puerperal fever is the subject of this particular science lesson. Ignaz Semmelweis's *Etiology, Concept, and Prophylaxis of Childbed Fever* was one of the first studies to establish protocols for avoiding this disease (chlorine hand washing), but his work was largely ignored until germ theory was adopted as a plausible medical paradigm. Like Pasteur, Semmelweis did not speak within the truth of the pathological discourse of his age and was only vindicated (became part of the medical discursive formation) by later studies in antiseptics (primarily Joseph Lister's). In Dieterle's depiction of Pasteur, hand washing and childbed fever loom large and provide the film's family drama.

In a discussion of Robert Hooke that directly relates to Shapin's argument, Barbara Stafford observes that the early microscopists "confused epistemology with perception. The scientific seer believed he would come to *know* things stably and eternally by *seeing* them proximately and minutely."[15] The fixing of the universal and the eternal is a crucial element of the rhetoric of scientific objectivity as it relates to optical technologies. Visible truth is freed of the vicissitudes of disruptive or imprecise linguistic description; however, as *Pasteur* makes clear on several occasions, such truth and its attendant universality are not so easily had. For the general (undisciplined) audience, the exotic flora and fauna of a microscope slide require explanation. The film demonstrates that such images must be supplemented by a theoretical enframing and validated by a disciplined audience in order to produce meaning, and, even given such support, those representations often remain sites of bitter contention.[16]

The space where scientific knowledge is produced is pedagogical—one that defines procedures, practices, admission, and activity—and a theater for the display of newly formed objective data to that disciplined audience. As Shapin notes, "The physical and the symbolic siting of experimental work was a way of bounding and disciplining the community of practitioners, it was a way of policing experimental discourse, and it was a way of publicly warranting that the knowledge produced in such places was reliable and authentic."[17] Film re-creations of laboratory life echo that privileged pedagogical space and discipline their audience, teaching it how to recognize the "truth" of objective fact. Dieterle's audience also learns how simple objective observation, particularly when it involves advanced technology, is no guarantee of credibility. The opening furor created by Pasteur's pamphlet on childbed fever gains him an audience with Napoleon III, but even though the women at court find his microscope and slides a novelty (similar to the stereoscope and other optical parlor games),[18] his impatience with those who doubt his conjectures earns him exile from Paris rather than the honors accorded a scientist who has made a significant discovery. The implied connection between the microscope and parlor games reflects an anxiety about visual truth both caused and remedied by optical technologies.

In *Screening the Body*, Lisa Cartwright discusses the importance of visual discipline as it relates to the technological apparatus itself: "Although by the nineteenth century microscopy still involved a singular monocular observer, that observer's perception was nevertheless continually corrected and calibrated by the apparatus just as the observer supervised and calibrated the bodies whose fragments were observed."[19] The practice of microscopy created observers who were disciplined by the very technologies they used. Similarly, film cameras and projectors train an audience to certain cultural and technological forms of seeing. In the nineteenth century, a broad range of optical devices (stereoscope, phenakistoscope, and so forth) produced startling visual distortions and were associated with illusionists.[20] From this perspective, the microscope is not a clarifying technology; instead it creates the potential for visual deception and marks Pasteur as a conjuror. Clearly film technology suffers the same fate. As "documentary" it depicts what was "really there," but as drama, it is the product of the sound stages in Hollywood's dream factory. It is significant that Dieterle insists on the validity of Pasteur's visual pedagogy, since his film serves a similar function. It emphasizes the trope of direct observation, even as those observations are technologically mediated (both in and by the film) and supplemented by explanatory discourse and representations.

Pasteur's exile results in his anthrax research at Arbois and gives Dieterle the opportunity to represent other elements of scientific practice. Articles that report the results of years of laboratory research rarely detail the sheer drudgery and repetition involved in the production of such knowledge. Instead, laboratory time is compressed; false leads, failures, and readjustments are suppressed in order to produce clear results that create an effect of timelessness and universality. In *Pasteur*, intertitles and montage provide easy compression of many years of research. They also raise Pasteur's stature by paralleling his years at Arbois to the Franco-Prussian War. Quick cuts between the violence of the front and the serene peace of the lab invite a contrast, while the voice-over relates these disparate scenes, asserting the significance of the war and peace of microbes (the French title of Latour's *Pasteurization of France*) in the rebirth of the French economy. Dieterle explicitly equates scientific advances with political progress, frequently depicting Pasteur not as the doctor of a few sheep in Arbois but instead as the savior of France and, quite possibly, all humanity. The film ends with Pasteur lecturing triumphantly in an auditorium (which architecturally resembles a theater for scientific demonstration as well as the theater in which the film is witnessed) on the importance of such scientific progress. This is the commonsense position of most scientific realists, particularly those engaged in the current "science wars" version of the political correctness debate; however, such a position is clearly fraught with contradiction. Pasteur is rightly celebrated for the near-eradication of certain diseases, but, as the famous biologist Jacques Monod observed the day before his death, even that is a mixed blessing: "I used to laugh at physicists' problems of conscience, because I was a biologist at the Pasteur Institute. By creating and proposing cures, I always worked with a clear conscience, while the physicists made contributions to arms, to violence and war. Now I see clearly that the population explosion of the third world could not have happened without our intervention. So I ask myself as many questions as the physicists ask themselves about the atomic bomb. The population bomb will perhaps prove more dangerous."[21] Monod's point (one that still bears traces of a colonialist attitude) is that progress, or indeed, scientific truth, must always be measured from a particular perspective, and that even a minor shift can raise serious questions about such progress or the accuracy of current scientific views of the world. In this sense, Dieterle's films are (albeit naively) more interestingly accurate pictures of (or supplements to) scientific practice as the cinematic apparatus (both the technology and the socio-technical enframing called "Hollywood") marks the "dual structure of eventfulness" embedded in any observational/optical socio-technology.

The Franco-Prussian war also creates the opportunity for the most dramatic and public pedagogical event of Pasteur's career—the anthrax experiment at Pouilly le Fort. Because of its need to pay war reparations, the new French Republic turns to agriculture and faces an epidemic that appears to have struck everywhere except Arbois. One of the two health officials dispatched to discover the reason wants to know for himself the cause of the sheep's apparent immunity. He is a trained observer and is readily convinced by Pasteur's slide show on the etiology of anthrax. Significantly, the film audience also learns how to see scientifically—to see through the superstition of "infected fields" to the "truth" of Pasteur's microbiology. However, convincing the Academy of Medicine requires a larger stage and simpler explanations, so Dieterle restages the Pouilly le Fort experiment replete with its tension and carnival atmosphere.[22] Pasteur's private explanation of anthrax is recast in a dramatic and irrefutable public presentation of scientific truth. This scene emphasizes the importance of public demonstration and enables the now well-schooled film audience to participate in the validation of the facts being produced.

Latour has discussed the significance of Pasteur's demonstration: "The Pouilly le Fort field trial is the most famous of all the dramatic proofs that Pasteur staged in his long career. The major mass media of the time were assembled on three successive occasions to watch the unfolding of what was seen as Pasteur's predication. "Staging" is the right word because, in practice, it is the public showing of what had been rehearsed many times in his laboratory."[23] Latour expands this discussion of staging in *The Pasteurization of France*: "Pasteur's genius was in what might be called the *theater of proof*. Having captured the attention of others on the only field where he knew that he was the strongest, Pasteur invented such dramatized experiments so that the spectators could see the phenomena he was describing in black and white. Nobody really knew what an epidemic was; to acquire such knowledge required a difficult statistical knowledge and long experience. But the differential death that struck a crowd of chickens in the laboratory was something that could be seen 'as in broad daylight.'"[24] Or in a darkened movie theater. The importance of this discussion in light of Dieterle's film is in the rarefaction of images or the production of stark differentials. Film is ideally suited to reproduce such highly dramatized events, to warrant their having actually occurred, and to mark the relative absence/presence of such objective knowledge.[25]

Scientific realists can (and should) object that these films simply reenact past scientific events and are obviously not science themselves. However, that position ignores two significant elements of scientific practice. The first Latour mentions above: Pasteur was not gambling with his dramatized public experiment because he had performed that very experiment many times before in the laboratory. This decenters the moment of scientific discovery: one cannot privilege one moment in the process over another. In other words, the "eureka" phenomenon is vastly overrated. Instead, scientific proof rests in the repetition of the experiment to a satisfactory regularity (the much-vaunted reproducibility), and this repetition always already includes the last in the series—in this case, public demonstration (all of them). Of course this moment is never the last. Pasteur's anthrax vaccine becomes a standardized procedure—to use Latour's term, it is a "black box"[26]—and its theater of proof is staged through silent, statistical repetition in barnyards across the world.

Public demonstration is also directly tied to another, equally significant element of the production of scientific knowledge: dissemination. Public presentation (or, as an efficient substitute, the scientific article) is never simply an event that occurs after the fact. Instead, it is part of ongoing scientific practice. Nascent truth must find its way to the larger community; it must be picked up and validated by later practitioners.[27] For example, it is crucial that Joseph Lister be present at Pouilly le Fort as authoritative witness and as someone who will carry this practice to other geographical regions and in different scientific directions. Such later discoveries are reenactments in the broadest sense, variations on repetitions. From that perspective, Dieterle's films are not simply historical representations and polite entertainments. They are also an element of scientific practice because they serve a crucial pedagogical function—the establishment and maintenance of scientific authority in the popular imagination. Science's theater of proof *requires* ongoing restaging to reinforce both its epistemology and its power.

The pedagogical function served by *any* staging of scientific proof (in the high school biology lab or the Salk Institute) cannot be underrated, nor can it be temporally separated from scientific practice. Much of life in the laboratory is spent producing marks that all scientists must constantly learn to see and interpret. The clear connection between cinematic technology and the production of scientific (particularly biological and astronomical) knowledge underlines film's importance beyond the instrumental.[28]

The regime of vision in post-Enlightenment society is inflected by a rhetoric of objectivity defined by the theater. As Kaja Silverman argues, "The screen represents the site at which the gaze is defined for a particular society, and is consequently responsible both for the way in which the inhabitants of that society experience the gaze's effects, and for much of the seeming particularity of that society's visual regime."[29] The term "screen" serves a particular function in Silverman's argument that I will not explore here. Instead I would substitute the notion of supplementarity raised earlier. The objective reality presented by the camera (or the optical technology depicted in Dieterle's films) requires that the audience ignore the qualities of the screen in order to see through to the reality behind it, but, paradoxically, the screen is the supplement. It marks its own absence (it is that which must be ignored) even as it is overwhelmingly present (it is that which must be watched). The audience's relation to this supplementarity is equally problematic. The cinematic apparatus provides the possibility of identification—here not just identification *with* (the actors, characters, or situations), but also identification *of* (the validation of the truth of the objects presented). It is learning to participate in a particular society's visual regime.

Learning to Write

> By remaining steadfastly obstinate, our anthropological observer resisted the temptation to be convinced by the facts. Instead, he was able to portray laboratory activity as the organisation of persuasion through literary inscription.
> —BRUNO LATOUR[30]

As *Pasteur* makes clear, simple observation is enframed by various linguistic and graphical practices and the testimony of trained audiences, even though consciousness of such enframing is necessarily suppressed by the rhetoric of the theater of proof. Where *Pasteur* is concerned with the presentation of "natural" objects (living microbes, dead sheep), in *Dr. Ehrlich's Magic Bullet*, Dieterle focuses more closely on these texts, inscriptions, and testimony. Latour's early work raises pointedly how scientific practice depends on devices whose inscriptions could be used to produce evidence: "But [the scientist's] end result, no matter the field, was always a small window through which one could read a very few signs from a rather poor repertoire (diagram, blots, bands, columns). All these inscriptions, as I called them, were combinable, superimposable and could, with only a minimum of cleaning up, be integrated as figures in the texts of articles people were writing."[31] Such inscriptions invoke an absent object or mark a statistical regularity. But, as *Ehrlich* demonstrates, such images are not unmediated, nor are they exempt from interpretive and epistemological problems. Indeed, as Jameson's discussion of realism here applies, they invite "ontic misreading."

Instead of emphasizing Ehrlich's development of chemotherapy (generally considered his most important scientific contribution), Dieterle's film concentrates on the pedagogical function of starkly rendered scientific images. In an interesting spin on Marshall McLuhan's thesis, Pasteur learned to grow microbes on a certain medium in order to observe them with the naked eye. This agriculture included both fertilizer and the extermination of unwanted "pests" so that he could observe his animals "in the wild." Ehrlich takes this process a step further. His early research with aniline dyes (coal tar derivatives developed by the German dye industry during the nineteenth century) led him to attempt to stain specific portions of living tissue, a concept he explains in the film as "affinity": specific chemicals have an affinity for certain portions of cells or organs in larger organisms.[32] The notion of affinity laid the foundation for modern chemotherapy. Ehrlich's dedication to Heinrich Waldmeyer in *From the Theory and Practice of Chemotherapy* notes, "Although modern chemotherapy has established itself in science and medical practice, its origin goes back to the histological stainings; so it is no coincidence that the first chemotherapeutic experiments that ended so promisingly were conducted with dyes—methylene blue or trypan red. Initially, therefore, chemotherapy was 'chromotherapy.'"[33] In other words, Ehrlich's earliest interest was in "biogrammatology," rendering specific portions of organisms visible, or, more precisely, legible.[34]

In one of the film's early dramatic moments, Ehrlich risks both his career and credibility by rashly proposing an alternative method for seeing biological entities. At a meeting attended by the leading medical minds of the era, Robert Koch announces his isolation of the tuberculosis bacillus and invites his audience to see for themselves, lamenting the fact that "the bacilli do not stand out clearly. Only those of you whose eyes are trained in the use of the microscope will be able to see and that of course makes my discovery of little or no value to the average doctor in diagnosing cases of tuberculosis."[35] Ehrlich, who is supposed to be on duty at the hospital, raises the possibility of staining microbes with aniline dyes so they would be easily read. As a result he loses his position at the hospital and must push through with his researches in a makeshift laboratory in his own home, contracting tuberculosis in the process. Koch's comments at this meeting underline the importance of trained observation. Scientists do not simply make observations and then report their findings; they must continue to learn to see even as they practice their craft. In addition, Ehrlich's successful staining of the bacilli raises pointedly the problematic relationship between inscription and observation, words and things. While a dyed organism is clearly not a word, it is also not exactly a thing (Latour, following Michel Serres, would call it a quasi-object[36]). It occupies ground similar to that of a graph that marks a statistical regularity: it is both pure (noise is suppressed) and enigmatic (one cannot be sure whether essential information has been suppressed along with the noise). The organism might write an obscure or ambiguous text, perhaps even a poem.

Ehrlich's first test of his self-inscribing microbes is the diagnosis of his own tuberculosis. Here the film's black-and-white technology limits its impact, since, as Koch himself described it, the TB staining process was not only vivid, but also aesthetically pleasing: "Under the microscope all constituents of animal tissue, particularly the nuclei and their disintegration products, appear brown, with the tubercle bacilli, however, beautifully blue."[37] Ehrlich's new-found inscription technology has immediate pedagogical/economic effect because, when he shows Koch his results, the latter is able to present his financial backers with clear evidence of his isolation of the bacillus. Rather than depending on an unstable theater of proof or undisciplined observers, he is able to call on self-inscribing microbes to represent themselves to doctors, funding agencies, and the now well-educated audience watching the film.[38] This scene emphasizes the pedagogical importance of clear inscriptions, a point that is later missed by Ehrlich himself. When he uses his microscope to explain his work on trypanosomes (which was to lead to his work on syphilis) to his funding committee, it is clear that they (unlike the film's audience) are unable to understand what they are seeing. This is the cause of some humor in the film, but it also emphasizes the importance of the link between pedagogy and practice. Ehrlich neglects this role as teacher and risks losing funding for his research. He complains that explaining his work to those who do not understand chemotherapy is like arguing about colors with the colorblind, and then mixes two clear chemicals to produce what the audience must assume is a brilliant color. This not only is good drama, it designates the film's audience as competent observers, as part of the visual regime of scientific knowledge. This gesture also defines Ehrlich's work (and science in general) as the production of inscriptions purified to simple oppositions: absence/presence, clear/stained.[39]

The self-inscribing organism is not the only role of inscription in *Ehrlich*. He first lays out his strategy to find the chemical cure for syphilis (what came to be known as 606 or, in the United States, "Salvarsan") on a graph he draws with chalk on the laboratory floor.[40] The search for this chemical is then depicted by a rapid montage of laboratory gestures— manipulation of chemicals, etc.—on which is overlaid a progressive charting of the graph. The audience follows the process without living through the tedium of the laboratory practices. This strategy duplicates the use of graphic images in a scientific article to compress laboratory time and to purify results. R. C. Lewontin notes (with more than a touch of irony) that in the contemporary scientific article, "There is some leeway for speculation in the 'Discussion,' which is considered somewhat adventitious since, after all, the 'Results' speak for themselves. An excessively articulated and discursive 'Discussion' will be truncated by the editor in the interests of space. Indeed, a really good experiment would need no 'Discussion.'"[41] Results typically summarize the data, usually transferring numerous readings, measurements, assays, and repetitions into a few proportional marks framed in a graphic representation. The trained reader simply "sees" the results without having to live through the quotidian tedium, the numerous false leads, contaminated specimens, and misdirected analyses. Like Lister observing the anthrax experiment at Pouilly le Fort, the reader of the article sees only a few perceptual contrasts: up, down, left, right. As Lewontin argues, the truth of such articles seems to reside in such displays rather than in an "excessively articulated" discussion.[42]

However, graphic images are not simply a replacement for the theater of proof. They contain and compress time by instantaneously representing that which cannot be perceived in an instant, and they enable scientific knowledge to circulate well beyond the walls of well-equipped laboratories. Ehrlich's production of 606 clearly enabled his obscure laboratory to become an economic center, but he was also active in producing a system of immunological imagery that continues to circulate in immunological theory. Alberto Cambrosio, Daniel Jacobi, and Peter Keating discuss these images, noting that he invented them for his Croonian lecture as, according to some historians, a "last ditch effort . . . to make his theory intelligible."[43] They go on to discuss the pedagogical importance of these visual images, a point that should not be lost on the audience of *Dr. Ehrlich's Magic Bullet*. When threatened with the loss of funding, Ehrlich finds himself searching for a benefactor and seeks support from Franziska Speyer. At a dinner party, he first scandalizes the polite company by openly discussing syphilis,[44] but then successfully woos Speyer with his side-chain theory,[45] his "magic bullet" metaphor,[46] and numerous examples of his immunological imagery drawn over her fine linen tablecloth.[47]

Ehrlich's form of immunological imagery neatly fits Latour's categories of successful scientific representation: "In sum, you have to invent objects which have the properties of being *mobile* but also *immutable, presentable, readable* and *combinable* with one another."[48] But these "immutable mobiles" are not just simple representations. Cambrosio notes that, according to Heymann, "Ehrlich conceived of the diagrams as 'a convenient pedagogical resource.' This remark points to a very common theme among those who assessed the role and value of Ehrlich diagrams, namely, the stress on their (mere) pedagogic function." He goes on to qualify this point: "The term *pedagogic* is not to be understood as referring to textbook knowledge and teaching alone. It also refers to explanatory capacity and selection of relevant phenomena or features."[49] Ehrlich's diagrams are meant to represent in the abstract a range of chemical/immunological relations on the cellular level. In the film, they serve an obvious pedagogical function by communicating his theory to his colleagues and Franziska Speyer; however, these diagrams have explanatory capacity beyond convincing a partially educated audience. They embody a conceptual theory and consequently point toward an emerging research program. In other words, they are not black-boxed completed science. In the absence of material or organic inscriptions, they carry scientific inquiry in a specific direction.

This expanded sense of the pedagogical ties directly back to Silverman's sense of the audience learning to participate in a particular visual regime. While Ehrlich helped develop this particular regime, he was also disciplined by his own visual pedagogy. Dieterle makes this habit of seeing embedded in these inscriptions clear in the scene where Ehrlich, having just ejected his purblind funding committee, glances at a drawing of the newly discovered syphilis spirochete. He immediately marks its structural affinity with the trypanosomes he has learned to stain (and kill) with various chemical preparations. What he observes is not a fanciful drawing of abstract cellular/chemical reactions, but rather a schematic rendering of an observable organism. However, Ehrlich's quick conceptual slide from affinity on the micro level to an organism's structural affinity on the macro level marks an epistemology informed by a specific regime of vision. The abstractions of his immunological imagery are not merely a representation, but rather are constitutive of this regime.

Dr. Ehrlich's Magic Bullet's penultimate scene is a courtroom trial on the safety of 606. There, proof of the drug's efficacy and Ehrlich's innocence (as well as his slanderer's guilt) does not depend on microscopes, graphs, statistics, immunological images, or direct experiments on living patients. Instead he is acquitted (according to the film) by the testimony of a single witness, Emil Behring, who had doubted chemotherapy but had now come to see the world in Ehrlich's fashion. *Ehrlich's* courtroom does not depend on the same forms of proof as scientific knowledge, but this scene does bring us back to Shapin's initial postulate. Truth in scientific matters depends on the training and expertise of the witnesses who are willing and qualified to testify to the significance of the issue at hand. To become a spokesman for scientific truth is to appear to be the supremely free and competent observer, but also to have the ability to speak not just for science, but for the truth of the objects themselves, to help them speak their own truth in the *theater of proof*, and to present the inscriptions they form through a range of mediating devices. Latour asks, "How can all the strength that a spokesman musters be retrieved?" He then supplies the same answer as Dieterle: "by letting the things and persons represented *say for themselves the same thing that the representatives claimed they wanted to say*. Of course, this never happens because, by definition, such direct communication is impossible. Such a situation however may be convincingly staged."[50] What Latour does not emphasize but Dieterle makes clear is that staging is doubly pedagogical: it teaches the audience (scientific or cinematic) *about* the objects presented and, by occupying a node in a network of truth production, it teaches that same audience *how* to see. Neither is prior to the other. In science and in film, the real and the simulated are unstable (or even useless) categories, and pedagogy is necessarily indistinguishable from practice.

Returning briefly to the impish question that opens this essay, are these films actually science in action? The answer is both yes and no. Clearly if one's conception of science is restricted to the production of new (perhaps increasingly accurate) descriptions of the natural world, then these films merely depict science in action. They dramatize the heroic efforts of two significant figures in the history of modern medicine. However, as both Latour and Dieterle teach us (one intentionally, the other perhaps unconsciously), the above conception of scientific practice is unduly narrow if not outright naive. One cannot restrict the notion of science to that rhetorically effective but elusive moment of insight. A fuller understanding of modern science requires that the crucial pedagogical function of presentation and inscription be acknowledged. Without doubt, nature's actors play a central role in science's drama, but the success of science also depends on a particular epistemology, on the creation of a regime of vision that is constantly being learned and supplemented. What Dieterle and Latour both show is that the seemingly stable boundaries designated by the term "science" are a (probably necessary) fiction, but that scientific practice is crisscrossed by numerous socio-technical networks that can combine in a broad range of configurations to form different concepts of objectivity and reality.

Notes

Ludwik Fleck, *The Genesis and Development of a Scientific Fact* (Chicago: University of Chicago Press, 1979), 84.

1. Bruno Latour, "Give Me a Laboratory and I Will Raise the World," in *Science Observed*, ed. K. Knorr and M. Mulkay (London: Sage, 1983), 141–42.

2. Paul R. Gross and Norman Levitt, *Higher Superstition: The Academic Left and Its Quarrels with Science* (Baltimore: Johns Hopkins University Press, 1994). The NAS conference was held in Cambridge, Massachusetts, on November 11–13, 1994, featuring, among others, E. O. Wilson, Christopher Ricks, Larry Lipking, and Paul Gross.

3. As the flyer for the NAS conference notes, questioning what counts as science is "a matter of widespread concern: it undermines public confidence; it alters directions of research; *it affects funding*; it subverts the standards of reason and proof" (emphasis mine). Actually, Latour finds the terms of this debate impoverished on all sides, participating as they do in the "modern regime": "The next books I want to write without critique—without denunciation. I hope that will show the possibility of doing other social sciences which are not predicated on the task of unveiling and denouncing, as the [modern] critique always asked us to

do." See T. Hugh Crawford, "An Interview with Bruno Latour," *Configurations* 1, no. 2 (Spring 1993): 267.

4. *Edison the Man*, directed by Clarence Brown (Metro-Goldwyn-Mayer, 1940) and *Young Tom Edison*, directed by Norman Taurog (Metro-Goldwyn-Mayer, 1940). Mervyn LeRoy's *Madame Curie* (Metro-Goldwyn-Mayer, 1944) follows a pattern similar to Dieterle's.

5. Londa Schiebinger, *Nature's Body: Gender and the Making of Modern Science* (Boston: Beacon Press, 1993), 114.

6. For extended discussions of documentary realism, see Jeanne Allen, "Self-Reflexivity in Documentary" and Peter Ohlin, "Film as Word: Questions of Language and Documentary Realism." Both essays are found in *Explorations in Film Theory: Selected Essays from* Ciné-Tracts, ed. Ron Burnett (Bloomington: Indiana University Press, 1991).

7. Fredric Jameson, *Signatures of the Visible* (New York: Routledge, 1992), 186–87.

8. The all-to-familiar debate between realists and constructivists in social studies of science can be read across this register. Perhaps the embodiment of this debate was the 1989 conference "Realism and Representation" sponsored by the Rutgers Center for the Critical Analysis of Contemporary Culture. Papers from this conference were later published in *Realism and Representation*, ed. George Levine (Madison: University of Wisconsin Press, 1993).

9. Bruno Latour and Steve Woolgar, *Laboratory Life: The Construction of Scientific Facts*, rev. ed. (Princeton, NJ: Princeton University Press, 1986), 176.

10. The term "practical" objectivity here is meant to distinguish it from concepts of objectivity in laboratory practice. "Practical" marks an objectivity based on simple seeing-is-believing rather than the accumulation of multiple sightings or evidence with high statistical probability before the confirmation of belief. This is of course a problematic distinction, but it does help reveal the juxtaposition and frequent opposition of the temporal and spatial in the production of proof.

11. Latour and Woolgar, *Laboratory Life*, 240.

12. Steven Shapin, "The House of Experiment in Seventeenth-Century England," *Isis* 79 (September 1988): 375. Shapin expands the issues raised in this essay in his book *Social History of Truth* (Chicago: University of Chicago Press, 1994).

13. For a preliminary "history" of objectivity, see Lorraine Daston and Peter Galison, "The Image of Objectivity," *Representations* 40 (Fall 1992): 81–128. See also Simon Schaffer, "Self-Evidence," *Critical Inquiry* 18 (Winter 1992): 327–62, and Susan Bordo, *The Flight to Objectivity: Essays on Cartesianism and Culture* (Albany: State University of New York Press, 1987). On the importance of authority and testimony in science, see Michel Foucault, "The Discourse on Language," in *The Archaeology of Knowledge* (New York: Pantheon, 1972).

14. For a discussion of the emergence of microcinematography in "actual" science, see Lisa Cartwright, *Screening the Body: Tracing Medicine's Visual Culture* (Minneapolis: University of Minnesota Press, 1995), 81–106.

15. Barbara Maria Stafford, *Body Criticism* (Cambridge, MA: MIT Press, 1991), 352.

16. Jacques Derrida's notion of supplementarity is particularly useful in understanding this point not so much because it sharpens the argument, but rather because this nexus of objectivity, proof, and theater supplements Derrida's own use of the term. The theater—the social and mechanical technologies—is not merely ancillary to the revealed truth of the objective world. Instead, it actively constitutes it. Any discussion of optical technology in science involves an infinite regress of necessary supplementation—including the supplement of scientific truth provided by Dieterle, his actors, and his audience. In other words, the theater as supplement is not a mere addition but instead is that which must not be ignored. Dieterle's films supplement science in precisely this fashion. See Derrida, *Of Grammatology*, trans. Gayatri C. Spivak (Baltimore: Johns Hopkins University Press, 1976), 141–64, and Derrida, "The Supplement of the Copula," *Margins of Philosophy*, tran. Alan Bass (Chicago: University of Chicago Press, 1982), 175–205.

17. Shapin, "House of Experiment," 373–74.

18. On the scientific import of such devices as the stereoscope, phenakistoscope, and others as they relate to an emerging sense of the visual in the eighteenth and nineteenth centuries, see Jonathan Crary, *Techniques of the Observer* (Cambridge, MA: MIT Press, 1990).

19. Lisa Cartwright, *Screening the Body: Tracing Medicine's Visual Culture* (Minneapolis: University of Minnesota Press, 1995), 84.

20. One need only recall the uncertain status of the daguerreotype as a form of proof in Hawthorne's *House of Seven Gables*.

21. Quoted in Michel Serres and Bruno Latour, *Conversations on Science, Culture, and Time*, tran. Roxanne Lapidus (Ann Arbor: University of Michigan Press, 1995), 17.

22. On the surface this scene contrasts with the precision and security of laboratory knowledge but, as Katharine Park has noted regarding fifteenth- and sixteenth-century Italian anatomies, such public presentations of scientific information often had a carnivalesque quality (Park, "The Criminal and the Saintly Body: Autopsy and Dissection in Renaissance Italy," *Renaissance Quarterly* 47, no. 1 [Spring 1994]: 14). This scene can also be related conceptually to Paul Feyerabend's famous (or notorious) "anything goes" carnival concept of scientific innovation. See Feyerabend, *Against Method* (New York: Verso, 1978), chap. 1.

23. Latour, "Give Me a Laboratory," 151.

24. Bruno Latour, *The Pasteurization of France* (Cambridge, MA: Harvard University Press, 1988), 85.

25. Clearly film is also a site of contention and divergent interpretations, as the use of videotape in the courtroom or in medical imaging technologies makes evident. Regarding the latter, see T. Hugh Crawford, "Imaging the Human Body: Quasi-Objects, Quasi-Texts, and the Theater of Proof," *PMLA* 111, no. 1 (January 1996): 66–79.

26. Challenging the truth of the black box requires the construction of expen-

sive counter-laboratories where those boxes can be opened. See Bruno Latour, *Science in Action* (Cambridge, MA: Harvard University Press, 1987), 79–94.

27. In *Science in Action*, Latour argues that the fate of scientific facts always rests in the hands of later practitioners who must take up those facts and use them. Ibid., 108–21.

28. Etienne-Jules Marey, who is generally considered if not the father, then at least the uncle of cinema, was a physiologist, as were the Lumière brothers.

29. Kaja Silverman, *The Threshold of the Visible World* (New York: Routledge, 1996), 135.

30. Latour, *Laboratory Life*, 88.

31. Bruno Latour, "Visualization and Cognition: Thinking with Eyes and Hands," *Knowledge and Society* 6, no. 6: 3–4. Film itself is also a form of inscription, as it is a technology that enables the camera to isolate and purify (through lighting, camera angle, focus, selection) the objects/subjects it presents. The roots of this inscriptive impulse in film can be traced to Etienne-Jules Marey's experiments with his "photographic gun." On Marey and inscription, see François Dagognet, *Etienne-Jules Marey: A Passion for the Trace*, trans. Robert Galeta with Jeanine Herman (New York: Zone Books, 1992).

32. *Affinity* is Ehrlich's, not Dieterle's term. In his 1913 lecture at the 17th International Congress of Medicine (in London) on Chemotherapy, Ehrlich explained his approach: "The whole field is governed by a simple, I might say natural, principle. If the law is true in chemistry that *Corpora non agunt nisi liquida*, then for chemotherapy the principle is true that *Corpora non agunt nisi fixate*. When applied to the special case in point, this means that parasites are killed only by those substances for which they have a certain affinity, thanks to which the substances become anchored to the parasites." Quoted in Ernst Bäumler, *Paul Ehrlich: Scientist for Life*, trans. Grant Edwards (New York: Holmes and Meier, 1984), 188.

33. Ibid., 8. There is a close link between dyes and many therapeutic compounds developed in the twentieth century. Between the wars, IG Farben (the company that funded Ehrlich's research on Salvarsan) and its subsidiaries developed a range of chemicals they tested as both dyes and drugs. They were later accused of hiding significant medical research under dye patents. For a postwar American account of these activities, see Richard Sasuly, *IG Farben* (New York: Boni & Gaer, 1947), 28–32.

34. François Dagognet uses the term *biogrammatology* in a different context (Marey's concern with marking the motion of living beings). See Dagognet, *Etienne-Jules Marey*, 86.

35. Ehrlich describes this lecture (March 24, 1882, in Berlin "On Tuberculosis"): "It was in a small room of the Physiological Institute that in simple, clear words, Koch convincingly described the causative organism of tuberculosis; his address was supported by numerous preparations and items of evidence. Everyone who attended that lecture was deeply moved, and I must say I always recall that evening as my greatest scientific experience." Quoted in Bäumler, *Paul Ehrlich*, 28.

36. See Bruno Latour, *We Have Never Been Modern*, tran. Catharine Porter (Cambridge, MA: Harvard University Press, 1993), 51–55.

37. Robert Koch, "The Aetiology of Tuberculosis," in *A Source Book of Medical History*, ed. Logan Clendening (New York: Dover, 1960), 394.

38. The circumstances regarding the production of films in this era provide an ironic perspective on some of these issues. At the time, three-color Technicolor was becoming the most widely used process, but, because of limited production capacity, the laboratories could not accommodate all the films being produced. Consequently, *Ehrlich* is in black and white even though its subject—visualization through color staining—begs for color production. More poignant is that the dyes Ehrlich used created the very possibility of color film and that AGFA (a subsidiary of IG Farben, Ehrlich's primary source of research funds) produced an alternative to Technicolor, though at the time it was experimental and later, because of World War II, was generally unavailable. On the development of color technology in film, see Dudley Andrew, "The Post-War Struggle for Colour," *The Cinematic Apparatus*, ed. Teresa De Lauretis and Stephen Heath (New York: St. Martin's Press, 1980), 62–72.

39. One of the more obvious manifestations of this phenomenon in contemporary society is the home pregnancy test, a device that enables a fetus to represent his or her presence.

40. The source of this image is probably Paul DeKruif, *Microbe Hunters* (New York: Harcourt, Brace & Co., 1926), 339.

41. R. C. Lewontin, "Facts and the Factitious in Natural Sciences," *Critical Inquiry* 18 (Autumn 1991), 141.

42. For a historical discussion of this issue, see Peter Dear, "*Totius in Verba*: Rhetoric and Authority in the Early Royal Society," *Isis* 76 (1985): 145–61.

43. Alberto Cambrosio, Daniel Jacobi, and Peter Keating, "Ehrlich's 'Beautiful Pictures' and the Controversial Beginnings of Immunological Imagery," *Isis* 84 (1993), 682.

44. A point of some consequence at the historical moment of the film. Not only was syphilis not discussed in polite company in 1940, because of the war, Ehrlich's cure was unavailable and antibiotics had not yet been introduced into general practice.

45. For an explanation of this concept, see Ehrlich, "On Immunity" (reprinted in Clendening, *A Source Book of Medical History*) and Arthur M. Silverstein, *A History of Immunology* (New York: Academic Press, 1989), 64–66, 94–107.

46. This image appears in a number of places, e.g., "for antitoxins and antibacterial substances are, so to speak, charmed [magic] bullets which strike only those objects for whose destruction they have been produced by the organism" (Ehrlich, "On Immunity," 409).

47. Bäumler relates a similar anecdote: "Once in Laubenheimer's house, at Number 3 Leverkuserstrasse in Höchst, Ehrlich is said to have pulled out of his waistcoat one of the colored pencils he always carried with him to illustrate his

side-chain theory on Frau Laubenheimer's spotless white tablecloth." See Bäumler, *Paul Ehrlich*, 81.

48. Latour, "Visualization and Cognition," 7.
49. Cambrosio, Jacobi, and Keating, "Ehrlich's 'Beautiful Pictures,'" 682.
50. Latour, *Science in Action*, 73.

chapter seven

Optical Constancy, Discontinuity, and Nondiscontinuity in the Eameses' *Rough Sketch*

Michael J. Golec

> [Science's] underlying ideas are not difficult and not at all extraordinary. They can be understood and enjoyed by everyone.
> —JACOB BRONOWSKI

Found Education

The Eames Office builds models. Charles Eames says, "In practice, we think of ourselves as tradesmen." The historian of science Owen Gingrich writes that craftsmen staff the Eames Office.[1] The Eames Office and its principle partners, the husband and wife team of Charles and Ray Eames, are best known as furniture designers. From the 1940s until the 1980s, the office produced a steady stream of modern iconic furnishings, most of which are still in circulation today. In addition to their furniture design work, Charles and Ray worked with the United States government, premier cultural institutions, and some of the country's most successful corporations. They made films, designed exhibitions, built houses, and made toys. In every instance, Charles understands that the work of the office is to design and construct models. He refers to the furniture and architecture end of the office's production as "models before the fact," ideas for things not yet built. The films and exhibitions are, according to Charles, "'models after the fact'—like a scientist's model of a giant molecule or a galaxy." Models before facts and models after facts are things the office makes in order to "communicate about a structure that interests you."[2] In the case of Eames films, structure is everything, even to the extent that it "takes the place of plot."[3] If a thing interests Charles, then it might interest others who can recognize the structure and learn from his model.

An Eames model is a multiplicity—film, chair, house, screen, exhibition, mask—that makes and is made from connections. Eames's models make up the texture of, what Charles calls, "found education." By this he means a process of learning where people find out about things "piece by piece, in pursuit of their own needs, loves, and curiosities."[4] He locates the origins of this type of learning in the lives of Benjamin Franklin and Thomas Jefferson, the subjects of the exhibition and film *The World of Franklin and Jefferson*, an example of a "model after the fact." The connective texture of found education—an assemblage of discrete pieces—possesses a quality of "nondiscontinuity." According to Charles, Americans have forgotten the value of found education with its emphasis on learning through discovery and observation. Indeed, as he observes, there exists a tremendous amount of data to find, but there are very few models that make sense of all this data and that can facilitate found education.[5] As he explains, "Considerable resources, complex institutional structures, and a whole range of techniques, go into producing these models, these found objects. The criterion is that these things are out in the world at large—where people can come upon them, without having enrolled in a course—without having chosen to become an initiate."[6]

As model-makers, the Eames office produces projects that are meaningful "for a non-specialist" but that are not "trivial or embarrassing for the person who knows most about the subject."[7] In a near paraphrase of Marshall McLuhan, Charles states that the film *Rough Sketch for a Proposed Film Dealing with the Powers of Ten and the Relative Size of the Universe* (1968) was "conceived in a way that conveys meaning to a distinguished scientist as to a small child."[8] The structure of nondiscontinuity—a continuous chain of distinct things—is such that connections are potentially profuse while at the same time manageable, hence his belief in a model possessing a broad range of receptions and multiple meanings. Yet, for all its multiplicities, a profusion of meaning within a nondiscontinuous structure requires a fixed point of reference.

In order to contend with the cosmic variables of the universe and the human body, *Rough Sketch* and the later *Powers of Ten* (1977) represent the scientifically known, but generally unfamiliar, at every scale and magnitude of viewing. Two figures occupy a picnic blanket in almost identical opening scenes of the films. The image of the picnic delimits the boundaries of the social within the confines of the domestic. Exactly, what do I mean by "the social" and by "the domestic," and, more importantly, how do both films represent these two spheres? I take social to mean connections, ties, assemblages, associations, networks, and systems. The social is thus a map or tracing of these relationships; it is not, following Latour, a substance or material added to another substance or material nor is it a background, frame, or other sort of container.[9] As I use the term, domestic designates a distinct sense of the social as the organization of human and nonhuman relationships pertaining to the home and/or the household, hence "the social within the confines of the domestic."

The picnic scene has led Beatriz Colomina to remark on how the films create an image of "[i]ntimate domesticity [that] is suspended within an entirely new spatial system."[10] The domestic produces a social space that, according to the logic of both films, relates to science. The sequence of views, or powers of ten — 10^{24} in both films and 10^{-14} in *Rough Sketch* and 10^{-16} in *Powers of Ten* — link to the establishing shot of the man and woman having a picnic. An invisible line extends to the limits of human knowledge, drawing from and connecting to this well-known and too often taken-for-granted location. From the very beginning of both films one can detect an attempt to situate science within the familiar — to find (or *found*) science in the midst of social relations framed by the domestic sphere and a mid-century ideology of normative pairings, as exemplified by the male and female couple on the blanket.[11]

The image of the picnic implies a mode of seeing, of taking a view of the world, it is a visual technique for coming to terms with the heterogeneity of science. But to what end? Why the picnic, not just in the first, "rough" version of the film but also in the second, more refined version of the film? On the one hand, the familiar scene of a couple sharing a meal out-of-doors in *Rough Sketch* and *Powers of Ten* introduce the lay public to science, underscoring Jacob Bronowski's admission that the "underlying ideas" of science are ordinary and easily understood. In addition to being a mathematician, a biologist, and a historian of science, Bronowski was devoted to the popularization of science, from writing for the catalog to the *Exhibition of Science* section of the *Festival of Britain* in 1951 and hosting the *Ascent of Man* series for BBC/PBS in 1973. His daughter Judith Bronowski worked at the Eames Office in the late 1960s and researched, wrote, and filmed *Rough Sketch* for the office. In her preliminary script for the 1968 film, sometimes referred to as *Cosmic View*, Bronowski wrote, "We begin with a familiar view — the scale is one that we are quite used to." In the film, however, Bronowski narrates, "We begin with a scene one meter wide, which we view from one meter away." At this point, the familiar scene has already passed without comment, yet it remains as the anchor image for the entire film. Beginning with a familiar view, quite literally a *common place*, is crucial to how the Eames Office conceived of its practice of model making and to the accessibility of its meanings, especially when it came to their film work for education.

Rough Sketch is a model in service of "found education." The film both opens with a scene of the everyday and *opens up* the everyday, here penetrating the colloquial ease of the picnic to reveal the vast and interwoven coordinates of the unseen, extraterrestrial context of ordinary human existence. Within the social-domestic relations, the exchange of plates of food and of conversation, *Rough Sketch* locates science in the everyday and the everyday in science. There exists no immediately perceivable discontinuity between the picnic blanket, the far reaches of outer space and the immensity of inner space. Uniformity and homogeneity seem to prevail as the film progresses through its powers of ten, rendering the world both extraordinary and ordinary at every scale. The film achieves what Charles identifies as nondiscontinuity through a technical virtuosity that results in a near seamless thread of images that reproduce all orders of sublime magnitude the same. It is this seamless animation that constitutes the "optical constancy" at the heart of the nondiscontinuous film. First introduced to describe the homogenization of visual representation through the use of one-point perspective, the term also implies uniformity along a chain of linked images.[12]

The picnic blanket, therefore, is a fixed point—a nodal fixture—that organizes the visual continuity of the rapid movement of multiple images that make up the film. Yet, so much more is active in *Rough Sketch*'s representation of galactic and subatomic spaces. In what Paul de Man might identify as a Derridian "articulation at the source," the film makes apparent discontinuity within its system of visualized order.[13] In doing so, the materiality of media technology or the materiality of film work, made manifest by the roughness of *Rough Sketch*, disrupts the seeming hegemony of visual experience at the heart of optical constancy. The film's articulation at the source points to anterior textures of visual, optical, graphic, and typographic technologies that add to and support the film. In the case of *Rough Sketch*, it is not that Bronowski and the Eames Office suppressed discontinuity in favor of the optical constancy in order to achieve nondiscontinuity. There is an unseen depth to the film's optical constancy where additional sources are interwoven and traceable but hardly perceivable.

In his unpublished Eliot Norton Lectures, Charles Eames notes the following: "Film as model; sketch model that gives specific insight. Model doesn't have to be total theory of field, or a golden thread through the labyrinth. Can be a tentative walk, after which you retreat and reconsider. Powers of Ten [*Rough Sketch*] was a model, a thin line through the universe."[14] In *Rough Sketch*, celluloid material—film itself—is literally threaded through an apparatus so as to make a *line* of scientific knowledge visible to an audience. In what follows, I will explore the lines of discontinuity that interrupt the "thin line" of optical constancy. The pathways of discontinuity—flights and deterritorializations—indicate an accumulation of source material that precedes the final film. For my purposes, discontinuity is akin to Paul Schrader's observation that *Rough Sketch* is an example of "information overload," where the film delivers "more data than [the viewer] can possibly process."[15]

Information overload in *Rough Sketch* results from a combination of the discontinuity of film as a material artifact—its threading together of discrete images—and the heterogeneous nature of science—the multiple and competing regimes of scientific study and publication.[16] Both discontinuity and heterogeneity saturate the film such that a reversal takes place so as to make science appear continuous and homogeneous. As Marshall and Eric McLuhan observe of media in general, "data overload equals pattern recognition;" anything pushed to its limit reverses its characteristics and becomes its complementary form.[17] The film's physical character of discrete images rapidly moving through the projector creates a perception of continuity. *Rough Sketch*'s gathering of multiple sources embedded in the film likewise creates a perception of homogeneity. What is of interest to visual culture and education in the life sciences is how *Rough Sketch* reverses back to discontinuity and heterogeneity so as to reveal and to school its viewers in the tacit connections that underscore scientific knowledge. Because the film is discontinuous and overloaded, I want to explore the possibility of divergence embedded in the early stages of research for the film in order to understand the multiple pathways inscribed into each power of ten.

In order to achieve this goal, I will attend to several breaks in the film that not only mark its *roughness and sketchiness* but that also inscribe a disruption in the easy flow of the structure of the animated images. The science library stores the data that produces an overload of information coming from the massive and ever-massing sources of science. It is the visual and textual archive of scientific knowledge, what Latour calls the "paper work" that precedes but is not materially distinct from the film.[18]

It too requires an apparatus—the library—to order its knowledge and to maintain constancy. The convergence between the film and the library, I will argue, is what constitutes the "sketch model" structure of *Rough Sketch*. This media convergence founds the educational purpose of the film. It is at this break that two media technologies converge—the science library and the film. Convergence happens at every power of ten within the film, but, because of limited space, I will focus on 10^{-4}—the power of blood cell mitosis—and 10^{-5}—the power of the leukocyte—to reveal a "viral" option that, if chosen, would result in "cellular takeover" of the structure of the film and its investment in the everyday and the familiar.

Making a Visual Model

Judith Bronowski began her work on the film in 1967 when she first began working at the Eames Office:

> The idea for the film was given to me by Charles, in the form of a small book, Kees Boeke's 1957 *Cosmic View*. I was asked to begin to think about and develop a film outline based on the concept powers of ten. I struggled with this assignment as conscientiously as an English girl at a new job would, thinking that if I didn't come up with something good I would lose my job. Little did I know, until much later, that new employees at the Eames Office were often given this assignment, almost as a test, and moved to another area when they failed to bring the project alive. But I wanted to make the film.[19]

There are several little-known facts revealed and significant points made here. First, while we know that Kees Boeke's *Cosmic View* was the starting point for *Rough Sketch* and *Powers of Ten*, it is not generally known that new employees were given the challenge to turn the book into a film. That Bronowski was the first Eames employee to meet this challenge is significant since Bronowski herself admits that Charles's willingness to employ a graduate in English and History of Science in a design studio had something to do with "his admiration" for her father, the distinguished theoretical biologist and mathematician.[20] It was Bronowski's enthusiasm and ingenuity that helped to realize the *Rough Sketch* model. It was not merely that she had solved the problem (something at which her peers had failed), but also that she possessed the ambition to make the film. She explains that her interest in the project arose from her commitment to "making science relevant and accessible."[21]

An immense effort went into the planning and the production of the film. A script had to be written that would convey reliable scientific information. Existing images were gathered from multiple sources and had to be drawn or photographed where visual sources were nonexistent. Camera tracking moves were calculated, images were enlarged and reduced, and everything was precisely registered in order to create the final animation. Music had to be scored and supervised. And, finally, the voice-over had to be added. Bronowski did all of this work. She wrote the script, drew all the storyboards, calculated the tracking moves (with the help of her father), and edited the sound and the final film. She even costarred in the film. Bronowski is the woman sitting at the man's feet in the opening picnic sequence. The only component of the film that Bronowski was not responsible for is the picnic scene. The office sometimes had picnic lunches in the back of the studio. As Bronowski tells it, "Probably one day at a picnic Charles just said, 'Rent a cherry picker, and let's get a shot we can use to start the film at a picnic and go from there.' That's how things happened around the office."[22]

The planning and production aspects of the films were well established by the time Bronowski started to work for the Eames Office. As Esther McCoy observes, an "affection for objects and love of facts" motivated the couple to make films.[23] From the very beginning, the Eames Office integrated the familiar into their films. In 1953, Ray and Charles produced *A Communications Primer*, a short film that was intended to familiarize architects and urban planners with a theory of communications developed from Norbert Wiener's *Cybernetics: Or Control and Communication in the Animal and the Machine* and formalized by Claude Shannon in the pages of the *Bell System Technical Journal*.[24] Discussing the film with Schrader, Charles explains,

> I had the feeling that in the world of architecture they were going to get nowhere unless the process of information was going to come and enter city planning in general. You could not really anticipate a strategy that would solve the increase in population or the social changes which were going on unless you had some way of handling this information. And so help me, this was the reason for making the first film, because we looked for some material on communications.[25]

Charles made contact with Bell Labs, hoping that its administrators would make suitable resources at their disposal, but they had only a few photographs of a "man with whiskers inventing the telephone."[26] This paltry showing held little interest for Charles and, he thought, his audience, so he and Ray rectified the situation and made A *Communications Primer*, "essentially for architects."²⁷

Communications Primer begins with what appears to be a static image of three horizontal lines at very slight angles to each other. Very soon, however, the audience realizes that these lines are multiplying and that they are in fact telephone wires as telephone poles rush by. The image moves. Over this scene, Charles observes in a voice-over, "The era we are entering might well be characterized as an era of communication." As Charles narrates, "If anything acts on the signal so as to vary it in an unpredictable and undesirable way in the communication system, it is noise." One of the many goals that occupied information theorists in the immediate postwar era was to reduce noise and to increase undisturbed and accurate transmission. In order to exemplify this concern, *A Communications Primer* continues through several examples of successful communications and unsuccessful communications as a result of noise. Charles explains,

> In speech, the brain is usually the information source. From it the message is selected. The message is a thought, not the words [I LOVE YOU]. The vocal mechanism codes the words into vibrations, and transmits them as sound across the communications channel, which is, of course, the air. The sound of the word is a signal. The ear picks up the signal, and with the associated . . . nerve decodes the signal [I LOVE YOU] and delivers the message to the destination. This time, noise can originate in the transmitter or in the vibrations that disturb the channel. Or, it could be a nervous condition on the part of the receiver. And it could change the message from 'I LOVE YOU' to 'I HATE YOU.' How do you combat it? One way is through redundancy. 'I LOVE YOU. I LOVE YOU. I LOVE YOU. I LOVE YOU.' Another is increasing the power of the transmitter. This combats noise, as does the careful beaming of the signal or duplicating of the message via other signals.

The choice of an intimate message is one attuned to the particular complexities of communications and its potential derailing. "I love you" (or "I hate you") as an instance of ordinary language is not merely a poetical or prosaic device, rather its articulation in the film draws on an audience's stock of forms and apperceptions of social relations. Ray and Charles had intended these familiar words to mitigate noise, to break through the clutter of communications because their contextual meaning is already stored in the mind. As such, the general structure of *A Communications Primer* follows Charles's desire to pursue "information that uses the medium for model making—trying to reduce the information to a form in which it will come through as a direct, primary experience."[28] It is the reservoir of apperception—what is established, familiar—that affords such an immediate responsiveness in the audience and their archive of social interactions. And yet, according to the film, the repetition of three words creates a pattern that when overloaded, because too emphatic, can switch into its perverse other. In this case, information overload transforms "I LOVE YOU" into "I HATE YOU."

The picnic blanket in *Rough Sketch* does the same work as "I LOVE YOU" does in *A Communications Primer*. *A Communications Primer* is a model for subsequent films. It offers a theory of communications that serves as a template for direct delivery of complex information. The familiar image as an instance of redundancy appears in many of the Eames films. For example, in the multiscreen presentation *The Information Machine*, unveiled at the IBM Pavilion at the 1964 New York World's Fair, the Eames Office relies on the image of everyday activities to explain the complex nature of "computer consciousness."[29] The multiscreen film uses prosaic examples like planning a dinner party and playing a game of football to exemplify, what the film presents as, "the method used in solving even the most complicated problems is essentially the same method we all use daily." Like images of dinner parties and football games, the picnic as an image of social interaction and domestic bliss draws on an already existing iconography so as to insure its reception and to assure the film's audience that the meaning of the film is firmly grounded in the familiar day-to-day activities.

The layering of complexity in communications implies that the routing of information is fraught with obstacles. In *A Communications Primer*, complexity is synonymous with noise, whether in the case of the mechanical duplication and distortion of text or in the case of disruptions to the human nervous system. In either case, the communications circuit is overloaded with extra information and multiple possibilities, making it more difficult, but not impossible, to maintain the fidelity of message. As Niklas Luhmann observes,

> Communicating something in particular involves selection from among a number of different possibilities. The motivation to transmit and receive selected choices becomes more improbable, and it thus becomes more difficult to motivate receptivity by means of the form of selection taken. It is, however, the function of media of communication to achieve precisely this.[30]

Luhmann continues to explain that information is the result of selection based on the projection of a system of experience onto a horizon of possibilities set within the limited range of "this and not that." For Luhmann, the arrival of potential disruptions is narrowly based on apperception and expectation. From different perspectives, *both* Luhmann and Eames propose that a system of experience results from a certain programming whereby the reception of a message is already anticipated in its transmission through media. In the case of love as in the case of the picnic we can say that both are able to communicate without any specific meaning to the message. Thus primarily indirect in its force, both the words "I love you" and the image of the picnic rely on receivers having already understood their meanings.

On the one hand, we can say that the everyday nature of the picnic makes for good drama that anticipates the "real" content of the film. The transition from the sequence of still images that constitutes the picnic scene to the animation that makes up the science sequences—from the ordinary to the extraordinary—is all the more apparent when the audience quickly realizes that most mundane activities, like having a picnic, are within the purview of the scientific. On the other hand, science directly relates to the domestic trappings of the picnic, as Beatriz Colomina observes, and its accompanying hierarchies, labor distributions, compensations, and negotiations.[31] In this case, the same transition from the ordinary to the extraordinary in *Rough Sketch* can be construed as a sly move from the domestic to the scientific where the organization of science as an institution intermingles with the domestic sphere. The transposition of science and the modern household dates back to the household engineering movement and reform-era discourse on hygiene for the home, both of which were by-products of the home economics movement and its adoption of modern scientific principles. According to this history, the domestic sphere is a scientifically controlled environment.[32]

The picnic blanket marks an intersection where domestic advice meets scientific discovery and late-1950s and -1960s discourses on the benefits of scientific research.[33] Also, the picnic scene in *Rough Sketch* inscribes the coordinates of normative heterosexual behavior for postwar scientists. The pairing of male and female picnickers and the arrangement of the two figures on the blanket is exemplary of immediate postwar domestic relations. The woman on the blanket is not at the center, but rather she occupies a place to the side of the action in the film. Because of her location away from the central focus of the film, she is effectively assigned to a support position.[34]

A theory of "the gaze" in cinema and its implications for gender identification is helpful here. Laura Mulvey's theory of the command of the masculine gaze indicates that the visual image of the male body on the picnic blanket in *Rough Sketch* determines, or reinforces, the social or gender identity of the male-self of the viewer.[35] The film's focus on the male body on the picnic blanket visually anchors scientific knowledge in the male self. It is the familiar image of the picnic that works, almost imperceptibly, to hedge against discontinuity introduced by convergence. When delivering his fourth lecture in the Norton series, Charles linked the importance of the familiar to the production of visual models. The presence of the picnic blanket as the starting point for an exploration of scientific knowledge allows the audience to find coherence in a chaotic situation. The image orders the disorder of the cosmos. As Charles might claim of this ordinary scene: "At that level there is no discontinuity."[36] In *House of Science* (1961), the office worked to create a sense of nondiscontinuity "between rewards of science and the rest of life," a quality evident in *Rough Sketch* and further emphasized in the later *Powers of Ten*.

Optical Constancy

Whatever direct senses, impressions, or truths that *Rough Sketch* delivers is due to optical constancy—the presumed antidote to discontinuity. As Charles explains, "Powers of Ten [*Rough Sketch*] had a feeling but also a structure."[37] The structure of the film is best articulated by Schrader where he describes the sequencing of the film's images as "skin becomes a wrist, wrist a man, man a beach, beach a peninsula, and so on, each change the square of the previous change, and each faster than the viewer can adjust his equilibrium."[38] The film's cuts and splices result in a series of rapid afterimages and make possible Schrader's chain of "becomes" or becomings. There are some forty roughly prepared images, repeatedly photographed and awkwardly registered to produce the semblance of an unbroken animation. The film's overall design is, however, never deceptive.

Rough Sketch conveys its truths through what the Renaissance scholar William Ivins calls its "optical constancy" at the level of the film's constructed, continuous tracking shot. In his book *The Rationalization of Sight*, Ivins discusses how the "internal invariances" or optical constancy of pictorial perspective, or what Bruno Latour calls a "regular avenue through space," prepared modern science and technology to acknowledge "homogeneity of space" and "uniformity of nature."[39] Influenced by Ivins, McLuhan observes, "Visual space is uniform, continuous, and connected."[40] For McLuhan, rationality and logic depend on the constancy of the "presentation of connected and sequential facts or concepts."[41] The film's breakthrough is in its animation of one-point, pictorial perspective whereby the picnic blanket is a vanishing point that organizes the disparities of the cosmic and the biological at every scale into the uninterrupted logic of scientific knowledge. McLuhan would say that the nodal picnic blanket not only determines the logic of representation all the way up and all the way down but it also fixes perceptions of the visual sequencing in the film.[42]

The perceived uniformity and scientific veracity of *Rough Sketch* are products of the mediating technology of film and its perceptual affects, thus we cannot attribute the film's continuity to scientific research and knowledge. Indeed, films are the accumulation and organization of so many cuts, splices, edits, and discontinuous frames. When humans view a film they do not see its celluloid materiality—the stuff that makes the projected images possible. The quick and imperceptible succession of these cuts, splices and frames as they run off of a reel make film possible as a convincing medium of representation, a medium that is all imaginary. As Friedrich Kittler remarks, "As phantasms of our deluded eyes, cuts reproduce the continuities and regularities of motion."[43] In other words, the technological successes of film render humans blind to its techniques. Second, there exists an argument that film is a technology originated in blindness. As Kittler further explains, the physiology of perception first explored by Gustav Fechner as a result of his literal blindness and the aesthetics of ancient Greek tragedy described by Friedrich Nietzsche in *The Birth of Tragedy* are formative inscriptions of an "applied perceptual practice" in film—the exploitation of the afterimage.[44] In film, rapid-fire images produce a sequence of afterimages that create perceptions of continuity. As Schrader remarks, "The pictorial area of the screen in itself has more information than the mind can assimilate."[45] Constancy is *in* the perception of the sequence of running images; it is, therefore, not a material character of the film itself.

The perception of unbrokenness—or visual coherence and logical order—results from a sequence of broken images, or, to put it more accurately, croppings. The entire film, as short as it may be, is made up of approximately forty-two discrete images. The parts of the film are discontinuous; they are connected by the film's animation. Each image is singularly composed, framed, and cropped to correspond to the dimensions of the one-meter square. It is this format—this frame within a frame structure—that provides the film with its visual continuity *and* its discontinuity. The film achieves its visual continuity of the frame through the technique of cropping, which is also the discontinuity at the center of *Rough Sketch*. To crop is to cut off or conceal parts of an existing scene or existing image. Cropping, according to Meyer Schapiro, "brings out the partial, the fragmentary and contingent in the image, even where the object is centered. The picture seems to be arbitrarily isolated from a larger whole and brought abruptly into the observer's field of vision."[46] Each cropping contributes to the thin line that connects all of the images that constitute *Rough Sketch*.

The technical nature of photography and cinema requires that the photographer and cinematographer crop or frame the visual field in order to produce static and moving images. The lens and aperture of the camera necessarily crop out all that exists beyond what can be seen beyond the viewfinder. The film underscores a commonplace in photography and film—that is, there is always something left out of the frame and of my field of vision—that structures the formal and narrative aspects of *Rough Sketch*. While the technology of film exploits discontinuity in the service of optical constancy, it can easily function to cause a film to reverse back to the discontinuous where there are breaks or inconsistencies in image, registration, and continuity so that the connective tissue of film work is unconcealed.

Model and Multiplicity

There are unseen worlds that occupy the gaps or splices in *Rough Sketch*'s chain of becomings. Each slice at the editing table strips away a potential or alternative becoming so that what appears on or as film is always partial, edited, and inconsistent. Yet, when I view *Rough Sketch* I do not nor do I expect to penetrate through the film back to the science lab where heterogeneity and complexity reside, but rather I linger in the visual-perceptual zones of astronomical and biological bodies.[47] There are, however, flashes of the discontinuous character of the film such that *Rough Sketch* does not withdraw into the merely visual. In addition to the breaks and misregistrations in the animation, there exists Bronowski's voice-over and the film's use of language to organize the visual perceptions of its audiences. For example, between 10^5 and 10^{-2} the images—from sky to skin—that flash before the audience are easily recognizable. As the camera ascends to 10^{24}

in both films and descends to 10^{-14} in *Rough Sketch* and 10^{-16} in *Powers of Ten* the images reveal almost nothing on their own. The passage from the familiar to the unfamiliar indicates an almost imperceptible break in the continuity of each film.

The "fix," as Roland Barthes might observe, happens when the voice-over acts as a running set of aural captions. Each spoken segment informs the audience as to what they are seeing at any given power of ten, thereby countering "the terror of uncertain signs" present in the unfamiliar realms of astrophysics and microbiology.[48] Fixing, stabilizing, or framing of visual perception, or, to put it more bluntly, how it is that I can see this array of marks at 10^{-5} as a cell requires that they are accompanied by the bibliographies of science. In following the flow of information between text and image, image and text, and the *writing* and *imaging* of territories the science library forms an information network with the *Rough Sketch*. The kind of information management involved in the ongoing project of producing a comprehensive catalog of all things known to science is a project of organizing the interaction of multiple pictures and diverse texts.

While the voice-over stabilizes the diverse images that flow before an audience, it also disengages the image from its sequential form. Like the captions that accompany all manner of visual material, the voice-over opens the image it attends to beyond itself. It is not that the voice-over/caption merely takes hold of an image or sequence of images, but, that it also, and perhaps most importantly, takes hold of other images and texts. Thus, the voice-over is a crucial link in a chain of visual signifiers that exceed the film. The voice-over marks and disturbs the morphological relationship between images, in a pure-formal sense, as neither merely perceptual nor merely formal. In the case of *Rough Sketch*, Bronowski's narration throws the sequence of discontinuous images into fields beyond themselves. The disruption of the film's pure self-presence, the violation of subjectivity, extends to that pure flow of interiority, pure image, that *would have been* had it not been for the exterior frame of the voice.[49] The voice interrupts the flow of optical constancy and reveals the massive amount of information—the unseen masses—required to create the semblance of nondiscontinuity.

The diverse texts that make up the multiplicity of the film model are too numerous to inventory here. There are, however, two instances that, while not universally applicable, do exemplify the disruptions embedded within *Rough Sketch*. At 10^{-5} *Rough Sketch* shows an illustration of a leukocyte, a white blood cell. It appears as if it were taken straight from a student biology notebook, but this is no expert illustration found in a textbook. The quality of the drawing is rough, but its main features are recognizable as nucleus and chromosomes, etc. It is worth noting that recognition of these two features is crucial to the transition that the film makes to the next power of ten where the focus is on the "donut"-shaped genes that make up the chromosome structure. A notation made to a thumbnail storyboard sketch of 10^{-5} indicates the emphasis: "leukocyte, genes in nucleus." While not showing the fine structure of cell and tissue, the illustration reproduced in the film is easily recognizable to any student of histology who has access to a light microscope and a biology textbook. But just in case, Bronowski's voice-over explains, "This cell is a leukocyte. It is self-reproducing and one of those which contains, in its nucleus, all the genetic material, the twenty three pairs of chromosomes that humans carry."

Yet, the film frame at 10^{-5} in *Rough Sketch* belies a network of interrelations where images and texts move freely from source to source. Invariance and transportation between distinct spatial locations—from the science classroom and lab to the animation stand to the screening room—are at play in exchanges that predate the appearance of the leukocyte image in the film. On the storyboard for power of 10^{-4} there is a rough drawing of a blood cell at a stage of potential division, or mitosis (figure 7.1). The same frame in the film is far less specific in its representation of cellular structure, instead showing a capillary that carries many blood cells. The sketch indicates Bronowski's research and her deliberations. In addition to the cell drawing one can see an accumulation of citations, evidence of her presence in the science library.

This debris from the planning stages of the film points to multiple sources: John McLeish's *Looking at Chromosomes* (1958), John Kendrew's *The Thread of Life: An Introduction to Molecular Biology* (1966), and Lemuel Diggs's *Morphology of Blood Cells* (1956). Each citation leads to a book in the science library. Information circulates from the sources to the sketch, making them all available for fixing the meaning of the film at that particular power and the unfolding narrative of the film. While they are spatially distinct, the books in the library and the drawings in the storyboard sketch maintain information constancy through the transposition of the data in the library to the data in the sketch. Here the heterogeneity of science edges into the film to reverse the consistency of visual perception.

Because of space restraints, I will discuss only one of the sources cited in the sketch to give a sense of the transpositions and fixes that are in play in *Rough Sketch* at the 10^{-4}. As I mentioned above, Bronowski cites John

Optical Constancy in *Rough Sketch* · 177

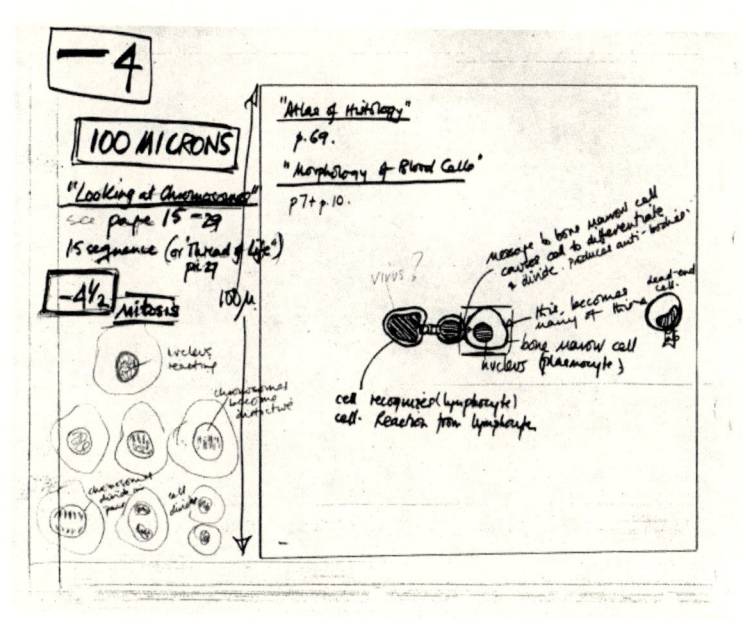

7.1 Sketch for storyboard, *Rough Sketch* (1968). Courtesy of and © Eames Office, LLC. www.eamesoffice.com

McLeish's *Looking at Chromosomes*. In 1958, at the time it was published, McLeish's book was praised for its skillful handling of mitosis and meiosis (two types of cell division).[50] The book was intended for the biology student in the early stages of her studies, so the relationship between the images, captions, and text are crucial for an accurate accounting of each stage of cell division.

The storyboard sketch for *Rough Sketch* transcribes seven of the thirteen images in *Looking at Chromosomes* and makes notations from several of the captions, which supplement the fourteen pages of text that explain root tip mitosis, microporogenesis, and megasporogenesis in the Regal lily. While the editing of texts in the process of transposing data to the sketch does not bear marks of absolute fidelity to the source (lily to human), the transposition of critical images and text performs the requisite task of indicating possible narrative directions for *Rough Sketch*. The edits to image and text, however, can be described as parts of the replicating network from library to film in terms of how seeing-as in the library compares to seeing-as in the film. The voice-over in the film remedies the break in continuity in the chain of replications from photomicrographs in *Looking at Chromosomes* to the drawings in the storyboard sketch to the animation

178 · The Educated Eye

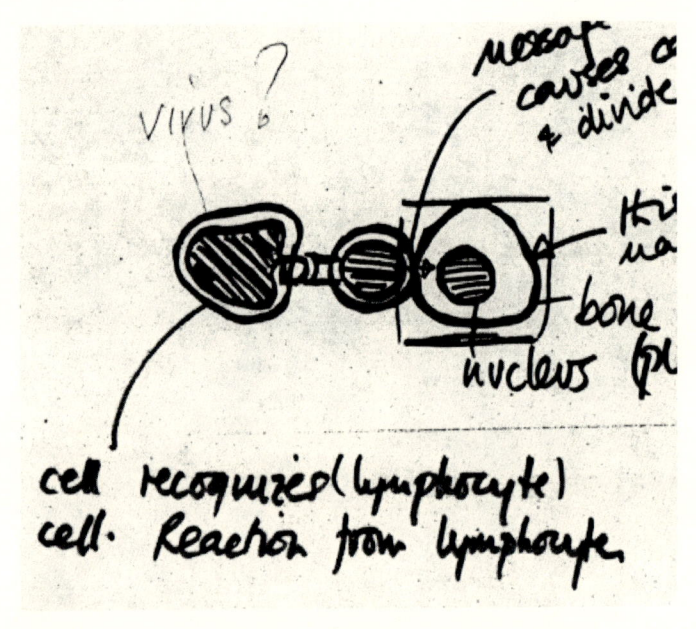

7.2 Detail of sketch for storyboard, *Rough Sketch* (1968).
Courtesy of and © Eames Office, LLC. www.eamesoffice.com

cell in the film by pruning mitosis to this: "Within the skin, the dark area we see is a tiny bloo[d] vessel—a capillary. The bloo[d] it carries is made up of different types of single cells." The voice-over instructs the audience to ignore all the discontinuities embedded in the image so as to see it from this general point of view.

A more damaging discontinuity exists where there appears a notation, "virus?" just to the left of the image that will transition into the drawing of the leukocyte at 10^{-5} (figure 7.2). The question mark after "virus" asks whether or not the film should take on the theme of "cellular take over." Cited to the left is John Kendrew's *The Thread of Life*, which offers a basic introduction to molecular biology, explaining in layman's terms "molecular architecture" and mutation.[51]

A note found among the storyboard sketches for *Rough Sketch* indicates further consideration of and resources for the inclusion of a virus into the film. The most suitable frame appears at 10^{-6}, where the chromosome is pictured as donut-shaped chromosomes. A fragment discusses two alternatives for the proposed viral narrative: introducing it from outside or identifying it inside the genetic structure. A draft of the voice-over script includes the text that would, if used, fix this particular image: "Many serious defects and abnormalities in humans have been found to be due to defective chromosomes which have either been broken or have failed to divide correctly." Additional materials found in the Eames Office files point to the complexities of introducing the viral line and defective chromosome into the narrative, namely "Mongoloid idiotism" resulting from the over reproduction of chromosome 21. The document goes on to explain how "a small mistake like this upsets the working of the body." And parenthetical instructions are given to "insert a panel of mongoloid children and chromosome layout" into the frame. The parenthetical instruction, like the gaps in the material film, could disrupt the normative narrative of the film and its emphasis on the familiarity of domestic bliss. Indeed, the viral encounter potentially derails the film, sending it on an alternate course with very different consequences for the master narrative of the all-knowing camera eye that dominates *Rough Sketch*.

The seemingly benign question mark at the end of "virus" draws on the science library with its offerings of multiple options for the film's narrative direction. But, if *Rough Sketch* were to take the viral route, then a host of complexities would present themselves. These complexities would surely overload the film more than it already is, adding even more unsettling paths to follow. Admittedly, not all branches of possibility can be taken up and followed in the nine minutes that it takes the film to move between the macro and the micro. On the one hand, it is fairly easy to see how the viral event breaches the continuum that is optical, structural, and narrative in *Rough Sketch*. On the other hand, the question, "virus?" introduces a greater degree of multiplicity in the model.

Rough Sketch makes it appear *as if* this world of the film exists apart from the overabundant bibliographies of science, and, in an interesting but neglected turn, as if the bibliographies themselves—their texts and images—were just mere piles of paper rather than being world-makers in their viral spread (to borrow from William Burroughs) of information. For whatever reasons, *Rough Sketch* does not include the *upsetting* narrative of the virus, leaving overload in the wake of a normative direction. There exists overload for sure, but it is at the source—in the bibliography of science. For the film to engage with scientific representation, which is hardly universally intelligible, it would have to demonstrate a greater understanding of what, for example, a virus or a leukocyte is in order to "bring the conviction" of what the frame at 10^{-4} or at 10^{-5} shows.[52] If such a representational conviction were introduced into *Rough Sketch* it would edge out the overriding narrative of normal science that undergirds the familiarity of the picnic scene and maintains what Freud calls the "germ-cell of civilization."

Conclusion: Dark Discontinuities

The overloading of the film with unseen discontinuities reverses into the perception of optical constancy and an emphasis on nondiscontinuity. In the process of the reversal in *Rough Sketch* there exists a surplus of dark discontinuities, the viral divergence at the power of 10^{-5} being just one of many. I say "dark" here because *Rough Sketch* is in the grip of a peculiar darkness in, to paraphrase Paul de Man, its groping toward a certain degree of insight.[53] Nondiscontinuity overshadows critical insights, relegating discontinuity to the shadows. Yet, in *Rough Sketch*'s overloading of information there exists the possibility to destabilize the film's "ontology of unmediated presence" in a reverse of the reversal.[54] This is why roughness is so crucial to the educational nature of the film. Indeed, as Charles Eames admits, "Non-discontinuous teaching is pretty hopeless; more exacting than teaching art. Not only new curriculum structures and a mass of people who don't exist—e.g. secondary school math teachers who don't hate math—few of them exist."[55] Roughness in *Rough Sketch* brings the dark uncertainties of science to light.

In his article on the Eameses' films, Paul Schrader writes, "The interstellar roller-coaster ride of *Powers of Ten* [*Rough Sketch*] does what the analogous sequence in *2001: A Space Odyssey* should have: it gives the full impact—instinctual as well as cerebral—of contemporary scientific theory."[56] Some forty years later, a better comparison might be a similar sequence in David Lynch's *Blue Velvet* (1986). In this scene, a middle-aged man is watering his lawn on a spring day in a small town. All is bright and sunny—the world is in bloom. But not all is well. Soon the water hose becomes entangled in some shrubbery. The man pulls at it, creating a kink that stops the flow of water. He immediately suffers a heart attack, causing him to fall to the ground. The hose spurts wildly as the man writhes on his lawn. The camera frames his now still body. An infant approaches from the driveway, and a terrier plays in the hose's shower. The camera cuts to the suburban lawn; it descends past the green grass blades into the dark earth. There it settles on a teeming colony of black beetles churning the soil. The movement of Lynch's camera connects the above ground activities of small-town existence to the underground stirrings of beetle existence. Because the camera magnifies the scene, the black hole of nonhuman life seems close to eruption, threatening the familiar and the normal that exists in the light of day.

Rough Sketch skins the male body, penetrating its dark interior to reveal a cascade of micro scales of nonhuman palpitation. Here a multitude of unfamiliar bodies work tirelessly to produce the horizons of normative behaviors that can transpire on the picnic blanket and beyond. That is, of course, until there is a breakdown in or rerouting of the causal chain established by the film's optical constancy. The viral element—"virus?"— proposes such a threat to the film, the body, and to the germ-cell of civilization. There is much to learn and a good deal to find at the moment of dark discontinuity.[57] But the question mark will always linger at the point of visualization. What is it that we do not see? My argument is that glimpses of disruptive discontinuity are there to grasp by way of an intense and intractable effort to produce nondiscontinuity. It is only by way of a process of an examination of the film's antinomies and dislodgings in the viewing of *Rough Sketch* that its modeling—of domestic bliss?—is visualized or made visual as such.

Notes

Jacob Bronowski, "The Story the Exhibition Tells," in *Exhibition of Science, South Kensington, a Guide to the Story It Tells* (London, 1951), 7. Passage quoted in Sophie Forgan, "Festivals of Science and Two Cultures: Science, Design and Display in the Festival of Britain, 1951," *British Journal for the History of Science*, 31, no. 2 (June 1998): 231.

1. Owen Gingrich, "A Conversation with Charles Eames," *The American Scholar* 46, no. 3 (Summer 1977): 326.
2. Charles Eames, "On Reducing Discontinuity," *Bulletin of the American Academy of Arts and Science* 30, no. 6 (March 1977): 24.
3. Charles Eames, "Interview" in Paul Schrader, "Poetry of Ideas: The Films of Charles Eames," *Film Quarterly* 23, no. 3 (Spring 1970): 15.
4. Eames, "On Reducing Discontinuity," 32.
5. Ibid., 33.
6. Ibid., 34.
7. Ibid.
8. Charles Eames, "Language of Vision: The Nuts and Bolts," *Bulletin of the American Academy of Arts and Sciences* 28, no. 1 (October 1974): 25. When referring to a film adaptation of historical events, like *Henry V* or *Richard III*, McLuhan observes, "Here extensive research went into the making of the sets and costumes that any six-year-old can now enjoy as readily as any adult." See Marshal McLuhan, *Understanding Media: The Extensions of Man*, 2nd ed. (London:

Routledge, 2005 [1964]), 314–15. It should be noted that there exists no evidence of any correspondence between the Eames Office and McLuhan in the Eames Papers held in the Library of Congress.

9. Bruno Latour, *Reassembling the Social: An Introduction to Actor-Network-Theory* (Oxford: Oxford University Press, 2005), 21–26.

10. Beatriz Colomina, "Reflections on the Eames House," in *The Work of Charles and Ray Eames: A Legacy of Invention*, ed. Donald Albrecht (New York: Harry N. Abrams, 1997), 12.

11. On the presence of the familiar in *Rough Sketch* and *Powers of Ten*, also see Michael J. Golec, "*Powers of Ten* (1977) and *Rough Sketch* (1968), Office of Charles and Ray Eames," *Design and Culture* 1 (July 2009): 201–4.

12. William Mills Ivins, *On the Rationalization of Sight, with an Examination of Three Renaissance Texts on Perspective* (New York: Da Capo Press, 1973), 9.

13. Paul de Man, "The Rhetoric of Blindness: Jacques Derrida's Reading of Rousseau," in *Blindness and Insight: Essays in the Rhetoric of Contemporary Criticism* (Minneapolis: University of Minnesota Press, 1983), 120.

14. Charles Eames, Lecture Notes for "I CARE," Eliot Norton Lectures, 1970–1971 (unpublished), in Charles and Ray Eames Papers, Box 217, File 10, Library of Congress, Washington, D.C. The Eames Office made the film more commonly referred to as *Powers of Ten* in 1977. In the Norton Lectures, Charles refers to *Rough Sketch*, which was made in 1968.

15. Paul Schrader, "Poetry of Ideas: The Films of Charles Eames," *Film Quarterly* 23, no. 3 (Spring 1970): 8.

16. For an account of how visualization streamlines the disjunctive nature of the sciences, see John Law and John Whittaker, "On the Art of Representation: Notes on the Politics of Visualization," in *Picturing Power: Visual Depiction and Social Relations*, ed. John Law and Gordon Fyfe (London: Routledge, 1988).

17. Marshall McLuhan and Eric McLuhan, *Laws of Media: The New Science* (Toronto: University of Toronto Press, 1988), 107.

18. Bruno Latour, "Drawing Things Together," in *Representation in Scientific Practice*, ed. Michael Lynch and Steve Woolgar (Cambridge, MA: MIT Press, 1990), 44.

19. Judith Bronowski, e-mail correspondence with author. May 2, 2006.

20. Ibid.

21. Ibid.

22. Judith Bronowski, e-mail correspondence with author May 4, 2006.

23. Esther McCoy, "Interior Design: The Charles Eames Office, An Affection for Objects," *Progressive Architecture* 54 (August 1973): 66.

24. See Norbert Wiener, *Cybernetics: Or Control and Communication in the Animal and the Machine* (New York: Wiley, 1948) and Claude Elwood Shannon, "A Mathematical Theory of Communication," *The Bell System Technical Journal* 27, no. 3 (July 1948): 379–423.

25. Schrader, "Poetry of Ideas," 9.

26. Eames, Lecture Notes for "I CARE," n.p.

27. Schrader, "Poetry of Ideas," 9.

28. Eames, Lecture Notes for "I CARE," n.p.

29. John Harwood, "Imagining the Computer: Eliot Noyes, the Eames and the IBM Pavilion," in *Cold War Modern: Design 1945–1970*, ed. David Crowley and Jane Pavitt (London: V&A, 2008), 197. Harwood makes a compelling historical case, however brief, for the human embodiment of media technologies. For Harwood, the Eames Office's work on the 1963 IBM Pavilion demonstrates how technologies impact human affairs. As he observed of the 1961 Mathematica exhibition at the California Museum of Science and Industry, "The Eames Office sought to impregnate the human body with mathematics, fitting the display technique to the task" (195). I would go further, however, to say that the mathematical body had long been established such that the display merely reinforced an already existing mode of being in the world. For an early source and a more recent source on the topic, see Ian Hacking, "Biopower and the Avalanche of Printed Numbers," *Humanities in Society* 5 (1982): 279–95 and Sarah Elizabeth Igo, *The Averaged American: Surveys, Citizens, and the Making of a Mass Public* (Cambridge, MA: Harvard University Press, 2007).

30. Niklas Luhmann, *Love as Passion: The Codification of Intimacy* (Cambridge, MA: Harvard University Press, 1986), 18–19.

31. Beatriz Colomina, "Reflections on the Eames House," in *Eames Design: The Work of Ray and Charles Eames*, ed. Donald Albrecht (New York: Harry N. Abrams, 1997), 12.

32. See Nancy Tomes, "Spreading the Germ Theory: Sanitary Science and Home Economics, 1880–1930," in *Woman and Health in America: Historical Readings*, ed. Judith Walzer Leavitt (Madison: University of Wisconsin Press, 1999), 596. For further discussion of the transposition of science and home, see Adrian Forty, *Objects of Desire* (New York: Pantheon Books, 1986), 156–57, and Sarah A. Leavitt, *From Catherine Beecher to Martha Stewart: A Cultural History of Domestic Advice* (Chapel Hill: University of North Carolina Press, 2002).

33. See James R. Killian, "Science and Public Policy," *Science* 129, no. 3342 (January 16, 1959): 130. On the impact of science policy on print media during the Cold War, see Michael J. Golec, "Science's 'New Garb': Aesthetic and Cultural Implications of Redesign in a Cold War Context," *Design Issues* 25, no. 2 (Spring 2009): 29–45.

34. The woman's supporting role reflects the increase of funding to science so that scientists could have the economic means to marry and have children. See David. Kaiser, "The Postwar Suburbanization of American Physics," *American Quarterly* 56, no. 4 (2004): 874–77.

35. Laura Mulvey, "Visual Pleasure and Narrative Cinema," *Screen* 16, no. 3 (Autumn 1975): 6–18. The crucial text for this generation of film studies theorists is Jacques Lacan, "The Mirror Stage as Formative of the *I* Function as Revealed in Psychoanalytic Experience," in *Ecrits: A Selection* (New York: Norton, 1977). Lacan's "mirror stage" is critical for his explanation of the role of "identification" in the formation of the "Ideal-I," what he regards as the *image* of the self that

structures the development of human subjectivity. Lacan explains that the significance of the "visible world" is in its role in the ongoing transformation of the self. He writes, "Indeed, for the *imagos*—whose veiled faces it is our privilege to see in outline in our daily experience and in the penumbra of symbolic efficacy—the mirror-image would seem to be the threshold of the visible world." Kaja Silverman takes up the role of situational apperception in Lacan's "mirror stage." See Kaja Silverman, *The Threshold of the Visible World* (New York: Routledge, 1996). For a view from the other side, as it were, see E. Ann Kaplan, "Is the Gaze Male?" in *Women and Film: Both Sides of the Camera* (New York: Methuen, 1983), 30. Kaplan remarks, "The gaze is not necessarily male (literally), but to own and activate the gaze, given our language and the structure of the unconscious, is to be in the 'masculine' position."

36. Eames, Lecture Notes for "I CARE," n.p.

37. Ibid.

38. Schrader, "Poetry of Ideas," 11.

39. William Mills Ivins, *On the Rationalization of Sight, with an Examination of Three Renaissance Texts on Perspective* (New York: Da Capo Press, 1973), 9. Also see Latour, "Drawing Things Together." *Rough Sketch* manages to consolidate the two spheres of what Martin Kemp refers to as anatomy and biology's "rhetoric of reality" and astronomy's "rhetoric of irrefutable precision." As Kemp explains, these two modes of representation were irreconcilable in the science of the Renaissance. See Martin Kemp, "Temples of the Body and Temples of the Cosmos: Vision and Visualization in the Vesalian and Copernican Revolutions," in *Picturing Knowledge: Historical and Philosophical Problems Concerning the Use of Art in Science*, ed. Brian S. Baigrie (Toronto: University of Toronto Press, 1996), 40–48.

40. Marshall McLuhan and Quentin Fiore, *The Medium Is the Massage* (New York: Bantam Books, 1967), 45.

41. Ibid.

42. Marshall McLuhan, *Understanding Media: The Extensions of Man* (London: Routledge & K. Paul, 2001 [1964]), 314.

43. Friedrich A. Kittler, *Gramophone, Film, Typewriter* (Stanford, CA: Stanford University Press, 1999), 119.

44. Ibid., 119–21.

45. Schrader, "Poetry of Ideas," 11.

46. Meyer Schapiro, "On Some Problems in the Semiotics of Visual Art: Field and Vehicle in Image-Signs," in *Theory and Philosophy of Art: Style, Artist, and Society* (New York: George Braziller, 1994), 7.

47. Law and Whittaker, "On the Art of Representation," 178.

48. Roland Barthes, "Rhetoric of the Image," in *Image, Music, Text* (New York: Hill and Wang, 1977), 39.

49. On sound as an "insidious means of affective and semantic manipulation, see Michel Chion, *Audio-Vision: Sound on Screen* (New York: Columbia University Press, 1994), 34.

50. Of special interest is its use of drawings and photomicrographs, prompting

one reviewer to claim, "It can be stated unequivocally, that no better exists in the biological literature." C. P. Swanson, "Review: Looking at Chromosomes," *Quarterly Review of Biology*, 34, no. 3 (1959): 241.

51. John C. Kendrew, *The Thread of Life: An Introduction to Molecular Biology* (Cambridge, MA: Harvard University Press, 1966), 93. A key early figure in molecular biology, Kendrew published *The Thread of Life* from lectures he prepared for a 1964 BBC television program of the same name.

52. Ian Hacking, *Representing and Intervening: Introductory Topics in the Philosophy of Natural Science* (Cambridge: Cambridge University Press, 1983), 205.

53. De Man, "The Rhetoric of Blindness," 106.

54. Ibid., 116.

55. Eames, Lecture Notes for "I CARE," n.p.

56. Schrader, "Poetry of Ideas," 11.

57. Dark discontinuity does not suggest an undoing of the early twentieth-century project of the unity of the sciences, as proposed by Carnap, Neurath, and Morris. As Ian Hacking remarks, the unity of science never claimed that the scientific method is an "impenetrable lump." Indeed, diverse styles of reasoning remain in the discovery of dark discontinuity. See Ian Hacking, "The Disunities of the Sciences," in *The Disunity of Science: Boundaries, Contexts, and Power*, ed. Peter Galison and David J. Stump (Stanford, CA: 1996), 74.

chapter eight

Educating the High-Speed Eye
Harold E. Edgerton's Early Visual Conventions

Richard L. Kremer

> Don't make me out to be an artist. I am an engineer. I am after the facts, only the facts.
> —HAROLD E. EDGERTON

By 1932, a young MIT electrical engineer, Harold E. Edgerton, had developed a portable, inexpensive light source, capable of generating flashes of extremely short durations, approaching one microsecond (one-millionth of a second) at regular intervals of up to several thousand pulses per second. Unlike earlier spark sources that could produce only a single flash at a weak intensity, requiring their use in a dark room, Edgerton's flashes were so intense that his stroboscope could expose photographic film even in relatively bright ambient light. And unlike earlier electronic flash lamps (commercially available from several French and American firms by 1930), Edgerton's lamps produced actinic light, weighted toward the violet end of the spectrum, which nicely suited the sensitivity of film then commercially available. Furthermore, after experimenting with standard movie cameras (twenty-four frames per second [fps]) and high-speed cameras with mechanical shutters (several hundred fps), Edgerton modified a shutterless camera manufactured by the General Radio Company so that it could record up to several thousand fps, with the film moving continuously at speeds reaching seventy-five feet per second. By synchronizing his stroboscopic flash lamp to the sprockets of the camera, he created a robust system for making high-speed motion pictures. Edgerton's technological innovations are well known.[1]

But how should high-speed photographs or motion pictures look? What knowledge, information or affectation should they impart to their viewers? If Edgerton's new technology did indeed allow one to "see the unseen," as he entitled one of his early films, would the content of the images be immediately recognized, or even recognized at all, by their viewers? Must viewers be instructed on how to look at the images, either by captions or by supplemental markings added directly to the prints? Could high-speed images rely on visual conventions established by previous photography or cinematography, which captured considerably longer chunks of time in their images and hence could not record crisp outlines of rapidly moving phenomena? Could Edgerton freely arrange the conventions of his images, or did his technological system of flash lamps and cameras or his adherence to earlier visual vocabularies constrain his freedom to innovate? Even if Edgerton's photographs presented "only facts" (and we shall see that they presented rather more than that), what conventions might be implicitly required before such facts could be "understood?" How, in other words, would one educate the high-speed eye?

In a now classic study of scientific illustrations printed on paper, Michael Lynch argued that laboratory researchers deploy normalizing conventions to construct their images, conventions that "reflect the disciplinary organization of scientific labour as much as they do the organization of natural objects and relationships."[2] His examples came from an electron microscopy laboratory that studied brain tissue. More recently, Luc Pauwel has proposed a general typology for visual representations in science, suggesting that the production of such images is invariably shaped by three types of constraints. Physical or material processes constrain the "transcription" of the referent as representation; structural processes constrain the selection of samples, of resolution or tonal range, of available human expertise or skill; and cultural processes of professional ethics, disciplinary practices, or broader scientific traditions constrain what a "good" image should look like. Given these constraints, Pauwel argued, visual representations always reflect a "style of execution." Such styles "have to do with genre conventions, cultural schemata, scientific traditions, specific circumstances of the production process, skill, preferences and idiosyncrasies of the maker, as well as the specific purposes the representations need to serve."[3]

This chapter seeks to explore such processes by considering the introduction of a new image-making technology to both scientific/technical and wider public audiences, that is, by examining Edgerton's first decade of making high-speed images. In 1932 he published his first photographs in *Electrical Engineering*, one of the most prominent journals in his chosen discipline. By 1937, his photographs would hang in the Metropolitan Museum of Modern Art. In 1939, he would publish a book of high-speed photographs and short strips of movie film. The next year, Edgerton and his high-speed equipment would appear in a Pete Smith short that won the 1940 Academy Award for best one-reel film. Many of Edgerton's commonplace motifs—splashes, animal movements, athletes in action, breaking soap bubbles, projectiles smashing into barriers—resemble those in earlier high-speed photographs made by Étienne-Jules Marey, Eadweard Muybridge, Arthur Worthington and their successors.[4] Yet Edgerton's visual conventions (his "style of execution") were shaped more immediately, I shall suggest, by the venues in which he presented his early work. The strong lines of the machine aesthetic of *Technology Review*, MIT's alumni magazine, and Henry Luce's new magazines, *Fortune* and *Life*; the vaudeville campiness and austere filmography of Luce's *March of Time* film series; the deliberate, pedantic tone of early educational movies prepared by MIT's Photographic Services and Harvard's Film Service; the gee-whiz, goofy entertainment of Pete Smith's shorts; and the celebration of celebrities by Madison Avenue advertising all come together in the photographs and movies made by Edgerton during the 1930s. By the end of that decade, his images have a distinctive look; when published in traditional science and engineering journals, they look quite different from the other images on those pages. Edgerton's look helped establish what it meant to see time "frozen"; his look presented the high-speed world as comfortable, domesticated, and visually accessible to everyone, not just the electrical engineers.

A "New Philosophy of Perception": Domesticating the High-Speed Eye

In January of 1940, MIT's *Technology Review* announced the publication of *Flash! Seeing the Unseen by Ultra High-Speed Photography*, a "unique volume of high-speed photographs" that introduces viewers to a "split-second world" not visible to the "slothful eye." Before Professor Edgerton had recorded them on photographic plates, "no eye had seen stop-motion pictures of a hummingbird in flight, . . . the beauty of falling drops of liquid, the inflation and collapse of bubbles, the incredibly fast thrusts of a fencer, and the flight of an arrow." Edgerton's work, the magazine gushed, "is an exquisite example of man's ability through science to improve upon, and extend, the function of his own being. Thus, stroboscopic light, brighter than the sun, flashing in rhythm with obedient motes of time, bespeaks through photography a new philosophy of perception."[5]

Educating the High-Speed Eye · 189

This new philosophy involved more than simply freezing motion. Already in 1851, William Henry Fox Talbot had captured, by means of a spark photograph, a crisp image of a rapidly spinning newspaper. But his viewers, of course, could also "see" the same image simply by holding the stationary newspaper before their eyes.[6] Edgerton's "high-speed eye" does not merely see things already known to the "slothful eye" but things moving at speeds too rapid for the latter to see. The "high-speed eye" sees things never accessible to the "slothful eye." An arrow in flight, captured on Edgerton's photos, bends as it whips past the bow; it assumes a different shape than does the familiar stationary arrow.[7] The "high speed eye" sees a different ontological world. This philosophy Edgerton promulgated during the 1930s not by writing prose on aesthetics but by making and publishing high-speed photographs of commonplace events. The education of the high-speed eye would be a domestic, comfortable experience for Edgerton's viewers. Like the austerity, purity, and simplicity of design in Machine-Age America in the 1930s, the high-speed eye would inhabit a safe, comprehensible world even if filled with visual surprises.[8]

Appearing at the end of his first decade of making high-speed photographs, *Flash!* was published by the Boston fine arts publishing company of Hale, Cushman & Flint. Yet many of the more than 200 images in that volume had previously been published in two rather different venues, MIT's *Technology Review* (*TR*) and the Royal Photographic Society's *Photography Journal* (*PJ*), published in London. The visual conventions of both journals would shape the high-speed look, the "new philosophy of perception," that Edgerton would develop.

Founded in 1899, *TR* began as a rather stodgy news organ for MIT's Alumni Association to communicate with its members. The quarterly publication included class news, obituaries, reports about campus activities, and essays of topical interest to scientists and engineers. Illustrations, photos, or nontextual elements of design appear rarely, if at all, in the early decades of the journal. In 1922, Harold Lobdell, MIT 1917, and staff member in the Institute's Office of the Dean of Students, assumed editorship of *TR* and massively revised its look. A larger magazine format with two columns of type, glossy paper, highly designed advertisements, dozens of photographs and engravings per issue, and covers showing drawings of architectural motifs, usually MIT buildings, provided Lobdell with a visually rich palette on which to paint a Machine-Age aesthetic. James Rhyne Killian, Jr., MIT '26, edited *TR* from 1930 to 1939, before moving on to become executive assistant to MIT president Karl Compton and then tenth president of the Institute (1948–59) and special science advisor to President Eisenhower. Deeply interested in the graphic arts, Killian brought in skilled typographers such as W. A. Dwiggins (who in 1922 coined the term "graphic designer") and Daniel Berkeley Updike to redesign the magazine again. Even more than Lobdell, Killian made *TR* a showcase for an industrial, modernist aesthetic. By 1931–32, Killian's covers featured three-side-bleed photographs by Margaret Bourke-White, Russell Atkins, and William M. Rattase (better known for their heroic pictures of machines and industrial landscapes in Henry Luce's new magazines *Fortune* and *Life*). The striking camera angles, severe cropping, deep contrast, and strong lines in these pictures present clean, crisp worlds of steel, concrete, and machines, without context, people, or the Great Depression.[9]

However, Killian also opened *TR*'s pages to Edgerton's photographs. Among the edgy photographs of steel and hard lines—soaring skyscrapers, massive dams, bridges, ore barges, industrial smokestacks, gears—came the high-speed prints, somewhat blurry at first, but ever more crisp and dramatic over the years. Brief captions accompany Edgerton's pictures, but usually not descriptive articles. Killian presented the images as stand-alone visual material, often along the margins of the pages and totally unrelated to their textual content. Edgerton's first *TR* photo, in the March 1931 issue, depicts the rotor of a large electric motor "turning at a rate corresponding to a linear speed of 95 miles per hour," an image from the laboratory context in which Edgerton initially had developed his stroboscopic technology.[10] Next year came the first domestic image, "a falling drop of milk caught in the act by a new method of high speed photography." Four strips of film, containing forty-four separate, successive images, show a milk globule falling onto a puddle of milk, forming a crater with a geyser erupting from its center. "This phenomenon is not observable by the eye, since it takes place too rapidly," intones the caption.[11] Never before had such a commonplace event, parsed by a high-speed eye, appeared in the pages of *TR*.

Although the caption added that Edgerton's stroboscopic method "is finding a great variety of uses in industry, particular for observing the operation of high-speed machinery and in making motion studies," it would not be industrial photos that Killian and Edgerton presented in *TR*. The high-speed eye, rather, would be educated by images of water squirting from a faucet, a golf ball compressed by a swinging club, a football being kicked, a hammer breaking an electric light bulb, a bullet shattering a pane of glass, a bursting soap bubble, a raw egg smashing against the floor, a swimming clam's cilium, a flying fish, a sneezing man's head, and house flies, mosquitoes and birds in flight. The captions emphasize the domesticity of these common events. "Plunk! When a cup filled with coffee is dropped by the clumsy waiter, this is a split-second photograph of what happens on the floor."[12] The accompanying image, strongly lit from one side, shows the cup disintegrating into six pieces, coffee starting to ooze out through the cracks, and a twisting finger of liquid extending above the rim. Presented without visual context or clues, this image, unlike the flying birds or the sneezing head, appears abstract and non-figural; without a caption, viewers might not have recognized its content, for the slothful eye had never seen such an image. But under Edgerton's tutelage, the high-speed eye could comprehend and enjoy.

For each year from 1933 to 1940, Edgerton submitted similar photographs to the annual exhibition of the Royal Photographic Society (RPS) of Great Britain in London. Founded in 1853, the RPS from its beginning had sought to bring together a wide variety of photographic enthusiasts and to bridge the scientific and aesthetic sides of the field.[13] Its official organ, *Photographic Journal* (*PJ*), had by the 1930s become one of the world's leading publications for technical articles on equipment and photochemical processes, discussions of technique and different types of photography, book reviews, and presentations of members' portfolios and photographic travelogues. With more than 2,000 members and nearly that many prints or cinematographic films submitted to each annual exhibition in the 1930s, the RPS by then represented traditional photographic conventions, not the avant-garde. The catalogs of its annual exhibitions do not feature the social realism of Dorothea Lange or Lewis Wickes Hine, the abstract forms of Edward Weston, the dynamic, ambivalent work of Henri Cartier-Bresson, the experimentation of Moholy-Nagy, or the Machine Age lines of Bourke-White. RPS photographs in the 1930s tend to be understated, carefully composed, static, and comfortable.

Edgerton's high-speed photographs were among the earliest such images to be exhibited at the RPS. Like his TR photographs, the content of Edgerton's RPS submissions remained distinctly domestic and commonplace. In addition to the dropped coffee cups and milk drops that had appeared in the TR, images of flying bats, firing revolvers, flipped playing cards, smashed tennis balls and a boxer hitting a punching bag were presented to the London viewers. When Edgerton occasionally included high-speed industrial images (spindles on textile looms, water flowing through Venturi tubes), the RPS critics were quick to express their disapproval.[14] In the traditionalist setting of the RPS, Edgerton's high-speed eye remained fixed on the domestic.

In 1937, six of Edgerton's prints appeared in New York's Museum of Modern Art's first exhibit of photography, curated by Beaumont Newhall. Again depicting commonplace events—the breaking coffee cup, water gushing from a faucet, a smashed electric light bulb, a kicked football, a struck golf ball and the elegant milk coronet (reproduced in the catalog)—these prints hung in the section for "scientific photography" as the only entries in a subsection entitled "stroboscopic photography." In his introduction to the catalog, Newhall referred to Talbot's 1851 spark photograph of the newspaper clipping. Yet the MIT photos, Newhall explained, present images "no eye has ever seen."[15] A decade later, in the first of the many editions of what would become a classic work on the history of photography, Newhall would include Edgerton's high-speed prints (including the milk coronet) in a section on "Experiments in Abstraction," along with work by the Zurich Dadaist, Christian Schad, a solarization by Man Ray, a photogram by Moholy-Nagy, and a three-hour exposure of the Horsehead Nebula in the constellation of Orion, taken at the Mt. Wilson Observatory.[16]

For the art critics of the 1940s, Edgerton's high-speed images of commonplace events had become abstract. But James Killian, in the introduction to *Flash!* on "the meaning of the pictures," made quite the contrary point, emphasizing the domesticity of the images.[17] He described in considerable technical detail Edgerton's stroboscopic methods and their value, in science and engineering, for measurement and analysis of motion. Yet only a few photos in *Flash!* depict high-speed photography "in the service of science and industry." Like the 1930s images in *TR* and the RPS exhibits, most of the *Flash!* pictures present the world of "unseen rapid motion" as it "surrounds us in our everyday life, and we are unable to penetrate it just as, before the telescope and the microscope, we were unable to break through all the barriers of space and see the mountains and craters of the moon, or the germs that make us ill and the micro-organisms that make our wine." Killian wrote of Edgerton's high-speed images that they "appeal because they bestow comprehension and increase our awareness. They endow us with a new kind of sight that reveals new forms, subtle relationships of time and space, the essence of motion. In the main the esthetic aspects of science are discernible only to the scientist, but here they appear in a universal language for all to appreciate."

Killian placed Edgerton's work in an artistic context but not one of abstraction. "He has photographed drops, birds, bullets, and athletes over and over again to arrive at a satisfying transcription, just as Edward Weston has studied cypress trees and rocks; Edward Steichen, sunflowers; and Alfred Stieglitz, clouds and hands." Edgerton's pictures, Killian concluded, offer "not only facts to help us in seeing and doing but new esthetic experiences"; they provide a "fresh aspect to the commonplace."[18]

The Self-Reflexive Photograph

Unlike Marey and most of the early twentieth-century high-speed photographers, Edgerton early developed the conceit of photographing himself making high-speed photographs. These self-reflexive photographs, in which either the image-making technology or the image-maker himself appears in the frame, feature prominently in Edgerton's public presentation of his work, both in scientific and engineering journals and in more popular venues. Edgerton not only showed viewers previously unseen worlds; he also showed them how he did it. Transparency, like domesticity, would ease access to the high-speed world.

In one of his first articles in the *Journal of the Society of Motion Picture Engineers* (March 1932), Edgerton described his new stroboscope and presented two "stroboscopic motion pictures" (he also called them "strobograms," a term he soon would drop), short strips of film showing only several frames on a single plate on the printed page. He also included a "Photograph of the stroboscope arranged to take motion pictures of the surges of an experimental engine," featuring Edgerton himself standing beside the flash lamps and an unidentified person tending to the movie camera. His first article published in a non-engineering journal, *American Golfer* (November 1933), also presents an eight-frame strip of movie film and a self-reflexive photograph of the laboratory with the golfer preparing to swing and the flash lamps suspended just above the teed-up ball. A massive engine in the background reveals that they are located in MIT's motor laboratory where Edgerton had begun his work with strobes and high-speed photography. Edgerton and two unidentified men stand to the side, readying the photographic apparatus. In his first appearance in *Fortune* magazine, Edgerton and ten other influential MIT faculty members, including several Nobel Prize winners, are featured. Only Edgerton is shown in his lab, standing over his stroboscope.[19]

Flash! also displays the self-reflexivity of the high-speed eye (figure 8.1). The volume opens with a full-page photograph as frontispiece, the only color print in the book, depicting the lower body of an unidentified athlete

194 • *The Educated Eye*

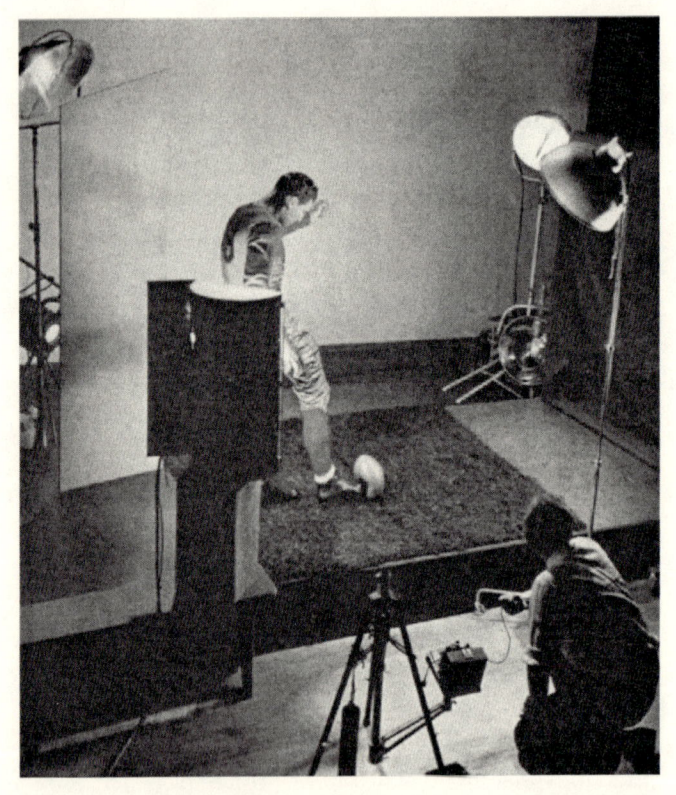

8.1 Self-Reflexive Photographs in Gjon Mili's studio. From Edgerton, *Flash!* (1939).

in football gear, with his booted toe sunk deeply into a football, frozen for a moment before the kicked pigskin will boom out of the frame. A black-and-white print, inserted amid the prose of Killian's introduction, depicts the same scene but with the lens drawn back to show the New York studio of photographer Gjon Mili, with the football kicker standing on a patch of artificial grass surrounded by three strobe lamps, screens, and Edgerton himself, seen from the back, kneeling to release the shutter of the camera recording the kick. "High speed picture of a high-speed picture being taken," reads the caption. In the introduction is another self-reflexive image of Edgerton standing before a massive motor in the MIT laboratory, the print usually denoted as the earliest photograph (1931) showing Edgerton's first stroboscope. With these two self-reflexive photographs in *Flash!*, Edgerton presented two self-portraits, of the electrical engineer in the laboratory and the art photographer in the studio.[20]

Why the self-reflexivity in Edgerton's presentation of his images, a convention that would continue to mark his work throughout his career? Almost never do self-reflexive photographs appear in the proceedings of the early gatherings of the International Congress of High-Speed Photography and Cinematography (1952 in Washington, 1954 in Paris, 1956 in London). Images of milk splashes, hummingbirds, or bullets slicing through playing cards or apples also do not appear in those proceedings. When asked about his "artistic" images, Edgerton always replied: "Don't make me out to be an artist. I am an engineer. I am after the facts, only the facts." Yet engineering journals of the 1930s do not feature self-portraits of engineers at work. As is well known, Edgerton would devote much of his career to popularizing and promoting the virtues of high-speed photography and the commercial lines of stroboscopes he had developed that were manufactured and marketed by the General Radio Company of Cambridge, Massachusetts. Always the hawker, salesman, and showman, Edgerton developed a legendary reputation for his performances. Walking on the street he would pass out signed postcards featuring his photographs and the telephone number of his MIT lab, inviting passersby to visit.[21] Surely his self-reflexive photographs figured in this showmanship. Yet they undoubtedly also contributed credibility to his images of worlds unseen by the slothful eye. Like the magician who shows the audience how he or she does the tricks and thus brings them into the guild, so too did Edgerton welcome his viewers into his world of high-speed photography. Unlike the mechanical objectivity of the nineteenth century, as described by historians Peter Galison and Lorraine Daston, in which scientists sought to remove themselves from the recording of nature, Edgerton always inserted himself into the process.[22] The high-speed eye would be educated not simply by photographic images but also by Edgerton's personal appearance in those images. The viewers' comfort, evoked by the commonplace events being presented to the high-speed eye, was further enhanced by the credibility of the self-reflexive engineer demonstrating his tools.

We might also note that, by the 1930s, self-reflexivity could be found in Hollywood. From Edwin Thanhouser's *Evidence of the Film* (1913) or Charlie Chaplin's *The Masquerader* (1914) through Buster Keaton's *The Cameraman* (1928), Clyde Bruckman's *Movie Crazy* (1932) or Harvey Lachman's *It Happened in Hollywood* (1937), movies about movies and movie-making increasingly entertained American audiences. It is not surprising that Hollywood filmmakers wanted to make movies about Edgerton's high-speed cameras and flash lamps; and perhaps it is not surprising that Edgerton himself so frequently made and published self-reflexive photographs.[23]

The Graphed Photograph

In one of his early articles touting the potential uses for stroboscopic photography, Edgerton described how quantitative measurements of velocity and acceleration of rapidly moving objects could be obtained from film exposed at regular intervals. Yet only rarely did he himself derive such data from his photographs (he did compute the speed at which cracks travel through plate glass—nearly a mile per second), and even more rarely did he display such measurements directly in his images.[24] Among its hundreds of images, *Flash!* presents only two examples of what I shall call graphed photographs. In Edgerton's high-speed conventions, the visual image and the quantitative analysis (if offered at all) would be presented separately.

In the late 1930s, the Chicago sporting goods company A.G. Spaulding and Brothers, commissioned Edgerton to make a series of high-speed multiexposure photographs of celebrity golfers and tennis players, all leading professionals in those sports.[25] Golfers such as Bobby Jones, Denny Shute, Jimmy Thomson, and Horton Smith and tennis stars John Bromwich, Virginia Wolfenden, and Bobby Riggs came into Edgerton's studio to expose their swings to his stroboscopic flashes. Some of Edgerton's images freeze the celebrities' bodies in mid-swing; other multiexposure images blur their visages but trace the motion of their swings frozen at regular intervals of time. Yet Edgerton also sought to submit these swings to quantitative analysis.

His first article published in a non-engineering journal, *American Golfer*, had featured nine frames from a high-speed motion picture of a golf swing, from which he separately computed the absolute velocities of the club head and the ball. For the Spaulding golf photos, however, Edgerton made multiflash exposures on a single negative. In *Flash!*, he presented side-by-side the multiflash photos of swings by professional golfer Bobby Jones and an "ordinary golfer." The caption under the latter urges readers to compare the images. Yet immediate visual inspection reveals no significant differences between the swings captured in the multiexposures. Hence, Edgerton reprinted both multiexposure images on which he superimposed graphs in polar coordinates showing the acceleration and velocity of the club heads as a function of time. This "graphical analysis," as Edgerton called it, now reveals differences between the skilled and

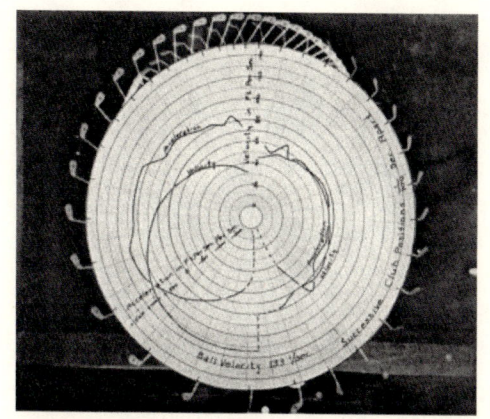

Educating the High-Speed Eye · 197

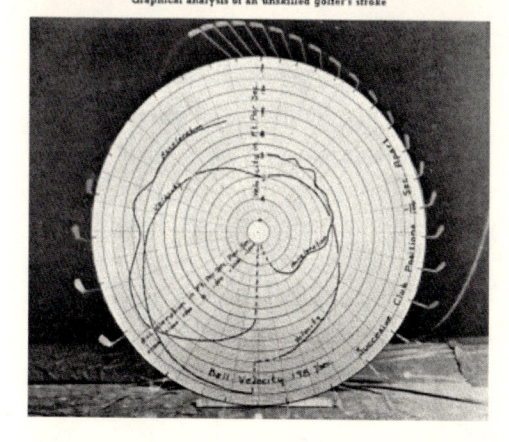

8.2 Graphed photographs.
From Edgerton, *Flash!* (1939).

UNSKILLED
Graphical analysis of an unskilled golfer's stroke

unskilled swings, even as it blocks out the central image of the golfer in the print.[26] Graphed photographs thus combine high-speed images and graphical representations added later, summarizing quantitative measurements of the photograph (figure 8.2).

The second example in *Flash!* might be called a graphed movie. To study the performance of an automatic tapping machine for the United-Carr Fastener Corporation, Edgerton made a high-speed motion picture of a rotating die as it travels in to cut a thread and then back out in 1.1 seconds. To make visible the rotational position of the die, he added to its top a white dial with its circumference divided into ten numbered sectors. Rather than showing the individual frames of the film, he reduced the data to a single complex graph, displaying several measured parameters on a single diagram. The "entire action of the automatic chuck," proclaims the caption, "may be studied at will and reduced to the graphical analysis shown in the plot."[27]

Under certain circumstances, Edgerton's high-speed images apparently did not "speak for themselves." Presenting images as "graphed photographs" enabled him to add quantitative analysis, the results of detailed measurements, directly to the photos. Yet it must also be emphasized that most of the hundreds of photographs that Edgerton made and published during the 1930s were not graphed. Like Marey and many other predecessors, Edgerton used high-speed photography more to freeze commonplace motions than to transcribe motion into quantitative measures.

The High-Speed Photograph as Analogy

Edgerton's 1939 book *Flash!* presents over 200 pages of single-exposure and multiexposure prints, plus several strips of movie film from ten to ninety frames in length. The photographs are divided into several titled sections: living motion (animals), bullets in action, sports (60 pages), drops and splashes, in the service of science and industry (only 20 pages), and people in action. As noted above, many of these photographs or their homologues had previously been published in *TR*, *PJ* or more specialized engineering journals. The sports photographs (the largest section in *Flash!*), however, generally had not been seen previously. Instead, corporate advertisers had commissioned many of the sports photographs, one example of which I shall here consider.

The captions for the football photographs in *Flash!* do not indicate that these images had been commissioned by the Madison Avenue firm of Batten, Barton, Durstine & Osborn (BBDO) for the Ethyl Gasoline Company.[28] One of America's earliest makers of leaded gasoline to fuel the new high-compression automobile engines of the late 1920s, Ethyl by the 1930s regularly advertised its product in heavily designed full-page color ads appearing in magazines such as *Life* (New York) and Luce's *Fortune* and *Life* (Chicago). Late in 1934, a representative of the A.D. Little consulting firm wrote Edgerton on behalf of BBDO, apparently asking whether Edgerton might supply some high-speed photographs of tennis balls and footballs being compressed. After clearing the idea with Vannevar Bush, MIT's dean of engineering and his longtime mentor, Edgerton took his apparatus over to Harvard and made his first photographs of that university's football coach kicking a football.[29]

BBDO's print ads for Ethyl appeared three or four times annually in the above magazines. Each ad was unique, although a given design and text might be carried through a series of ads extending over several years. In 1933–34, the ads played on fear, warning viewers of the dire consequences of not tanking with Ethyl's anti-knock leaded gas. Rainy weather might leave one stranded, cheaper gasoline might send unwanted signals about the socioeconomic class of the "cheap" driver, or heat in the summer might dramatically reduce engine performance and leave one unable to keep up with other cars on the road. In 1935, BBDO introduced a lighter theme, featuring large drawings of animals and comical text. Beside a monkey hanging in a tree, the text reads: "This daring young man on the flying trapeze motors through traffic with the greatest of ease . . . next time get Ethyl." Or accompanying a photo of two identical zebras: "Look much the same and so do many fuels. But Ethyl makes a difference between plain glass and jewels . . . next time get Ethyl."[30]

BBDO's ad for November 1935 reveals a dramatic change in focus.[31] Now Ethyl sought to educate readers about the inner workings of their automobile engines. Featuring a large photograph of a football being compressed by the kicker's toe, breathlessly described as "caught with stop-motion camera in 1/100,000th of a second by Prof. H. E. Edgerton and Mr. K. J. Germeshausen of the Massachusetts Institute of Technology," the ad presents an analogy, visually and in the text. "It's compression that puts power in the kickoff; it's compression that puts power in our car's engine." A sidebar at the right presents three smaller Edgerton photographs, showing the football just before, during, and after the kick, each paired with a cut-away diagram of the cylinder of an automobile engine, indicating the piston's position just before, during and after the compression and ignition of the gasoline vapor. Although the diagrams and photographs are paired, this ad is not a graphed photograph; rather it proposes a relation of analogy between the high-speed photographs and graphic representations of the engine.

For reasons I have not been able to discover, BBDO did not develop the football-engine analogy ads into a series. In 1936, they introduced a new format, in which color photographs of homey, rural scenes depict happy families tanking their automobiles with Ethyl. The first ad of this series kept the right sidebar, showing one diagram of the cylinder (Edgerton's photographs have disappeared) and filled with fine print offering details about anti-knock fluids containing tetraethyl lead. A title apologizes that the sidebar will take "only 37 seconds to read." The next happy family ad drops the cylinder diagram and eliminates the technical prose.[32] As far as I know, Edgerton photographs did not appear in subsequent Ethyl ads.

Apparently, someone decided that the abstraction, both visual and conceptual, in the analogy of the Edgerton ad would not serve Ethyl's needs.

The Normal and "What Really Happens"

A strobe light flashing at, say, 1,000 times per second, with each flash lasting only a millionth of a second, is "on" for only 0.1 percent of the time. To put it cinematographically, Edgerton's high-speed cameras of the 1930s captured on film only 0.1 percent of the total flow of time over which the camera operated. Nonetheless, one of Edgerton's most prominent conventions, especially in his movies, was to contrast views of processes shot at normal speeds with those filmed with his high-speed camera and then projected hundreds of times more slowly than shot. All of his movies in the 1930s offer such comparisons; and all of them emphasize the epistemic superiority of the high-speed eye in showing "what really happens."

Edgerton's first widely viewed film appeared in a newsreel, the second in *Time* magazine's newly inaugurated "March of Time" series that premiered in hundreds of American movie houses in the spring of 1936.[33] Only two minutes in length, the clip opens with the voice-over ebulliently introducing the professor and his camera: "In the Electrical Engineering Laboratory at Massachusetts Institute of Technology Professor H. E. Edgerton perfects a new high-speed movie camera which shows the eye of man things that happen too quickly to be seen." The film then pairs short clips, showing the same phenomenon, first as motion "too quick to be seen" by the slothful eye and then "what really happens" as seen by the high-speed eye in slow motion. An electric fan blade, a bird released into the air, a bursting soap bubble, and a light bulb smashed with a hammer—all very domestic scenes—appear, between shots showing in considerable self-reflexive detail the components of the camera and flash lamps as wielded by Edgerton and his assistant.

The convention pairing the normal against "what really happens" also appears in Pete Smith's *Quicker'n a Wink*, a short film that was screened in thousands of American movie houses. In 1940, the film won the Oscar (one of only two Oscars awarded to Smith, who made nearly 300 shorts in the 1930s–50s) for the best short film of the year. Directed by Sidney Smith, who would go on to become a distinguished maker of feature-length films, *Quicker* was shot at the MGM studios in Hollywood. Edgerton and his MIT colleague, Herald Grier, traveled to California to assist in the shooting and to appear in the film. *Quicker* remained a favorite for both Pete Smith and Sidney Smith. The latter once recalled that he had enjoyed working with Edgerton, although the MIT professor "only thought of the technical use of the stroboscope, and my thinking was visual entertainment and informational."[34]

Eight minutes in length, *Quicker* consists of eleven scenes, each demonstrating another example of "what really happens" (according to Pete Smith's voice-over) when motion is slowed down by Edgerton's camera.[35] The moving objects had all been presented in Edgerton's earlier still photographs and films of the 1930s—a cat lapping milk, hummingbird hovering, milk splashing into a hard surface, an electric fan blowing twirls of smoke, bullets shattering a light bulb, etc. The first four scenes are self-reflexive, showing a rather serious-looking Edgerton setting up and using his flash lamps and high-speed camera. Yet despite its simple, effective explanations of stroboscopy, the voice-over consistently expresses his "lay" awe at the complexity of the "engineer's" apparatus, not trying to explain Edgerton's "thyratron amplifier" and upon seeing the cat's tongue in action, marveling at "how smart you can get when you go to the movies." Sidney Smith's concern for the informational is indeed emphasized. Viewers learn that hummingbird wings beat seventy times per second, that eggs bounce before shattering, that soap bubbles can be deeply penetrated by a pencil without breaking, etc. The remaining scenes introduce other human actors. A pretty young woman blows a soap bubble to be burst, about which Smith's voice-over offers a sexual innuendo. A well-known professional golfer drives a ball through a phone book. And in perhaps the most visceral scene, a dentist drills the tooth of a goofy patient. All quite entertaining in Sidney Smith's sense.

Two conventions unify the film's scenes—humor and comparisons. In nearly every scene, the voice-over cracks a joke: quieter fans will no longer disturb sleeping employees; our milkman finds the splash experiments "very interesting at thirteen cents a quart," etc. More importantly, each scene juxtaposes a cut of the given phenomenon at normal speed, where the slothful eye "sees nothing," with Edgerton's high-speed shots run at slow motion. These latter cuts, the voice-over keeps repeating, allow the high-speed eye "to see what really happens." In each case, the voice-over humorously instructs viewers on what they should "really see" in the slowed phenomena. These comparisons of the "normal" and the "slowed down," for everyday phenomena well-known to viewers, would become a standard convention in Edgerton's films and the many documentaries about his work that would appear from the 1960s through the 1990s.

Taking the High-Speed Eye into Biology

As I have argued already, Edgerton was not only an engineer and a creator of a new "style of execution" for the high-speed eye but also a consummate showman who worked energetically to promote the use of high-speed photography in industry, science, and entertainment. He traveled widely, demonstrating his stroboscopic equipment to professional, public, and commercial audiences. He coauthored hundreds of articles with scholars in many scientific fields, illustrating how high-speed photography could serve as a research tool in these disciplines. And he propagated not only research apparatuses and methods but also his visual conventions for the high-speed eye. Although biology would never become a major site of research for Edgerton himself, in the 1930s it represented one of the first areas beyond engineering and industry where the electrical engineer deployed his new tool and visual conventions.

In his usual self-effacing way, Edgerton later described his move beyond the world of electrical engineering: "I suddenly realized, hey, there's a lot of things in the world that move. I looked around and there was a faucet right next to where I worked, so I just moved the strobe over and took a picture of water coming out of the faucet. And that was the first picture I ever took, except for a[n electric] motor."[36] Edgerton's detailed laboratory notebooks, however, record a richer story. The initial venue for Edgerton's strobe lamp and his high-speed movies was indeed the electrical engineering laboratory at MIT. News traveled quickly, however, and by mid-1931, other MIT engineering professors began bringing their high-speed phenomena to Edgerton's cameras. These faculty in turn introduced Edgerton to colleagues working in industry and by October 1931, manufacturing companies began hiring Edgerton to help solve problems with their high-speed processes and machinery. Box, textile, shoe, soap, and gyroscope manufacturers, printing and lithograph companies, and public power-generating companies were all fascinated by the high-speed eye. Edgerton carefully recorded all these travels and visits in his laboratory notebooks, which seamlessly combine business agreements, travel reports, and schematic writing diagrams. Then suddenly on 31 January 1932, while testing a new arrangement of flash tubes wired in parallel, Edgerton wrote in his notebook: "Germ[eshausen] came about 11 and we set up a milk drop experiment. Took movies at 250 frames a second of the drops just as they struck the surface. For these tests we used three tubes in parallel about 5 or 6 inches from the subject."[37] By mid-February, Edgerton had screened the milk drop movie for MIT's Visiting Committee; 44 frames of the movie film were published in the April 1932 issue of *TR*. And within months, Edgerton had turned his high-speed eye on all kinds of motion, including transient phenomena of life: birds and insects in flight, the flick of a snake's tongue, and the wink of a human eye.[38]

As commonplace events never seen by the slothful eye, these early images were intended to demonstrate the virtuosity of Edgerton's technology, not to contribute to zoological or physiological research. But by the spring of 1934, MIT's biologists, led by John W. M. Bunker, started bringing their problems to Edgerton's "Stroboscopic Laboratory."[39] Significantly, they began by trying to capture on film one of the classic objects of nineteenth-century experimental physiology, the rapidly contracting gastrocnemius muscle of the twitched frog leg.[40] They next turned to the mechanical motions of cilia on the gills of a clam, a topic then being investigated stroboscopically by the British zoologist, James Gray. Using a commercially available cinematographic camera shooting through a microscope at twenty frames per second and a rotating disk to stroboscopically illuminate the field, Gray had made a series of crisp photographs showing the effective and recovery strokes of an active cilium slowly beating at about 2 hertz. Since most cilia beat at 10 hertz or faster, Gray admitted that at least 100 exposures per second would be required to record the "true sequence" of the ciliary beat.[41]

Working with the biologists and MIT physicist Arthur C. Hardy, a leading expert on physical optics, Edgerton arranged for a 10-microsecond spark flashing at 200 hertz to illuminate the microscope's field with the images captured directly (i.e., not stroboscopically) on his high-speed movie camera mounted above the eyepiece. To measure and document the positions of the cilium, they printed small grids directly on the 35mm film, prepared enlarged prints, and drew the successive stages of motion on graph paper. "As an aid in interpreting results," they also projected the movies in slow motion. The published article, by MIT biologists Bunker and Maurice W. Jennison (Edgerton is not listed as an author), does not present what we above called "graphed photographs." Instead, the graphed drawings appear in figures without photographs. A separate plate (figure 8.3) presents twenty-five "photomicrographs" selected at various intervals from the movie film, showing blurry images of a cilium slowly recovering (here the images are at 50-millisecond intervals) and then snapping forward much more quickly (images at 5-millisecond intervals). Although Gray's photographs were crisper, Edgerton had increased the frequency

8.3 Stages in the beat of a cilium from gill of clam. From the *Journal of Cellular Physiology*, M. W. Jennison and J. M. W. Bunker (1934), courtesy of John Wiley and Sons.

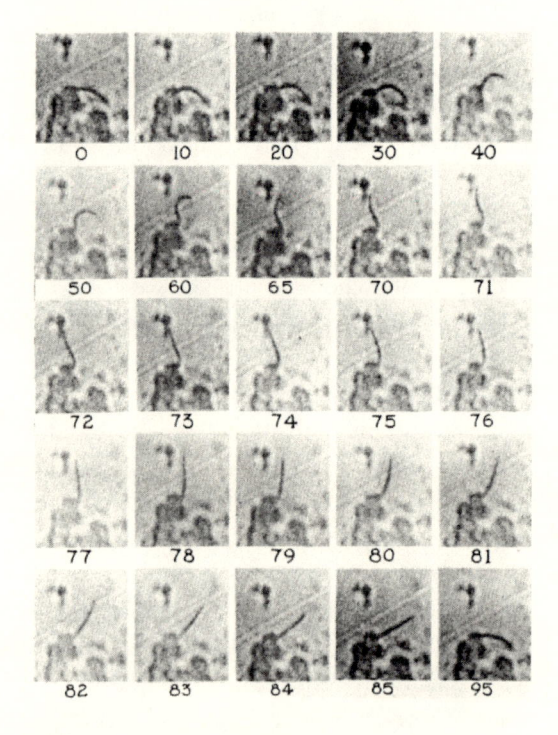

by a factor of ten, offering, in the words of the biologists, "a new type of high-speed photomicrography . . . [that] may prove a valuable tool for the study of certain biological phenomena." After shooting fifteen films of the beating cilia, Edgerton recorded in his notebook: "We worked a lot upon the biology work the last week!"[42]

Edgerton's work on ciliary motion would serve as an example of film's pedagogical potential. For example, Harvard's Film Service, established in 1934, sought to promote the use of film as a pedagogical and research tool. As an example of the latter, wrote the HFS director in 1935, "one has only to look at the splendid work of Dr. Edgerton of the Institute of Technology. He can take pictures at the rate of 6,000 feet [*sic*] a second [he must have meant frames a second] and has kindly offered to cooperate with us in matters of high-speed photography. The application of this form of research to various problems in Biology, Physics, and Chemistry presents important possibilities. Much could be learned from a slow motion analysis of ciliary action, from a similar study of wave forms, or from a photographic record of the rate of diffusion in liquids."[43]

Yet, although neither Edgerton nor the MIT biologists would continue this study of cilia, photographing animal locomotion would soon become a leitmotif for the Stroboscopic Laboratory with the zoological images often reflecting the visual conventions of the high-speed eye. For example, *Quicker'n a Wink* featured hummingbirds hovering with their wings beating at seventy strokes per second, distinctly visible to Edgerton's high-speed eye but only a blur to the slothful eye of the normal movie camera. Edgerton would continue to photograph hummingbirds in flight, but the effort never yielded research publications.[44] Collaboration with zoologist C. M. Breder, Jr., a leading student of fish locomotion then at the New York Aquarium, proved more productive, yielding in 1941 the first sharp photographs of flying fish in action. The article, coauthored by Breeder and Edgerton, features handsomely produced plates showing single-exposures (10 milliseconds) of fish in flight; the opening image of the series is a self-reflexive shot of Edgerton sitting on the deck of the boat, silhouetted against the black night sky, with two assistants holding the high-speed flash equipment.[45] The next year Breeder and Edgerton took 300-fps movies of swimming seahorses, a small fish that maintains a vertical posture as it moves forward by oscillating its dorsal and pectoral fins as fast as thirty-five beats per second. They prepared enlarged tracings of the individual frames from which they could measure the movements of particular points on the fish's body. Although they did not "graph" the individual photos, they did convert their measurements into vectors moving in orthogonal planes and presented the results of their mechanical analysis of the fish's various motions in vector diagrams. The only photographs published were single exposures, freezing the vertical position of the seahorse and its rapidly moving fins.[46]

A final example of Edgerton's early biological collaborations occurred in 1937, when he indicated in his lab notebook: "Mr. Paul Moore of Northwestern University School of Speech was here Feb 1 and 2. We took pictures of his vocal cords by both the stroboscopic and the high speed motion picture methods."[47] Moore had just completed his doctoral dissertation at Northwestern on "a stroboscopic study of vocal fold movement." Working with a Chicago vacuum tube company, he had developed a short-duration flash lamp not unlike Edgerton's light, which could be directed down a laryngoscope to illuminate the working vocal cords. How do the vocal folds move, as they produce tones of different pitches and intensities? The question was old, Moore admitted, "but satisfactory solutions had not been found."[48] By flashing the light at nearly the same frequency as the sound being produced, Moore had been able to make direct observations and stroboscopic photos with a conventional movie camera of the wavelike movements of the tissues. Similar stroboscopic laryngoscopes, using electronic flash lamps, had been introduced previously; that is, Moore and Edgerton were not the first to realize that rapidly moving vocal tissue could be slowed down stroboscopically. But the earlier researchers had not published any films or photos of the active larynx.[49]

At MIT, Moore and Edgerton also began by observing stroboscopically the wavelike movements of the vocal folds. But the difficulty of keeping the voice vibrating at exactly the same frequency made it impossible to observe the changing behavior of the glottis; hence they took high-speed movies of the vocal folds, exposing the film at 1,200 fps at durations of less than 1 millisecond. For a voice singing at 220 hertz, this method yielded about five or six exposures per cycle of the larynx. In a single-authored article, Moore printed several series of nine to fifteen consecutive exposures, identical in format to the milk-drop film strips that Edgerton had published in TR several years earlier. The images are grainy, not sharply focused, and, as Moore admitted, do not present self-evident visual information: "The extraneous parts of the pictures have been trimmed away to conserve space, leaving scarcely more than the image of the larynx as reflected in the guttural mirror. This reduction makes it difficult to orient the image; therefore, the structures will be pointed out before an analysis of the pictures is attempted." Moore then offered lengthy verbal descriptions of the images, instructing viewers how to "read" the photographs before concluding (without irony) that "it is clear from these high speed pictures that the vocal folds perform a complex opening and closing wavelike movement."[50]

Moore also published a strip of twenty-four consecutive stroboscopic photographs of the vocal fold motions, presumably made in Edgerton's lab with a standard movie camera at 24 fps. Showing a wider field and sharper details, these pictures project "very well," Moore indicated, and thereby give a better "pictorial and illustrative" sense of the wavelike movements of the vocal folds than do the individual frames of the high-speed movies. Yet Moore downplayed the epistemic value of the stroboscopic images. Given the unavoidable pitch variations and the fact that each exposure superimposes several sequential positions of the tissue (to yield the required intensity, the flash frequency was greater than the fps), the composite stroboscopic photo "might produce a picture of a fold position not actually existing."[51] That is, Moore in his 1937 article clearly contrasted, both in words and photographs, the "normal" (i.e., the stroboscopic summaries of many cycles of the vocal folds) with "what really happens" (a single cycle). Although Edgerton never coauthored any work with Moore, he did publish in *TR* a self-reflexive picture of Moore's laryngoscope. Moore would go on to become a leading high-speed photographer of vocal fold movements. Although he apparently never again collaborated with Edgerton, he nonetheless employed Edgerton's visual conventions for the high-speed eye.[52]

Conclusion

Edgerton's visual conventions for the high-speed eye—domesticity, self-reflexivity, graphed photographs, high-speed analogies, and contrasts of the normal and "what really happens"—were to a large extent shaped, I have argued, by the venues in which he presented his images during the 1930s and early 1940s. His images appeared in engineering journals, biology journals, photographic and motion picture journals, MIT's alumni journal, cinemas, and art museums. Edgerton himself screened his movies for hundreds of audiences, ranging from rotary clubs and church groups to professional societies and industrial organizations. In the decade when the new talkies were marketed as "electrical entertainment" based on "scientific progress" and when Machine Age–designed lives enveloped Americans' material worlds, Edgerton's high-speed eye found wide appeal.[53] His images of previously "unseen worlds" astonished their viewers; but they also delighted audiences who found Edgerton's visual conventions easily accessible.

While Edgerton may have professed to be after "only the facts," he obviously realized that cultural work is required, even for images that present "only the facts." As many critics, museum curators, and historians of photography have argued, Edgerton's images did much more than present facts. They also taught viewers from science, engineering, and many varied publics how to see with high-speed eyes.[54]

Notes

Quoted in Harold E. Edgerton, Estelle Jussim, and Gus Kayafas, *Stopping Time: The Photographs of Harold Edgerton* (New York: H.N. Abrams, 1987), 18.

1. Cf. Richard L. Kremer, "Inventing Instruments and Users: Harold Edgerton and the General Radio Company, 1923–1970," in *Who Needs Scientific Instruments?* ed. Bart Grob and Hans Hooijmaijers (Leiden: Museum Boerhaave, 2006), 253–62; Richard L. Kremer, "Harold E. Edgerton's High-Speed Cameras," in *East*

and West, the Common European Heritage: Proceedings of the XXV Scientific Instrument Symposium*, ed. Ewa Wyka, Maciej Kluza, and Anna Karolina Zawada (Cracow: Jagiellonian University Museum, 2006), 175–80.

2. Michael Lynch, "Discipline and the Material Form of Images: An Analysis of Scientific Visibility," *Social Studies of Science* 15 (1985): 38.

3. Luc Pauwels, "A Theoretical Framework for Assessing Visual Representational Practices in Knowledge Building and Science Communications," in *Visual Cultures of Science*, ed. Luc Pauwels (Hanover, NH: University Press of New England, 2006), 12–13.

4. For some recent studies: Marta Braun, *Picturing Time: The Work of Étienne-Jules Marey (1830–1904)* (Chicago: University of Chicago Press, 1992); François Dagognet, *Étienne-Jules Marey: A Passion for the Trace*, trans. Robert Galeta (New York: Zone Books, 1992); Simon Schaffer, "A Science whose Business Is Bursting: Soap Bubbles as Commodities in Classical Physics," in *Things That Talk: Object Lessons from Art and Science*, ed. Lorraine Daston (Cambridge, MA: MIT Press, 2004), 147–94; Peter Geimer, "Fotographie als Wissenschaft," *Berichte zur Wissenschaftsgeschichte* 28 (2005): 114–22; Jimena Canales, *A Tenth of a Second: A History* (Chicago: University of Chicago Press, 2009).

5. *Technology Review* 42 (1940), 106–7.

6. William Henry Fox Talbot, "On the Production of Instantaneous Photographic Images," *Philosophical Magazine* ser. 4, no. 3 (1852): 73–77.

7. Harold E. Edgerton and James R. Killian, Jr., *Flash! Seeing the Unseen by Ultra High-Speed Photography* (Boston: Hale, Cushman & Flint, 1939), 109.

8. A full study of Edgerton and Machine-Age aesthetics would require more space than available here. See Richard Guy Wilson, Dianne H. Pilgrim and Dickran Tashjian, *The Machine Age in America, 1918–1941* (New York: Abrams, 1986); Jeffrey L. Meikle, *Design in the USA* (Oxford: Oxford University Press, 2005).

9. Silas W. Holman, "The Function of the Laboratory," *Technology Review* 1 (1899): 13–35; C. Zoe Smith, "An Alternative View of the 1930s: Hines' and Bourke-White's Industrial Photographs," *Journalism Quarterly* 60 (1983): 305–10; James R. Killian, Jr., *The Education of a College President: A Memoir* (Cambridge, MA: MIT Press, 1985), 12–17.

10. *Technology Review* 33 (1931): 290.

11. *Technology Review* 34 (1932): 278.

12. *Technology Review* 36 (1933): 104.

13. "The object of the Photographic Society is the promotion of the Art and Science of Photography, by the interchange of thought and experience among Photographers, and it is hoped that this object may, to some considerable extent, be effected by the periodical meetings of the Society." *Photographic Journal* 1 (1853): 1.

14. Although the Venturi tube photo was a "remarkable achievement, this example is somewhat difficult to appreciate and is certainly less likely to impress visitors than the more homely subjects—a coffee-cup breaking on hitting the floor and an electric bulb disintegrating under a hammer blow—exhibited last year,"

intoned the judges in *Photographic Journal* 74 (1934): 491. For Edgerton's other entries in the Royal Photographic Society annual exhibitions, see *Photographic Journal* 73 (1933): 435; 75 (1935): 545; 76 (1936): 532–33; 77 (1937): 585; 78 (1938): 625; 79 (1939): 595; 80 (1940): 395. Edgerton's prints appear in the annual catalogs of the Royal Photographic Society exhibitions for 1933, 1934, 1939, and 1940, each published as a separately paginated issue of *Photographic Journal*.

15. Beaumont Newhall, *Photography, 1839–1937* (New York: Museum of Modern Art, 1937), 88, 125, Pl. 87.

16. Beaumont Newhall, *History of Photography from 1839 to the Present Day* (New York: Museum of Modern Art, 1949), 201–18.

17. Both Edgerton's and Killian's names appear on the title page of *Flash!* But they divided the profits from sales of the volume, which retailed for $3.50 in 1939, at a ratio of 75/25 between them, respectively. See Notebook 10 (Jun 1939 to Sep 1940), 67, Box 51, Harold Eugene Edgerton Papers, 1889–1990 (MC 25), Institute Archives and Special Collections, MIT Libraries, Cambridge, Massachusetts (henceforth cited as HEE Papers). Edgerton's laboratory notebooks are also available at http://edgerton-digital-collections.org/notebooks (accessed 1 December 2010).

18. Edgerton and Killian, *Flash!* 10, 21–22.

19. Harold E. Edgerton, "Stroboscopic and Slow-Motion Moving Pictures by Means of Intermittent Light," *Journal of the Society of Motion Picture Engineers* 18 (1932): 356–64; Harold E. Edgerton and Kenneth J. Germeshausen, "Catching the Click with a Stroboscope," *American Golfer* 17 (November 1933), 37–39; *Fortune*, November 1936, 111.

20. Edgerton and Killian, *Flash!* frontispiece, 15, 20.

21. Many colleagues who personally knew Edgerton at MIT have told me this story, which is also repeated in J. Kim Vandiver and Pagan Kennedy, "Harold Eugene Edgerton (1903–1990)," *Biographical Memoirs, National Academy of Sciences* 86 (2005): 102.

22. Lorraine Daston and Peter Galison, *Objectivity* (New York: Zone Books, 2007), chap. 3.

23. See Christopher Ames, *Movies about the Movies: Hollywood Reflected* (Lexington: University Press of Kentucky, 1997).

24. Harold E. Edgerton and Kenneth J. Germeshausen, "The Mercury Arc as an Actinic Stroboscopic Light Source," *Review of Scientific Instruments* 3 (1932): 535–42; F. E. Barstow and Harold F. Edgerton, "Glass-Fracture Velocity," *Journal of the American Ceramic Society* 22 (1939): 302–7.

25. Notebook 9 (Apr 1938–Jun 1939), 39, 49, 73, 76, 118, 124–25, and back cover (unpag.), Box 51, HEE Papers. Edgerton also set up a stroboscopic camera at a Spaulding retail store in New York City, so that potential buyers could photograph themselves swinging the clubs.

26. Edgerton and Killian, *Flash!* 63–66.

27. Ibid. 162.

28. Ibid., 197, acknowledges the Ethyl Gasoline Corporation for two photos of smashed golf balls, but not for the football images.

29. Notebook T-5 (Oct 1934–Aug 1936), 8, 27, 114, Box 162, HEE Papers.

30. *Fortune*, May 1935, rear cover; *Life* (New York), August 1935, 3; *Fortune*, September 1935, rear cover.

31. *Life* (New York), November 1935.

32. *Life* (New York), April 1936, 33; May 1936, front cover verso; June 1936, front cover verso.

33. Time Inc., *Four Hours a Year: A Picture-Book Story of Time Inc's Third Major Publishing Venture—The March of Time* (New York: Time Inc., 1936); Raymond Fielding, *The March of Time, 1931–1951* (New York: Oxford University Press, 1978). I thank the Edgerton Center at MIT for providing me with a copy (perhaps not the final version) of the Edgerton clip, with their shelfmark DVD-Advanced Tech D.

34. Quoted in Leonard Maltin, *Great Movie Shorts* (New York: Crown Publishers, Inc., 1972), 143. For Edgerton's autobiographical comments about the making of the film, see "M.G.M. invitation—Pete Smith short," typescript, Box 104, Folder 12, HEE Papers.

35. The complete short is available at http://www.youtube.com/watch?v=jUZwFrGzQGw (accessed December 1, 2010).

36. Quoted in *Strobe in Industry, 1931*, a film clip available at http://edgerton-digital-collections.org/docs-life/strobe-in-industry (accessed December 1, 2010).

37. Notebook T-3 (Jan 1932–Jul 1933), 14, Box 50, HEE Papers.

38. Notebook T-4 (Jun 1933–Nov 1934), 42, 99, Box 50, HEE Papers; *Technology Review* 36 (December 1933): frontispiece; 36 (July 1934): 344–45; 36 (October 1934): cover, 24.

39. By November of 1931 Edgerton had so named his MIT laboratory. See Notebook T-2 (Jul 1931–Jan 1932), 90, Box 50, HEE Papers. Bunker, a professor of biology who published X-ray studies on gastrointestinal motility and the therapeutic use of ultraviolet radiation, would after 1936 play a leading role (with Vannevar Bush) in reconfiguring MIT's Department of Biology and Public Health into a "Department of Biology and Biological Engineering."

40. Notebook T-4 (Jun 1933–Nov 1934), 102–6, Box 50, HEE Papers. Apparently neither Bunker or his student Fuller ever published these researches. For nineteenth-century fascination with the contracting frog leg, see Henning Schmidgen, *Die Helmholtz-Kurven: Auf der Spur der verlorenen Zeit* (Berlin: Merve-Verlag, 2009).

41. James Gray, "The Mechanism of Ciliary Movement, VI: Photographs and Stroboscopic Analysis of Ciliary Movement," *Proceedings of the Royal Society of London* B, 107 (1930): 321. Also, see James Gray, *Ciliary Movement* (Cambridge: Cambridge University Press, 1928).

42. Notebook T-4 (Jun 1933–Nov 1934), 122–23, Box 50, HEE Papers; M. W. Jennison and J. W. M. Bunker, "Analysis of the Movement of Cilia from the Clam (Mya) by High-Speed Photography with Stroboscopic Light," *Journal of Cellular and Comparative Physiology* 5 (1934), 189–97.

43. Harold J. Coolidge, *Annual Report*, 5 October 1935, Harvard Archives, Conant Papers, UAI 5.168, Box 56, Folder "Film Service, 1934–36."

44. *Technology Review* 42 (July 1940): frontispiece (on Kodochrome film); Harold E. Edgerton, "Hummingbirds in Action," *National Geographic* 92 (1947): 220–32; Harold E. Edgerton, R. J. Niedrach, and Walker Van Riper, "Freezing the Flight of Hummingbirds," *National Geographic* 100 (1951): 245–61. For an overview of studies of the mechanics and aerodynamics of bird flight, in the decade before Edgerton began his work, see Oskar Proschnow, "Die Verfahren zur Erforschung des Tierfluges," in *Handbuch der biologischen Arbeitsmethoden*, Abt. IX/4: *Methoden der vergleichenden Physiologie mit besonderer Berücksichtigung der Wirbellosen*, ed. Email Abderhalden (Berlin: Urban & Schwarzenberg, 1930), 215–94.

45. Harold E. Edgerton and C. M. Breder Jr., "High-Speed Photographs of Flying Fishes in Flight," *Zoologica* 26 (1941): 311–14; C. M. Breder Jr. and Harold E. Edgerton, "Do Flying Fish Really Fly? One of Nature's Riddles Unraveled . . . in Startling Stroboscopic Flash Pictures," *Travel and Camera* 1, no. 1 (1946): 73–76. See C. M. Breder Jr., "Field Observations on Flying Fishes: A Suggestion of Methods," *Zoologica* 9 (1929): 295–312; C. M. Breder Jr., "The Perennial Flying Fish Controversy," *Science* 86 (1937): 420–22.

46. *Technology Review* 44 (January 1942): 116–17; C. M. Breder Jr. and Harold E. Edgerton, "An Analysis of the Locomotion of the Seahorse, *Hippocampus*, by Means of High-Speed Cinematography," *Annals of the New York Academy of Sciences* 43 (1942): 145–72. Edgerton's penchant for undersea photography would later become legendary in his collaborations with the undersea explorer Jacques-Yves Cousteau, his work at the Woods Hole Oceanographic Institute, and his fanciful search for the Loch Ness monster. See Harold E. Edgerton and Charles W. Wyckoff, "Loch Ness Revisited," *IEEE Spectrum* 15, no. 2 (1978): 26–29; R. H. Rines, Harold E. Edgerton, Charles W. Wyckoff, and Martin Klein, "Search for the Loch Ness Monster," *Technology Review* 78, no. 5 (1976): 25–40.

47. Notebook T-7 (Apr 1936–May 1937), 117, Box 51, HEE Papers.

48. George Paul Moore, "A Stroboscopic Study of Vocal Fold Movement," *Summaries of Doctoral Dissertations of Northwestern University* 4 (1936): 185.

49. See Marcel Clary, "La Phonoscope à cordes vibrantes," *Revue scientifique* 70 (1932): 462–67; Robert Pollack-Rudin and Leopold Stein, "Ein neues handliches Laryngostroboskop," *Wiener medizinische Wochenschrift* 82 (1932): 916–17; L. A. Kallen and H. S. Polin, "A Physiological Stroboscope," *Science* 80 (1934): 592.

50. G. Paul Moore, "Vocal Fold Movements during Vocalization," *Speech Monographs* 4 (1937): 50, 52.

51. Moore, "Vocalization," 54–55.

52. *Technology Review* 39 (May 1937): 276; G. Paul Moore, F. D. White and H. Vonleden, "Ultra High-Speed Photography in Laryngeal Physiology," *Journal of Speech and Hearing Disorders* 27 (1962): 165–71.

53. Donald Crafton, *The Talkies: American Cinema's Transition to Sound, 1926–1931* (New York: Scribner, 1997); Andrew Sarris, *"You Ain't Heard Nothin' Yet": The American Talking Film, History and Memory, 1927–1949* (Oxford: Oxford University Press, 1998).

54. For their stimulating questions and comments I am deeply indebted to the participants in the 2006 Humanities Institute at Dartmouth's Leslie Center for the Humanities. I also thank staff at the Institute Archives & Special Collections of MIT Libraries, the Edgerton Center at MIT, and the MIT Museum for their generous help with this research. Librarians at the Dartmouth College Library and Janice Smarsik at Dartmouth's Visual Resource Center also helped to secure needed materials.

chapter nine

On Fate and Specification
Images and Models of Developmental Biology

Sabine Brauckmann

> Scientific illustrations are not frills or summaries;
> they are foci for modes of thought.
> —STEPHEN JAY GOULD

Between *Animal Videns* and Craftsmanship

Studies of the history of biology and, in particular, of the field of embryonic development ponder many and diverse issues that either focus on the developmental history of a model organism, an innovative experimental method, or how evolutionary issues are connected to ontogenesis. When investigating the processes and mechanisms shaping an embryo, biological scientists produce images and decipher them with the standards that are specific for the discipline in question and with their set of tools. In doing so, they observe a specific embryonic shape attentively, observe it and assay it for hours, days, or weeks until their hands can trace the specimen in all its depth. For instance, they touch cells or embryonic material, then lay it between glass plates under a microscope, shake it in the test tube, and measure it to observe and study it. Then they order the experimental data and analyze them, organizing their observations in tables or lists, and visualize them through figures, material models, and/or computer simulations. In the next step they tweak the imagery and prepare their results for publication.

When leafing through biology textbooks, which either train students in embryology or offer insights into molecular genetics, many of these images catch our eyes immediately. It seems obvious that life scientists prefer reasoning in *pictoribus* rather than with mathematical equations, as we find in physics or chemistry. For some years, the historiography of the sciences has dealt with the imagery in the sciences in general, not least of all in the life sciences.[1] However, a concise study is still missing that does for the life sciences what Martin Rudnick accomplished for geology, namely to delineate the coming into being of a visual language for newly developing disciplines such as developmental biology, or molecular cell biology.[2]

My discussion on images and models of developmental biology complements the research on visualization in the biosciences by providing a more epistemic approach when arguing that scientists (and nonscientists involved) "see with their hands and talk without words."[3] In this essay, I will consider historical data on images depicting cell dynamics and morphogenetic movements for reasoning on the subtle distinction of embryonic fate and cell specification. These images were created by developmental physiologists working on amphibians and chicks from around 1900 through 1950s.

Fate mapping is closely connected to Walther Vogt who perfected the method of *in vitro* staining and depicted classical maps of folding and warping layers of germinal cells of amphibian and toad embryos.[4] Almost immediately, other developmental biologists working on gastrulation of the amphibian embryo took up his technique, including Ludwig Gräper, Robert Wetzel, and Jean Pasteels, among many others. However, these scientists voted for another model organism, the chick, as well as took up a slightly different research program, one that attempted to solve the enigma of the primitive streak and to predict to which organs cells become individualized, or specified. In the 1940s Dorothy Rudnick published two hand-drawings of specification maps of the chick embryo that persuasively summarized the knowledge about organ-forming stages and shapes. As it turned out, her work would be recycled in textbooks until the 1970s although, it needs to be pointed out, textbook authors during this time often did not distinguish between fate and specification in their composites.[5] It is important to note that the terms "fate map" and "specification map" were not yet coined at the time under consideration. For instance, Lester G. Barth described fate maps as "gastrular maps of presumptive value" in 1953, and even in 1955 Johannes Holtfreter and Viktor Hamburger spoke of "so-called fate maps" when reviewing amphibian development.[6]

As for my text here, after some cursory remarks about the history of fate mapping, I will focus on Vogt's and Rudnick's work to consider how scientists mapped or specified these fates by the means of hand drawings, camera lucida drawings, and material models. Both Vogt and Rudnick worked along the same crucial issues of classical embryology, namely: how the germ layers develop; which specific fates (potencies) cells permute during embryogenesis; whether development, once started, proceeds reversibly or irreversibly; and which organs are developed from which cells. Implicitly, both followed the pathway of potencies carved into the theoretical landscape of biology by Hans Driesch around 1900, whether depicting fate or representing specification.[7] In addition, I will look at a lab manual utilizing the in vitro staining technique and will show how these images were used for training students. My main argument will be that these scientists depicted a "mental imagery" in their figures or models of developing embryos, which suggests that perception relies upon a visual-tactile experience.

But as we analyze how fate and specification were distinguished, we can address another theme, namely, *tacit knowledge* ("we do know more than we know") and the significant role it played and still plays in transmitting the experimental "know-how" of laboratory practice.[8] For deeply embedded in the imagery of developmental maps rests an intuitive knowledge of the trade that results to a large extent from manual training at the workbench, and the manifestations of these skills are often distinguished between different laboratories in regard to styles of doing experiments and training students. However, I argue that in the case of fates and specification, *tacit knowledge* rested not so much on the reliability and "fineness" of the experimental tools in the laboratory, but rather on the fine-tuning of individual observational and manual skills in handling such delicate objects as living embryos.[9]

As Kathryn Olesko has elaborated in her study of German research schools in physics, we need to study the notion of tacit knowledge by undertaking more penetrating analyses of laboratory manuals, notes, and diaries.[10] In tracing the decade-long experimental work of Vogt and Rudnick, I will show that, due to their extensive training in observation and reflection on their visual data, both developed a profound tacit knowledge of embryonic fates pushed on by cell migration. This knowledge had a deep impact on their imagery of specifying cells, images that had a productive influence on students as well as colleagues. For example, Robert Wetzel received his training in vital staining at the same department of the University of Würzburg where Walther Vogt worked until 1924.

Aperçus on Fate and Specification

In the nineteenth century, comparative embryology and anatomy developed into research programs—first still focusing on standardized observation but then incorporating invasive experiment—that aimed to describe how embryogenesis proceeded to the fate of the undifferentiated germ and to predict the fate of the future embryo. The issues guiding this program pertained to the share of the three germ layers in organogenesis, the fate of undifferentiated cells, and the formation of the mesoderm in amphibian embryos (or the function of the primitive streak in chick embryos). The embryologists investigating the dynamics of the formation of germ layers strove to resolve whether the germinal cell (material) was responsible for development or whether morphogenetic movements of the layers (mechanism) played the determinate part. However, it was difficult to reconstruct embryogenesis visually by observing whole or sectioned embryos with the then available microscopes and sectioning instruments. To make the task easier, embryologists used the camera lucida to trace, again and again, the moving contours of these delicate objects, trying to identify the area inducing cell differentiation. An advantage of the camera lucida was that it aided in depicting the three-dimensionality of the specimens ostensibly flattened between glass slides.[11]

These drawings did not settle matters, however. Even if one agreed upon the validity of the observational data, the analyses of the data, theories and models created from the data, were often under dispute and controversy for years. For instance, many images that were drawn either without holding on, or with the camera lucida, displayed appraised values, or they provided idealizations and often functioned to demand further experiments and improvements of the experimental equipment. Thus, the issues mentioned above were not yet solved around 1910 when Vogt started his research on amphibian gastrulation.

In the early twentieth century a new method of tissue cultivation came along and improved the observation and experimentation on cell dynamics.[12] To visualize the presumptive fate of the germinal cells, one depicted their morphogenetic movements in images (fate maps), which mapped the early gastrula (or blastula) with reference to what its various parts would produce. A series of fate maps would then depict trajectories in the development of an organism from the fertilized egg. With the in vitro technique of tissue culturing, one schematized to which particular structure a tissue explant would be specified after isolation from the embryo.[13] These schemes were compared to histological stainings of differentiating cells in vitro (specification maps). For constructing the maps, one invented tools for disclosing the potency (fate) of a cell or cell clusters from a defined region of the embryo to the phenotypes. The hope was that the maps would answer some crucial questions regarding gastrulation and serve to demonstrate whether embryogenesis proceeds in regulative or in mosaic-like fashion.

Around 1900 Ludwig Gräper stained the chick blastoderm with neutral red to trace the formation patterns of the three germ layers during gastrulation. A few years later Hubert Dana Goodale refined the vital-staining method such that it could be applied to the amphibian embryo. At nearly the same moment, but, independent of Goodale, Vogt experimented with little pieces of agar that he saturated with dyes to mark the folding and warping germ layers from the blastula stage to neurulation.[14] At this point it became possible for experimental embryologists to depict the fate of the germinal cells at successive intervals during gastrulation.

To approach the problem of the intrinsic properties of the germinal areas, developmental physiologists relied mainly on grafting experiments, exchanging parts between embryos. This caused technical difficulties because of the internal tension of the blastodermal walls. Benjamin Willer surmounted this problem by depositing pieces of a blastoderm cultivated on the chorio-allantoic membrane of an older egg. (Later, plasma clots were also used.) The membrane then reacted to the local irritation by surrounding the implant with a richly vascularized tunic of indifferent tissue.[15] The scientists learned how to record the phenomena they observed, using camera lucida-drawings, cyanotype, and colored pencils. Thus, by the 1940s a solid body of knowledge had accumulated and was published in textbooks and taught in biological and medical courses.

There were disadvantages to these fate maps, however. They had a coarse consistency, partly due to the imprecision of the instruments, partly due to the isolated embryonic cells containing remnants of the three germ layers. Furthermore, experimental embryologists often stained germinal areas that were situated too closely to each other, and in these cases they would not be able to observe which embryonic cells invaded which germ layer and where they laterally left the embryonic body. In any case, it turned out that fate maps of amphibians and toads could not exhaustively answer these questions, and so many biologists changed the model organism and started to work with chick embryos, another classical model organism of comparative embryology of the nineteenth century. Employing the visual data of fate maps, which was the precondition of mapping cell specificity, the scientists sought the phenotypic future of the organism, namely, which blastodermal areas or blastodermal cells induced which organs. In other words, they wanted to know what the (organ) fate was for individual cells.

How to Map and Specify

Enter Walther Vogt and his fate maps, icons of classical developmental biology. Starting from the idea of depicting a visual topography of the "prospective meaning" of specific embryonic areas before gastrulation, he documented the dynamics of the germinal layers in his amphibian and toad maps. He baptized his descriptive research program *Gestaltphysiology*, synthesizing classical comparative embryology with physiology and aiming at a detailed study of embryonic *form*.[16] According to him, embryology meant serial form. One had to think in sequences of images and compare the altering shapes at different developmental stages. Without ever naming them as such, he circumscribed a fate map as *Abbildungsreihe des Anlagenplans* (image sequence of the primordial anlagen).[17] Numerous drafts preceded all final images (maps), drafts that Vogt made with the camera lucida from the living specimen. He called them *Photogramme*, although he took photographs with a small camera, using a small aperture and two-sided illumination. Evidently, the photomicrographs in Vogt's article are not photograms, that is, photographic images of three-dimensional objects (and photosensitive material without involving any optical system such as a photoapparat). Nevertheless, his usage of the word "photogram" was less a *lapsus linguae* than Vogt's way of describing an experimental method that left the embryonic object completely undisturbed as the microscopic eye hovered above it. He emphasized repeatedly that the photos were not retouched and showed solely the controlling of gastrulation movements.[18]

His basic objective was to reproject the embryonic anlagen to the germ regions by directly observing how the material moves to its final location, as well as to visually trace how these movements determine the anlagen of the organs. Vogt's point of departure was to reevaluate Ernst Haeckel's gastrea theory, and so he chose the blastula, the developmental stage that looks similar across different species. Further, the experiments studying the formation of the germ layers in amphibian and toad embryos relied upon the conviction that the germ layer starts off in a mosaic-like fashion; that is, its anlagen are directly canalized by the nucleus (genome) and, simultaneously, epigenetically induced by an organizer center.[19] What distinguished Vogt from the Spemann Group at the University of Freiburg was his strong argument for pattern formation before gastrulation, that is, before Spemann's organizer induces embryonic tissue to specific organs. Moreover, he had reservations about Spemann's transplantation experiments, as this method disturbed the course of normal development. Instead, he preferred a noninvasive style of doing experiments.[20] In several articles he repeatedly emphasized that the technique of vital staining went beyond just determining in a descriptive way the future specification of the germinal areas, or the reprojection of organ anlagen to all developmental stages during embryogenesis. According to Vogt, this tool extended beyond being a purely descriptive method and moved toward a causal-analytical one because it allowed the scientist to ascertain material movements and to decide whether these movements represent mere growth phenomena, or "true" morphogenesis of cell specification.[21]

Vogt published his first amphibian maps in 1923 and then a comprehensive study of all developmental stages six years later in the *Spemann-Festschrift*. Initially he showed how germinal parts that were stained in vivo could retain the color over days while development took place. In further experiments he placed pieces of agar saturated with dyes of different color (mainly Nile blue and neutral red) upon the surface of amphibian embryos. The dyes penetrated into the cell membranes and stained the surface of the late blastula. By the marking of particular regions with neutral red and Nile blue it became apparent which cell layers or groups inside the opaque structure of the embryo migrate to which location and which folding and invaginative mechanisms they trigger during their migration. Due to his observational experiments, the cell material moving into the germ was marked so clearly that particular segments of the chorda developing into the spinal column of vertebrates could be traced from the beginning of gastrulation.

During various developmental stages, Vogt placed red and blue markers (sometimes up to eighteen dots) along the dorsal median surface of an *Axolotl* embryo to indicate how the stained spots were sucked into a lunula-like split (blastopore) (figure 9.1e–f). Finally, one could see how the migrating spots come to rest in the foregut, where the round dots ended up as faded longitudinal stripes (figure 9.1m). The following schemes demonstrate how Vogt visualized the species-specific fate maps of anurans and urodeles during gastrulation and the formation of the mesoderm (figure 9.2a–d). The scheme of the beginning of gastrulation (figure 9.2a–b) has

220 · *The Educated Eye*

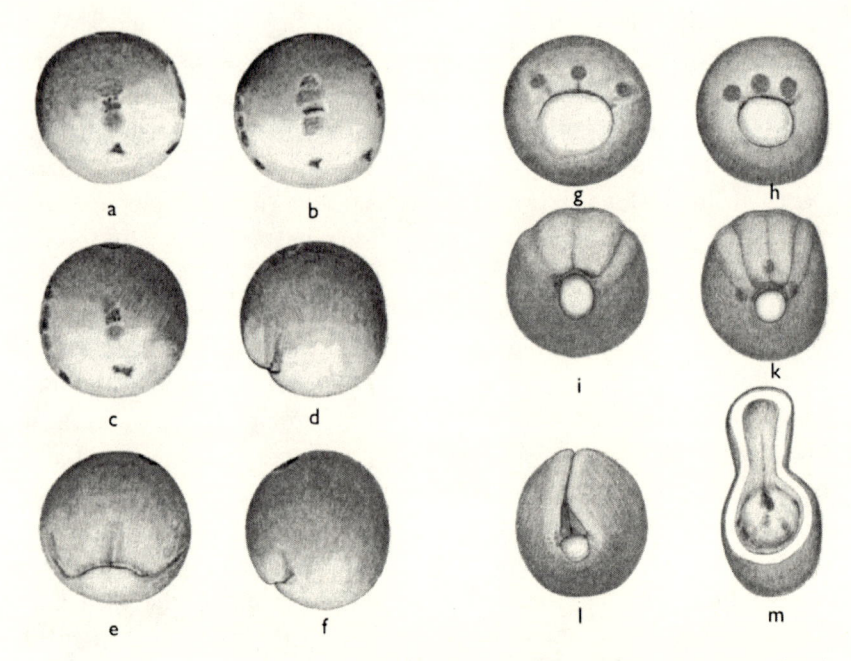

9.1 Axolotl, four-line marking of the marginal zone during different gastrulation stages, dorsal and lateral; *a–c* views after first marking; *d–f* invagination and stretching of the marginal zone with lunulus-like blastopore; *g* second marking; *h* third marking; *i* invagination of the red markers; *k* fourth marking; *l* closure of blastopore; *m* neurula, frontal view, the entodermal markers of the first marking are visible in the foregut. From Walther Vogt, "Gestaltungsanalyse am Amphibienkeim mit örtlicher Vitalfärbung, II Teil: Gastrulation und Mesodermbildung bei Urodelen und Anuren," *Wilhelm Roux' Archiv für Entwicklungsmechanik der Organismen* 120 (1929): 462, 464, figure 15.

been set against the complementary view of amphibian gastrulation (*Axolotl*). The lines in boldface indicate the stretching of external material into the marginal region, while the thinly drawn curves map colored streaks that are situated inside. They also demonstrate how the material moves from the blastoporal region and migrates into the lunula-like split of the blastopore (figure 9.2c–d).

As Vogt used more methods, besides the technique of vital staining tissue culture and sectional cuts with the micrometer, to confirm his data, he also employed different modes of visualization, for example, camera lucida drawings on paper (figure 9.3a), diagrammatic schemes (figure 9.3b), and wax models for teaching purposes.[22] Interestingly, in Vogt's case, his research group at the University of Munich projected a sequence of numerous marked images of earlier blastula stages to two wax globes of 20 cm diameter (Figure 9.3c). The wax globes were then photographed and

On Fate and Specification · 221

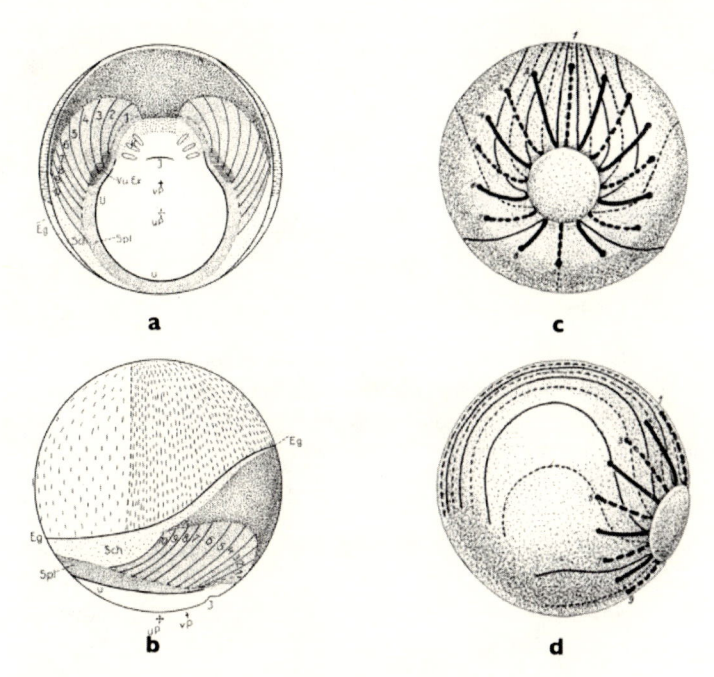

9.2 *a–b* scheme of the presumptive anlagen of urodele embryo at the beginning of gastrulation, the material migrating into the lunula-like blastopore is lightly punctuated (from Vogt, "Gestaltungsanalyse," 392, figure 1); *c–d* orientation scheme of movements during gastrulation and formation of mesoderm; frontal (*c*) and ventral (*d*) view, the thin lines lag inside and correspond to the stained streaks gleaming through the mesodermal walls and the archenteron (from Vogt, "Gestaltungsanalyse," 431, figure 8).

printed on wall charts that were used for teaching and training students. The globes visualized the most important transitory stages of development by a three-dimensional material model. A by-product of the modeling was that passive observation was transformed into active knowledge when students were able to mold the shapes with their own hands. By means of these globes Vogt rendered spatially the time sequences of embryogenesis, the stages of which he scaled to the microphotographs of the histological sections of the in vitro cultures. The drawings are two-dimensional, the wax globes have three dimensions, and the tissue culture provides the flat image.

The *differentia specifica* between images and models is their dimensionality representing the nonvisible structures, and their common feature is visibility. The notion of visibility, or *Anschauung*, means that between

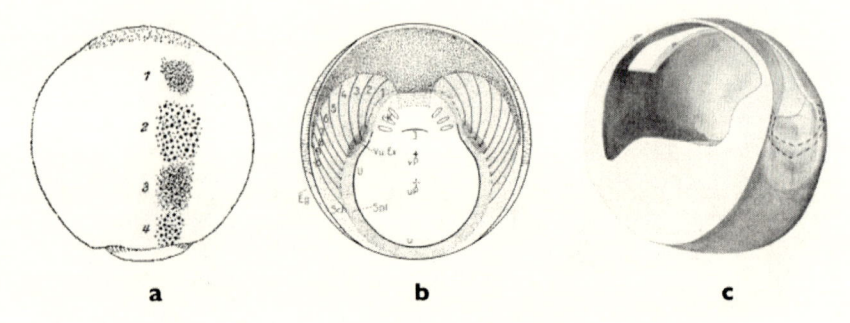

9.3 *a* Axolotl blastula with staining (from Vogt, "Gestaltungsanalyse," 568, figure 1c); *b* scheme of the presumptive anlagen of an amphibian embryo during beginning gastrulation (from Vogt, "Gestaltungsanalyse," 392 figure 1b); *c* part of a wax globe depicting a closed toad gastrula (from Vogt, "Gestaltungsanalyse," 476, figure 17).

the specimen (object) we are observing and the eye arises an invisible (imaginative) scheme serving as, in essence, the spatial knowledge about this specimen. What Vogt implicitly projected served as the retrospection of the embryonic anlage, which is hidden (as it were cocooned) inside the germinal areas and invisible to the naked eye. His accurate and aesthetic figures convinced him of the harmony of the movements as well as of the uniqueness and irreversibility of primitive development. Both dimensionality and visibility (*Anschauung*) disclose Vogt's style of reasoning about embryonic forms of space.

For nearly two decades Vogt built up these fate maps, drawing on his mental images, using the camera lucida, when modeling the time sequence of embryogenesis either from material models or by comparing (side by side) microphotographs of sectional cuts of embryonic tissue cultures. By reasoning that modified and enhanced the representation on paper, he improved the "evidence" of his data on the *Gestaltungsanalyse* of gastrulation over a period of fifteen years. Now the trained eye could see the fate of the marked cells through later developmental stages. All these material images served to make visible gastrulation movements, and thus reproduce embryonic data, which allowed other biologists and students to replicate the experiments to reveal for themselves the "new" knowledge on gastrulation. Vogt's meticulous descriptions and models became important teaching tools.[23]

In the second edition of his lab text, *A Manual of Experimental Embryology*, published in 1960, Viktor Hamburger presents classical experiments of experimental embryology, or developmental physiology, and assigns

Vogt's experiments with the vital-staining method a prominent place for several reasons. First, the published works, in particular the monographic review of 1929 with its nearly 100 figures (18 in color), outline an almost complete record of gastrulation movements and the mechanism of germ-layer formation.[24] Further, these experiments could easily be reproduced, not least of all because of the availability of embryos and their technical facility. Most importantly however, for the case here, is that these investigations are well suited to train students to coordinate eyes and hands and, in this way, internalize the standards of scientific observation for biological experimentation. Before the student can start at all to open the delicate body of an embryo in order to place the red or blue markers, he or she has to memorize the shapes of the developmental stages. The next step forces him to fix his or her eyes and hands on the object in order to learn how to observe carefully and to reconstruct and visually represent the observed forms by sketching them. Again and again, whether replicating Vogt's or others' experiments (e.g., Wetzel's staining of chick embryos), Hamburger emphasized the need to observe and sketch at the same time as a way to interpret experiments. Although he does not explicitly mention it, one could say that the repeated contemplation of embryonic shapes that recur with the reliability of normal development and the transformation of hands into tools for making visible what the naked eye has seen and the mental eye has imagined serves the purpose of building tacit knowledge.

Although a considerable number of experimental embryologists and tissue specialists worked on fate maps using transplantation and explantation experiments, or in vitro techniques, the focus here will be on Dorothy Rudnick and her research on organ-forming areas of the chick blastoderm. Her experimental program complements Vogt's by introducing another classical model organism of embryology—the chick. Rudnick excelled in that she was more proficient than her colleagues in memorizing and extracting the relevant information from the comparative imagery on embryogenesis stretching over nearly 150 years. In addition, she belonged to the minority of developmental biologists who openly announced the disadvantage of fate maps, which depicted cellular black boxes, thus concealing the internal processes inside the cell, namely, which particular cell component induces which organ. To disclose these mechanisms, intervention experiments were necessary that would either investigate the cell metabolism of the germinal areas known to condition embryogenesis or reveal the genetic properties forcing the cells to differentiate by modifying environments. For Rudnick, embryology allowed one to observe and describe accurately and in a standardized way the visible processes related to organ-forming germinal areas as well as opened up experimental studies on the morphogenetic potencies of these areas. The former had been undertaken extensively by embryologists since Baer's time, but the blind spot of embryology still remained how the germinal cells were localized during the cleavage stage.[25] Therefore, Rudnick's research program aimed to open up the embryonic black box in order to localize the presumptive anlagen and to act upon the future (virtual) embryo (and not to focus on the embryonic past as Vogt had done).

Unlike Vogt who, as already mentioned, had an aversion to transplantation experiments, Rudnick was not shy about cutting open a chick embryo. Also, her commitment to biochemical analysis distinguishes her from Vogt, although she conceded that this only makes sense when the cellular morphodynamics of the primitive streak is known (up to now an open issue).[26] Furthermore, she argued in favor of pluripotency, assigning to cells the ability to develop into different types of mature cells, and reversibility of the embryonic fates—an argument that opposed Vogt's conviction of the irreversibility of cell determination.

In her 1944 review article "Early History and Mechanics of the Chick Blastoderm" Rudnick reported Gräper's and Wetzel's vital-staining work, which had succeeded in interpreting the ontogenetic events leading to the formation of the primitive streak of the chick embryo in a new way.[27] Both Gräper and Wetzel experimented under the hypothesis that a metabolic gradient around the primitive streak conditioned mesoderm formation, and they argued that the cell migrations they depicted in their figures proved vortex-like development of the embryonic layers. Wetzel mainly adopted Gräper's ideas, but discovered that a double crossflow vortex formed these layers and, by forcing the cells to invaginate, developed the primitive streak. His long article indeed shows how painful experimental research can be. When discussing his over 950 staining and extirpation experiments, he explicitly conceded that the initial positions of the staining markers referred at the outset to the axis of the germinal area really were an instance of inferential reasoning.[28] His frank statement discloses that his figures are *realiter* mental images, a perceptive problem that troubled him. Also, the issue of three-dimensionality caused him great difficulties, as he took great pains to plot the surface view onto the two-dimensional cross section, that is, to relate the rough form to the cell clusters of tissue cultivation.[29] Wetzel's confession of how difficult it was to visualize embryogenetic processes with flat images provoked Rudnick to criticize him wrongly as the "artist-morphologist" who had a wealth of the experimental data but fell short in interpreting this data.[30] However, even nowadays it is an extremely difficult task to represent a three-dimensional specimen, notwithstanding the technology of computer simulation.

As had happened to Vogt's amphibian fate maps, Pasteels corrected Wetzel's scheme, rejecting the vortex image, which he interpreted as a polonaise. Pasteels's maps demarcate between a presumptive neural system, a mesoderm, and the streaming of epiblastic cells in a clean cut toward the midline, where the cells invaginate to form the primitive streak.[31] Rudnick evaluated Pasteels's figure as the most satisfactory, although she thought it still gave an incomplete idea of the embryogenetic processes taking place inside the embryo.[32]

Rudnick completed her research program on avian embryogenesis by focusing her experiments on the intrinsic properties of the germinal areas over two decades. Using the same needles as Hans Spemann, she cut the blastoderm and transferred the pieces by pipette to an embryonic plasma clots on cover slips.[33] By culturing in vivo sections or areas of the blastoderm at various stages during the formation of the primitive streak, she could show that the differentiation centers of the heart lie in the posterior half near the streak. However, what distinguishes her from her male colleagues is her openness toward still unsolved questions, such as the meaning and function of the primitive streak for the first stages of embryogenesis. In referring to the preparatory work of Elizabeth Butler and, in particularly, Mary Rawles, Rudnick utilized Butler's experiments and the figures of Rawles to construct her diagrammatic view of cell specification.[34]

A few years before Rudnick's work, Butler had shown that the whole unincubated blastoderm can induce practically all embryonic tissues when transplanted to a chorio-allantoic membrane of a nine-day-old chick embryo. However, the one and only schematic figure of that article does not indicate any substantial information about embryogenesis. The scheme looks more like a repetition of Wetzel's technical advice on how to section the blastoderm. Rawles, who was especially interested in the fate of cells that pigment the feathers, continued Butler's experiments on cell specification and presented in her diagrammatic figures the region that develops the heart area during head formation. In working closely along the lines of Wetzel's experiments, but using chorio-allantoic grafting instead of vital staining, she adopted his diagrammatic figures with some modification (cf. figures 9.4a–c).[35] Her figures map the localization of the organ-forming areas in the three embryonic layers separately and indicate, therefore,

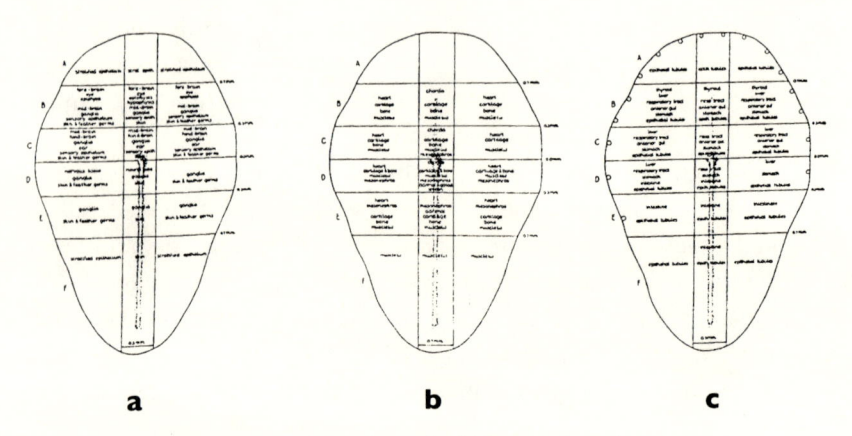

9.4 Diagram of a head-processes blastoderm of a chick; *a* ectodermal structures; *b* mesodermal structures, S striated muscle, U non-striated muscle; *c* endodermal structures, circles at margin indicate pieces from which primordial germ cells were obtained. From Mary E. Rawles, "A Study in the Localization of Organ-Forming Areas in the Chick Blastoderm of the Head-Process Stage," *Journal of Experimental Zoology* 72 (1936), 309–11, figures 7–9.

mental images rather than actually observed sequences of developmental stages. To gain the three-dimensional effect of the model, all three figures have to be envisaged as if stacked one above the other, starting from the ectodermal plate (figure 9.4a), and peeling off the mesodermal region (figure 9.4b) until the view looks at the entodermal zone, the most internal area (figure 9.4c). Her diagrammatic drawings intentionally reduce the three-dimensional object to two-dimensions for enhancing the visibility of organ-forming sectors, which were at the center of her research program. Although she did not make these drawings herself, the figures became classics of developmental biology and were reproduced in textbooks (e.g., in Witschi's *Development of Vertebrates*) until the 1960s.[36]

Rudnick's research question focused on the issue regarding what material migrates into the chick embryo and what happens to the primitive streak when this happens. To find an answer, she worked on locating the areas destined to become heart, kidney, limbs, and extraembryonic mesoderm, and on gaining knowledge about the histochemistry of the invaginating cells (cell respiration of blastoderm pieces). As previously mentioned, her objective was to project the future (virtual) embryo at the stage of the definite primitive streak whereby she could take advantage of the mapping images of Wetzel and Pasteels. By distinguishing the three stages of head, axis, and blastoderm formation, which are sequentially timed, she corrected the accepted claim that had depicted the developing blastoderm

On Fate and Specification · 227

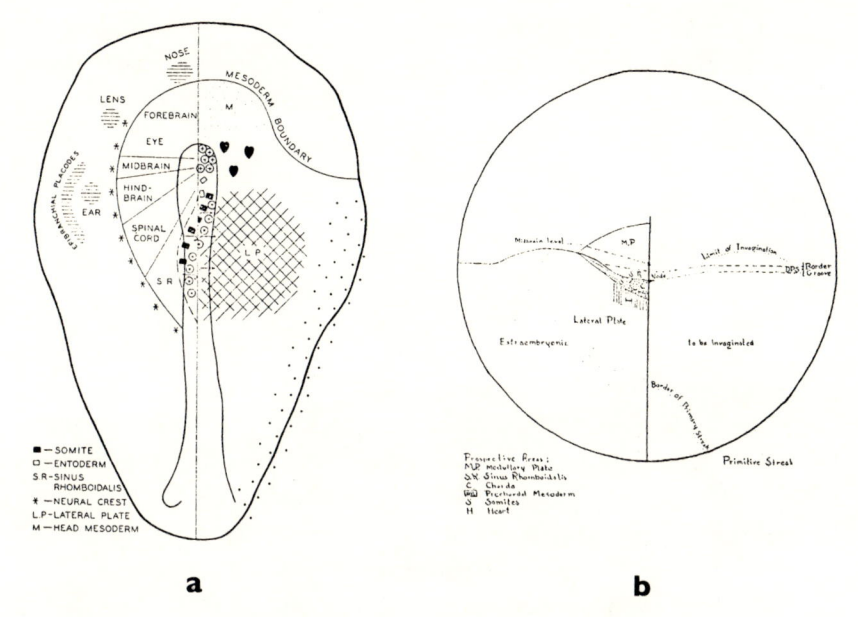

a　　　　　　　　　　　　　　　　**b**

9.5 *a* Prospective area map of the definitive primitive streak, boundaries of the primitive streak; *b* diagram of the prospective areas in the unincubated blastoderm. *a* From Dorothy Rudnick, "Early History and Mechanics of the Chick Blastoderm: A Review," *Quarterly Review of Biology* 19, no. 3 (1944), figures 9 and 10. Courtesy of University of Chicago Press. *b* From Dorothea Rudnick, "Prospective Areas and Differentiation Potencies in the Chick Blastoderm," *Annals of the New York Academy of Sciences* (1948), p. 763, figure 1. Courtesy of John Wiley and Sons

as if it were a jigsaw puzzle of transitory features and boundaries forming specific areas by growth mechanisms. Like Vogt, she opted for a regulative development of the blastoderm by which there were cells destined to induce the formation of the primitive streak.[37]

Rudnick visually compressed into two hand-drawings an overview of knowledge on chick embryogenesis, but was rather cautious about assigning to the images of differentiating cell potencies anything other than a hypothetical character.[38] Her schematic hand-drawing (figure 9.5a) represents the assumption that the degenerating primitive streak invaginates further, which she construed with the pluripotency of embryonic cells. She advised the attentive reader (or observer) to compare both drawings with the "potency maps" of Rawles (cf. figure 9.4).

Four years later Rudnick proposed a revised map of the areas of the blastoderm surface. In this figure she recreated the future (virtual) embryo during the primitive streak stage from Wetzel's and Pasteels's map and projected the embryo onto the unhatched blastoderm (figure 9.5b). She reiterated that the boundary between the mesodermal regions was merely a hypothetical construct and did not exist *realiter*.[39] She concluded that the cellular specificity is expressed by the morphogenetic relations between particular cells. Thus, she claimed the blastula material that occurred as an undifferentiated and amorphous mass to be a complex organized structure of specific cell compounds.[40]

What Is the Difference between Fate and Specification?

While the fate maps that are closely connected to Walther Vogt became icons of twentieth-century developmental biology, Rudnick's specification maps did not attain this status. In order to explain why this would be, it may be helpful to compare the highly complex fate map of Vogt to Rudnick's specification map. Both figures visualize the same fate of a particular developmental stage but are different in regard to the model organism (toad or chick), the experimental mode (observation or tissue culture), and in the dynamics of the representation. Concerning the specification map, biologists must calculate a diagrammatic projection from rotating three-dimensional structures. Therefore, to produce specification maps is a much more difficult task than to trace a fate series by following embryonic shapes with the camera lucida or redrawing the contours of microphotographs.

Already in 1828 Karl Ernst von Baer realized that mapping organic structures observed in their three-dimensionality to a two-dimensional plate flattens the object and turns a developing, growing organism into a still life. To compensate for the flattening and the invisibility of (physiological, or cell) dynamics, he made use of developmental trajectories (*Bildungsbogen*) that geometrized his typological reflections. These included arrows and dotted lines simulating pathways of two-dimensional plates to three-dimensional organs, just as Gräper, Wetzel, and Pasteels used them in their figures and schemes over a century later.[41]

Fate and specification maps differ perceptually because of the tool the researcher utilizes to observe the embryonic object, that is, either the eye aided by the microscope or the mind's eye, both of which are improved by professional training and study of the same model organism for years or decades. Rudnick claimed to perceive embryonic processes that she in reality could not yet have seen; that is, she made visible an epistemic construct that she mapped to the actual course of differentiation and induction. She assembled her prognosis of spatial cell fates representing the virtual embryo during the primitive streak from colleagues' images, whereas Vogt depicted the time sequence that his physical eye observed by looking through the microscope. What distinguishes both biologists further is their respective attitudes toward their own imagery. While Rudnick saw them as useful for future research instead of factual evidence of developmental reality, Vogt understood his fate maps as embryonic reality that could be reproduced by any biologist at any time.

I started with the statement that "we see with our hands and talk without words" as I proposed to describe how embryonic fate was mapped and cell specificities visually represented. It is evident that scientists have to coordinate eyes and hand each time they seek to produce images of cells colliding with each other. As this coordination is an ongoing, historical process, perhaps it confirms William Ivins's dictum that "representation is a visual language which has to be learned and which changes over time."[42]

To translate a microscopic view into a scientific representation scientists have to undertake different experimental procedures with the specimen, the technical equipment that traces the developmental processes of the organism, and the image itself; for example, specific cell migrations are visually enhanced when cells are stained. This implies a conscious selection, a theoretical attitude toward one's own style of doing experiments and of perceiving living organisms, which all flow together into the representational process. Further, to depict by hand is a way of experiencing knowledge about the specimen that one cannot achieve by passively looking at a photograph or a computer screen. When the hand draws what the "educated eye" sees and participates in the manual training, it serves as a lesson in reasoning about the embryo and plays a significant role in helping the researcher or student to build a memory of the developing patterns. By drawing the same lines, angles, dots, and circles for years, sometimes even decades, experimental biologists have literally embedded the know-how of doing biology into their bodies. In short, experimental reasoning proceeds in a visual-tactic way whereby the visual imagery entails the (epistemic) history of scientific observation and improves the "naked eye" to an educated one. Finally, it results in the know-how (or *tacit knowledge*) of the discipline in question.

Acknowledgment

My time as visiting fellow of the Fannie and Alan Leslie Center for the Humanities at Dartmouth College was a pleasure. Having spent over a year already at Dartmouth, the place was comfortably familiar to me. To meet new colleagues from disciplines like history of design, history of film, the neurosciences, literature studies, and philosophy broadened my own reasoning about images and their impact, whether in regard to inspiring students or to convincing professional audiences. When a family member became seriously ill and I felt desperate, the empathy and the aid of my colleagues helped me enormously. My special thanks to Nancy Anderson and Mike Dietrich for organizing this wonderful workshop on visual culture and pedagogy.

Notes

Stephen Jay Gould, *Bully for Brontosaurus: Reflections in Natural History* (New York: W. W. Norton and Co., 1991), 171.

1. For a selective overview, cf. Sabine Brauckmann, Christina Brandt, Denis Thieffry, and Gerd B. Müller (eds.), *Graphing Genes, Cells, and Embryos: Cultures of Seeing 3D and Beyond*, Preprint 380 (Berlin: Max-Planck-Institut für Wissenschaftsgeschichte, 2009); Sabine Brauckmann and Denis Thieffry, "Introduction to Picturing Eggs, Embryos and Cells," *History and Philosophy of the Life Sciences* (Special Issue) 26 (2004): 283–89; Horst Bredekamp, Angela Fischel, Birgit Schneider, and Gabriele Werner, "Bildwelten des Wissens," in *Bilder in Prozessen*, ed. Horst Bredekamp and Gabriele Werner, *Bildwelten des Wissens. Kunsthistorisches Jahrbuch für Bildkritik*, Bd. 1, 1. (Berlin: Akademie Verlag, 2003), 9–20; Peter Geimer (ed.), *Ordnungen der Sichtbarkeit: Fotografie in Wissenschaft, Kunst und Technologie* (Frankfurt: Suhrkamp, 2002); Daniel Guggerli and Barbara Orland (Hg.), *Ganz normale Bilder: Historische Beiträge zur visuellen Herstellung von Selbstverständlichkeiten* (Zürich: Chronos, 2002); Bettina Heintz and Jörg Huber (ed.), *Mit dem Auge denken: Strategien der Sichtbarmachung in wissenschaftlichen und virtuellen Welten*, Institut für Theorie der Gestaltung und Kunst (Zürich: Springer, 2003); Caroline A. Jones and Peter Galison (ed.), *Picturing Science. Producing Art* (New York: Routledge, 1998); Wolfgang Lefèvre, Jürgen Renn, and Urs Schöpflin (ed.), *The Power of Images in Early Modern Science* (Basel: Birkhäuser, 2003); Henri Michel, *Images des sciences: Les anciens instruments scientifiques vus par les artistes de leur temps* (Rhode-St. Genèse: Albert de Visscher, 1977).

2. Martin Rudwick, "A Visual Language for Geology, 1760–1840," *History of Science* 14 (1976): 149–95, 177.

3. Gottfried Boehm's interesting chapter about visualization in the sciences localizes the (produced) image between eye and hand, according to the Kantian tradition, but oversees the importance of the handicraft when fabricating these images. Gottfried Boehm, "Zwischen Auge und Hand: Bilder als Instrumente der Erkenntnis," in *Konstruktionen, Sichtbarkeiten*, ed. Jörg Huber and Martin Heller (Wien: Springer, 1999), 215–227; Sabine Brauckmann, "Seeing with Hands and Talking without Words: On Models and Images in the Sciences," *Biological Theory* 1 (2006): 199–202.

4. Although biologists continue to refer to his seminal fate maps even now, Walther Vogt (1888–1941) is among the famously unknown biologists; for a short biographical note, cf. Hans Spemann, "Walther Vogt zum Gedächtnis," *Wilhelm Rouxs Archiv für Entwicklungsmechanik der Organismen* 141 (1941): 1–13.

5. Like Vogt, Dorothy Rudnick (1907–1990) also deserves more attention; her life and scientific achievements are passim discussed in Edna Yost, *Women of Modern Science* (Westport, CT: Greenwood Press, 1959), and referred to in an article about Marcella O'Grady Boveri, whom Rudnick succeeded as professor of zoology at Albertus Magnus College in 1940 after obtaining her PhD under Ralph Lillie at the University of Chicago. The appointment also offered her the opportunity to do experimental research at the Osborn Zoological Laboratory of Yale University, cf. Margaret R. Wright, "Marcella O'Grady Boveri (1863–1950): Her Three Careers in Biology," *Isis* 88 (1997): 627–52, here 644, 646–47. For the textbooks, cf. Boris I. Balinsky, *An Introduction to Embryology* (London: W. B. Saunders, 1975); Aleksander L. Romanoff, *The Avian Embryo. Structural and Functional Development* (New York: MacMillan, 1960).

6. Lester G. Barth, *Embryology*, rev. and enlrg. ed. (New York: Holt, Rinehart and Winston, 1953), 110; and Johannes Holtfreter and Viktor Hamburger, "Amphibians," in *Analysis of Development*, ed. Benjamin H. Willier, Paul A. Weiss, and Viktor Hamburger (Philadelphia: W. B. Saunders, 1955), 230–96.

7. See Hans Driesch, *Die organischen Regulationen: Vorbereitungen zu einer Theorie des Lebens* (Leipzig: Engelmann, 1901). Also, Reinhard Mocek published a treatise about Hans Driesch's experiments and theory of biology up to 1899. See Reinhard Mocek, *Die werdende Form: Morphogenese, Evolution und Selbstorganisation in der zweiten Hälfte des 19. Jahrhunderts*, Acta Biohistorica, Bd. 3 (Marburg: Basilisken Press, 1998).

8. The classic account of *tacit knowledge* was written by Michael Polanyi, *Personal Knowledge: Towards a Critical Philosophy* (London: Routledge, 1958).

9. For the importance of the standardization of the experimental instrumentarium, see Kathryn M. Olesko, "Tacit Knowledge and School Formation," *Osiris* 8 (1993): 25.

10. Ibid., 27–28. Also, Harry Collins's discussion of the "Q of Sapphire," poignantly elaborated the epistemic value of *tacit knowledge* in laboratory practices by comparing Russian and Western measurements of gravitational radiation. See Harry M. Collins, "Tacit Knowledge, Trust and the Q of Sapphire," *Social Studies of Science* 31 (2001): 71–85.

11. Concerning the technique of drawing with the camera lucida and its "nonsubstitutability" for modern software systems being programmed for reconstructing developing embryos geometrically, cf. Erna Fiorentini, *Instrument des Urteils: Zeichnen mit der Camera Lucida als Komposit*, Preprint 295 (Berlin: Max-Planck-Institut für Wissenschaftsgeschichte, 2005).

12. The technique had several fathers who discovered it independently. To what extent "mothers" were involved gender studies of the life sciences has not yet researched exhaustively. For a short overview of the (female) history of tissue culture,

cf. Brigitte Hoppe, "Die Institutionalisierung der Zellforschung in Deutschland durch Rhoda Erdmann (1870–1935)," *Biologie Heute* 366 (1989): 2–9; Sabine Brauckmann, "Networks of tissue knowledge, 1910–1960," *Bulletin d'histoire et d'épistémologie des sciences de la vie* 13 (2006): 33–52. For Bridget Honor Fell, the head of the Strangeways Research Laboratories in Cambridge, cf. Janet Vaughan, "Honor Bridget Fell: 22 May 1900–22 April 1986. *Biographical Memoirs of Fellows of the Royal Society*, 33 (1987): 236–59.

13. Jonathan M. W. Slack, *From Egg to Embryo: Determinative Events in Early Development* (Cambridge: Cambridge University Press, 1983), 12. Slack's definition is still the most commonly used, cf. Andres Collazo, "Cell Determination and Differentiation," in *Keywords and Concepts in Evolutionary Developmental Biology*, ed. Brian K. Hall and Wendy M. Olson (Cambridge, MA: Harvard University Press, 2003), 30–35.

14. Ludwig Gräper, "Beobachtung von Wachstumsvorgängen an Reihenaufnahmen lebender Hühnerembryonen nebst Bemerkungen über vitale Färbung," *Archiv für Entwicklungsmechanik der Organismen* 33 (1912): 303–27. For his motion pictures, see Ludwig Gräper, "Die Methodik der stereokinematographischen Untersuchungen des lebenden vitalgefärbten Hühnerembryos," *Wilhelm Roux' Archiv für Entwicklungsmechanik der Organismen* 115 (1926): 523–41. For the technical description of the method of vital staining, cf. Hubert Dana Goodale, "The Early Development of *Spelerpes bilineatus*," *American Journal of Anatomy* 12 (1911): 173–247; Walther Vogt, "Ueber Zellbewegungen und Zelldegenerationen bei der Gastrulation von *Triton cristatus*, I. Untersuchung isolierter lebender Embryonalzellen," *Anatomische Hefte* 48 (1913): 1–64.

15. Benjamin H. Willier, "The Specificity of Sex, of Organization, and of Differentiation of Embryonic Chick Gonads as Shown by Grafting Experiments," *Journal of Experimental Zoology* 46 (1927): 409–65.

16. Walther Vogt, "Morphologische und physiologische Fragen der Primitiv-Entwicklung: Versuche zu ihrer Lösung mittels vitaler Farbmarkierung," *Sitzungsberichte der Gesellschaft für Morphologie und Physiologie (München)* 35 (1923): 30.

17. Walther Vogt, "Gestaltungsanalyse am Amphibienkeim mit örtlicher Vitalfärbung, II Teil. Gastrulation und Mesodermbildung bei Urodelen und Anuren," *Wilhelm Roux' Archiv für Entwicklungsmechanik der Organismen* 120 (1929): 391.

18. Walther Vogt, "Gestaltungsanalyse am Amphibienkeim mit örtlicher Vitalfärbung, Vorwort über Wege und Ziele, I Teil: Methodik und Wirkungsweise der örtlichen Vitalfärbung mit Agar als Farbträger." *Wilhelm Roux' Archiv für Entwicklungsmechanik der Organismen* 106 (1925): 542–610, see pages 578, 602.

19. Walther Vogt, "Mosaikcharakter und Regulation in der Frühentwicklung des Amphibieneies," *Sitzungsberichte der Gesellschaft für Morphologie und Physiologie (München)* 40 (1928): 28.

20. Vogt, "Gestaltungsanalyse am Amphibienkeim mit örtlicher Vitalfärbung, Vorwort über Wege und Ziele," 545–47, 605–7; for Spemann's and Hilde Pröschold's famous constriction experiment, cf. Hans Spemann and Hilde Mangold, "Über Induktion von Embryonalanlagen durch Implantation artfremder

Organismen," *Archiv für mikroskopische Anatomie und Entwicklungsmechanik* 100 (1924): 599–638.

21. Vogt, "Gestaltungsanalyse am Amphibienkeim mit örtlicher Vitalfärbung, Vorwort über Wege und Ziele ," 548–50.

22. In his article on Wilhelm His's embryo models of rubber and wax, Nick Hopwood discusses at length plate modeling in wax that was the preferred tool of anatomists around 1900. Nick Hopwood, "Giving Body to Embryos," *Isis* 90 (1999): 487.

23. Jessica A. Kaiser, Department of Biology of Swarthmore College, exemplifies this point quite well. See http://www.swarthmore.edu/NatSci/sgilber1/DB-lab/Frog/jess/tester.html

24. Viktor Hamburger, *A Manual of Experimental Embryology*, 2nd rev. ed. (Chicago: University of Chicago Press, 1960), 50.

25. Dorothy Rudnick, "Early History and Mechanics of the Chick Blastoderm: A Review," *Quarterly Review of Biology* 19 (1944): 187.

26. For an example of the many unsolved questions concerning the primitive streak, see L. Bodenstein and Claudio D. Stern, "Formation of the Chick Primitive Streak as Studied in Computer Simulations," *Journal of Theoretical Biology* 233 (2005): 253–69.

27. Ludwig Gräper, "Die Primitiventwicklung des Hühnchens nach stereokinematographischen Untersuchungen durch die vitale Farbmarkierung und verglichen mit der Entwicklung anderer Wirbeltiere," *Wilhelm Roux' Archiv für Entwicklungsmechanik der Organismen* 119 (1929): 382–429; Robert Wetzel, "Untersuchungen am Hühnchen: Die Entwicklung des Keims während der ersten beiden Bruttage," *Wilhelm Roux' Archiv für Entwicklungsmechanik der Organismen* 119 (1929): 188–321.

28. Wetzel, "Untersuchungen am Hühnchen," 260.

29. Ibid., 193.

30. Rudnick, "Early History," 191.

31. Jean Pasteels, "Analyse des mouvements morphogénétique de gastrulation chez les Oiseaux," *Bull. Proceedings de Societe Royale Belgique* (ser. 5) 22 (1937): 737–52.

32. For images on "gradient vs. polonaise," cf. Gräper, "Die Primitiventwicklung," table II, Wetzel, "Untersuchungen am Hühnchen," figs. 93–94, and Pasteels, "Analyse des mouvements," fig. 2; Rudnick, "Early History," 191–92; Jean Pasteels, "New Observations Concerning the Maps of Presumptive Areas of the Young Amphibian Gastrula (*Ambystoma* and *Discoglossus*)," *Journal of Experimental Zoology* 89 (1941): 255–81.

33. Hans Spemann had come to the University of Chicago in 1931 to visit Frank R. Lillie. Then he also showed the students, among them Mary Rawles and Dorothy Rudnick, how to make glass needles with fine points he had invented some years ago to make isolations on the amphibian material. Cf. Mary E. Rawles, "A Study in the Localization of Organ-Forming Areas in the Chick Blastoderm of the Head-Process Stage," *Journal of Experimental Zoology* 72 (1936): 271–315.

34. Elisabeth Butler, "The Developmental Capacity of Regions of the Unincubated Chick Blastoderm as Tested in Chorio-Allantoic Grafts," *Journal of Experimental Zoology* 70 (1937): 287–338; Mary E. Rawles, "The Pigment-Forming Potency of Early Chick Blastoderm," *Proceedings of the* U.S. National Academy of Sciences 26 (1940): 91.

35. Cf. Wetzel's figure 120 (Wetzel, "Untersuchungen am Hühnchen," 311) and compare to Rawles's figures 7–9 (Rawles, "A Study in the Localization of Organ-Forming Area," 308–9).

36. Emil Witschi, *Development of Vertebrates* (Philadelphia: Saunders, 1956), fig. 137b, 247.

37. Rudnick, "Early History," 206.

38. Ibid., 205.

39. Dorothy Rudnick, "Prospective Areas and Differentiation Potencies in the Chick Blastoderm," *Annals of the New York Academy of Sciences* 49, no. 5 (1948): 763–64.

40. Rudnick, "Prospective Areas," 771–72.

41. Karl Ernst von Baer, *Entwickelungsgeschichte der Thiere*, vol. 1 (Königsberg: Borntraeger, 1828), xix. For Baer's reflection on dimensionality and the difficulty of how to visualize a developing embryo on flat paper, see Sabine Brauckmann "Axes, Planes and Tubes, or the Geometry of Embryogenesis," *Studies in History and Philosophy of Biological and Biomedical Sciences* (forthcoming, 2011).

42. William M. Ivins, *Prints and Visual Communications* (Cambridge, MA: Harvard University Press, 1953).

chapter ten

Form and Function

A Semiotic Analysis of Figures in Biology Textbooks

Laura Perini

The explosive growth of the life sciences in the twentieth century poses significant pedagogical challenges for college-level education in biology. Life science majors must learn basic concepts from domains as diverse as biochemistry and biogeography. Majors are usually required to take one or more introductory-level courses that each covers a very broad domain, as preparation for more advanced courses, which focus on subfields or topics. Genuine understanding requires more than simply learning disparate concepts: students must grasp key connections among different topics, such as those between the structure of DNA and the theory of natural selection. The introductory courses thus play an important role in educating life science majors. Students must understand the overall landscape of biology in such a way that they are prepared both to study a variety of individual topics in depth, and to maintain (and ideally, strengthen) their grasp on how those different topics relate to one another. How, precisely, is all of this accomplished? How do students learn such diverse content? How do they achieve genuine understanding—in particular of the linkages between different topics they encounter in their introductory biology classes?

Textbooks are important educational tools in such courses. They are repositories of the bulk of the information presented. Their use outside of classroom settings gives individual students much greater control over the pace and sequence of information delivery. On the other hand, their content is static, in contrast to live instruction, which can be adjusted at any time to clarify, explain, and elaborate as needed in response to questions or blank stares. Textbooks also do not offer the social interactions and experiential feedback available in laboratory work. Unread books offer no advantages at all: in order to complement classroom instruction, textbooks must be engaging as well as comprehensible. As learning tools, both their greatest advantages and most serious liabilities stem from the fact that textbooks are individual student-managed resources.

These are the general advantages and constraints of textbooks. As noted, biology has undergone a dramatic increase in explanatory depth and breadth in the twentieth century. This growth in the discipline is matched by changes in college-level biology textbooks, indicating that textbooks are being used as one means of meeting the specific pedagogical challenges involved in teaching contemporary biology. Woodruff's 1926 publication, *Foundations of Biology*, has 411 pages in the body of the text.[1] Contemporary introductory textbooks are much larger: they are typically in the neighborhood of 1,100 pages long. But comparison in page numbers can give only a rough idea of the degree to which biology textbooks have changed, because the number of figures presented has increased far more than the number of pages. Woodruff contains 211 pictures—about one every other page. A midcentury text by G. G. Simpson et al. has many text-only pages.[2] Current editions of general biology textbooks designed for life science majors, on the other hand, are filled with images: a page without a figure is rare, and pages with multiple images are common.[3] The dramatic increase in the relative amount of pictorial content suggests that images are being used in response to the need to present information in a way that addresses the goals of an introductory course—which require students to comprehend and integrate many diverse concepts.

Could images serve such pedagogical goals in the life sciences? Do images really contribute to the cognitive goals of a biology course—or does their value stem only from their aesthetic appeal, serving to attract and hold student attention rather than convey essential content? To address this question, I will investigate three textbooks, which are all designed for introductory courses for life science majors, and together comprise the majority of the market share of college-level major's textbooks.[4] A quick look at a contemporary general textbook may arouse some suspicion, because some of the images do function primarily to draw student interest. Examples include pictures of famous scientists and (often beautiful) photographs of whole organisms when their visual appearance is irrelevant to the point under discussion. But the fact that *some* figures play such a limited role does not imply that all do. Can figures make a substantive pedagogical contribution? If so, just what do they have to offer? Are they an effective learning tool for particular concepts? Can they do even more—can they help students understand connections among different concepts?

These questions cannot be given general answers, because there is significant variety *among* the figures in biology textbooks. The growing literature on scientific images presents reasons to doubt that some figures—detailed pictures, like photographs or electron micrographs—can play such roles effectively. On the other hand, diagrammatic representations have been shown to convey theoretical content that is clearly cognitively significant, and so may have pedagogical value as well.[5] In this paper I will present an analysis of these different kinds of figures in order to clarify their potential pedagogical value. I will show that in spite of the genuine obstacles that have been identified with detailed pictures, they can indeed make important pedagogical contributions. This opens up the question of why both highly detailed as well as visually abstract images are presented in textbooks, but further analysis of diagrams provides an answer: different kinds of diagrams provide different sorts of pedagogical advantages.

Visual Representation

Textbook figures include drawings, diagrams, and images produced by various kinds of detection processes, like photography and electron microscopy. There are significant differences in the way these images look, how they are made, and the kinds of content they convey. In order to assess their pedagogical contributions, we need to understand what they have in common as visual representations, and how they differ.

In his analysis of images in a biology textbook, Myers approaches this diversity by applying Peirce's tripartite division of signs.[6] Indexical signs have a direct link to the thing referred to, so the form of the image is caused by features of the referent. Iconic signs have forms, which resemble their referents. Symbolic signs refer in virtue of convention; there is an arbitrary relation between their form and their content. While these factors all come into play in scientific visual representations, they cannot be used to sort images into distinct classes, because all visual representations involve both convention and resemblance.[7] Myers himself stresses the blurring of the categories, explaining that his usual motive for applying them is to "make readers more critical of the indexical and iconic end of the scale" by calling into question assumptions about the naturalness of such images;[8] in this paper, Myers's main concern is with highly conventional images and the pedagogical challenges they involve.

Here I offer an alternative semiotics as a means for analyzing the roles of images in biology textbooks. The problem with Peirce's conceptual apparatus is not that the categories are vague, but that they are too general to clarify important similarities and differences among images. It is not the distinction between whether or not an image is conventional or resembles its referent that explains the use of different kinds of images, but the different *kinds* of conventions and resemblance relations involved. Those are significant factors, because they jointly determine the relationship between the form of an image and its content. Differences in form-content relations provide the means to categorize images.[9] Furthermore, they ground explanations for why different kinds of images are suited for different kinds of communicative tasks. For this reason, I'll present my analysis in terms of a semiotic approach that is designed to clarify the relation between form and content.

All visual representations have one thing in common: they use spatial properties of the picture to convey information.[10] This can be done in a variety of ways; protein diagrams use two-dimensional spatial properties of the diagram to represent three-dimensional spatial features of proteins. Graphs use spatial relations to represent relations among properties. Other visible properties may also be used to convey information. For example, color photos use their spatial layout and the colors of the image to represent visible properties of the scene depicted. The use of spatial relations to convey content is, however, the defining feature of visual representations.

Comprehension of any visual representation requires interpretation on the part of the viewer: only by understanding the relation between the form of the image and its content can a person comprehend the image. This understanding is often tacit. We comprehend familiar kinds of pictures so readily that we are rarely aware that it is a cognitive achievement to understand a flat, marked surface as a representation of anything. Our ability to understand various kinds of pictures demonstrates that we can work with many different kinds of form-content relationships: we look at, and/or focus on, different visible features of the image and relate those features to different kinds of properties. Comprehending a black and white photograph requires relating the visible properties of the photo to visible properties of visible objects. But not all visible features of the photo convey information: the fact that brightness ranges on a white-to-black scale rather than through light-to-dark sepia tones is not relevant. Additionally, the visible properties of the image that are interpreted convey information about some, but not all, properties of the referent; specifically, the tones in the photograph are not understood as representing colors of depicted objects. Understanding a color photo depends on interpreting the image according to a different relation between form and content from that appropriate for black and white photographs. Comprehending a color photograph involves relating a different set of visible properties—now including the specific colors in the photo—to a different set of properties of depicted objects, including their colors. These differences in the form-content relations relevant to different kinds of images underwrite differences in representational capacity. Black and white photos can represent visible properties like relative brightness and spatial features; color photographs convey information about those features and color as well. In both cases, the visible form of the image is related to its content, due to the defining feature of visual representations—the fact that some spatial features are interpreted to represent some feature of the referent. Summing up, interpretation plays an essential role in determining the meaning of every picture, and the kind of interpretation involved varies considerably because the relations between visible form and content vary.

Because all visual representations involve a relation between form and content, they all (in a broad sense) resemble their referents, but those resemblance relations vary; they are not limited to relations of visual resemblance. The conventional aspect of visual representations—the fact that communication with visual representations depends on application of shared interpretive practices, and not just on the visible features of the image—grounds the fact that different visible features in images can be used to represent different kinds of properties. This in turn explains the broad expressive capacity of images, which can be used to represent many different kinds of things. Scientists have exploited this capacity: scientific images are frequently used to represent phenomena that are not visible at all. In such cases, the visible features of the image are interpreted as conveying information about *nonvisible* features of the referent. Some images represent phenomena that are simply too small to be seen, like the helical structure of DNA. Other images represent phenomena that are not even spatial, such as diagrams of mechanisms or graphs of relations between properties. For understanding the value of a particular image, clarifying the relationship between its form and its content sets the stage for articulating what makes the content conveyed by that type of image distinctive, and for explaining why the mode of representation matters.

Replete Pictorial Representations

Some visual representations stand out due to their detailed appearance. Such figures include "naturalistic" drawings or paintings, as well as photographs and other images produced through detection mechanisms, like electron microscopy. Despite the differences within this class of figures, their form-content relationships are all similar in one significant way: Most visible details of the picture are interpreted as conveying information about specific properties of the depicted scene or subject matter. I'll refer to this broad type of visual representation as "pictorial," since it includes familiar types of pictures such as photographs, drawings, and paintings done in naturalistic style. The examples just mentioned involve a form-content relation that correlates specific visible details in the image with specific detailed properties; a particular contour line used to represent a particular shape, for example, or a particular hue used to represent a particular color in the scene. There are two different ways in which detail is important in these images. First, *specific* visible features of the picture convey information about specific properties of the referent. This is the feature that all pictorial representations have in common. Second, *most* of the visible features of the image are interpreted in this way—specific visible features are used to represent specific properties. The fact that most visible features convey meaning in this way is a distinct feature, that of relative repleteness.[11] Visual representations with such a form-content relationship convey large amounts of very specific information.

The use of richly detailed visible forms to convey correspondingly detailed information is common in introductory biology. Textbooks frequently include photographs of medium-sized objects like plants and animals, and photographs made with light microscopes. The form-content relation that defines pictorial representations is also produced by imaging techniques that detect nonvisible features. For example, electron micrographs represent the form of the biological material in a particular specimen through the array of light and dark tones in the figure (figure 10.1, left panel). Images made as a result of mechanisms like electron microscopy are often presented as evidence in research publications. The pictorial nature of the visual representation offers a concise yet comprehensible way to convey information about very complex properties. Also, the visible form of the figure is produced by a mechanism designed to correlate the form of the image with properties of the sample.[12]

How do such images contribute to *learning* biology? These images have

Form and Function · 241

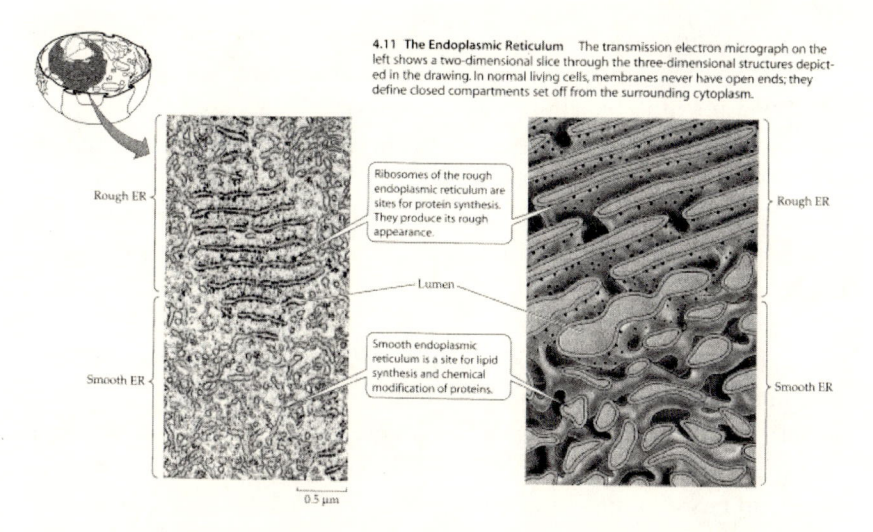

4.11 The Endoplasmic Reticulum The transmission electron micrograph on the left shows a two-dimensional slice through the three-dimensional structures depicted in the drawing. In normal living cells, membranes never have open ends; they define closed compartments set off from the surrounding cytoplasm.

Rough ER

Ribosomes of the rough endoplasmic reticulum are sites for protein synthesis. They produce its rough appearance.

Rough ER

Lumen

Smooth endoplasmic reticulum is a site for lipid synthesis and chemical modification of proteins.

Smooth ER

Smooth ER

0.5 μm

10.1 Electron micrograph (left panel) paired with a schematic diagram (right panel), with text bubbles and pointers, and linked to a schematic diagram of a cell interior (upper left) with an arrow. From *Life: The Science of Biology*, 7th edition, edited by Purves et al. Reproduced with permission from Sinauer Associations and Visuals Unlimited. Photographs copyright D. Fawcett/Visuals Unlimited.

a capacity for detailed representations of biological forms. While this might seem like a representational asset, the literature on scientific images has clarified two issues that present reasons to question their pedagogical usefulness.

First, the large amount of detailed information conveyed about the subject can impede learning, which usually depends on awareness of a particular part of an image, such as the facial expression on a particular chimp in a photograph that depicts that individual amid a group of conspecifics, in a natural setting: Myers identifies this as the problem of "gratuitous detail."[13] Replete pictorial representations fall into this category; photographs, for example, are very visually complex images. If learning requires understanding *which* visible features are significant in terms of the subject at hand, how do students identify those features out of all the detail a photograph presents? Law and Lynch analyze the use of different kinds of images in guidebooks for birdwatchers and find books with more replete images (photographs) less useful for species identification, which requires focusing on a few visible traits that matter for determining which of two similar species has been sighted.[14] The "extra" information about the bird's appearance presented by the more detailed pictorial representations was not helpful because the photos give the reader no guidance about which features, among all those depicted, matter for determining species membership. Less replete images, which have less information about how the birds look but which put the visual emphasis on a few traits that matter for identification, were more helpful in allowing birdwatchers to categorize the birds they saw. Lynch describes the problem with photographs as one of "too much reality," which can cause trouble even for researchers—experts—and thus reinforces the worries about using such images in pedagogical contexts.[15]

A second problem regarding pictorial representations is raised by Daston and Galison's research on the history of objectivity.[16] They look at atlas images, which are intended to provide information about classes of objects. They discuss changes in images over time, focusing on how the human skeleton was depicted. Their examples show that there was little change in terms of the drawing techniques; the skeletons are depicted in naturalistic style. These are pictorial representations: visible detail in the image is used to represent detailed features of the depicted individual. The visible differences among these pictures are due to different choices of *which* skeleton to depict, rather than to differences in *how* to depict a particular skeleton. The use of a type of representation in which visual details convey information about specific properties poses a problem for representing classes whose individual members vary in terms of those specific properties. The atlas authors must choose *which* individual should be depicted in order to best represent the class of individuals. Daston and Galison demonstrate that different atlas makers have embraced different views about which individual is the appropriate representative for the class, such as a typical individual, one with averaged properties, or an ideal.[17] Photographs, like the atlas engravings, are also relatively replete pictorial images: They depict individuals with a particular set of specific properties. The pedagogical value of a photograph requires that students do more than simply comprehend the picture; students must also grasp the relation between the information about that individual and a wider biological category, which includes cases that are similar to that depicted by the photograph, but not identical. The problem of using naturalistic images to represent a biological class is caused by the use of images that represent detailed visible features of referents. While the atlas authors aimed to resolve the problem of instructing about a class by choosing the right individual to depict in detail, contemporary biology textbooks use alternative tactics.

The pedagogical drawbacks of pictorial representations can be miti-

Form and Function · 243

gated. The problem generated by using a pictorial representation of an individual to represent a class, for example, is often resolved by pairing the pictorial representation with a different kind of visual representation. Since the resolution depends on the distinctive form-content relations of a particular type of diagram, I'll present my support for this claim in the discussion of schematic images.

There are also ways to get around the problem of gratuitous detail. Myers notes that it can be resolved textually. Figure legends are often used to direct attention to significant features, such as a reference to a "play face" expression in the caption of a picture of a group of chimpanzees.[18] Bastide shows how grouping images in clusters can help a reader focus on a significant detail in one that differs from the others.[19] As figure 10.1 shows, the use of lines and arrows superimposed on the image also directs attention to a particular part of an image, and placement of a textual label at the other end of the line indicates a relation between that part of the picture and a linguistically expressed concept.

The pedagogical problems involved with replete pictorial representations raise the question of whether there is any pedagogical value to these images. However, authors and editors are choosing to mitigate the drawbacks of these images by combining them with text, pointers, and diagrams, rather than eliminating them. This suggests that detailed pictorial representations have some distinctive advantages over other forms of representation such as text and diagrams.

What kind of advantages do detailed images offer? Myers claims that the detail is not informative, but that it does convey the impression that the picture provides immediate contact with reality.[20] If that is all the detail provides, then it might be useful in generating student interest by establishing a sense of personal connection to the depicted subject. If so, then pictorial representations would make a pedagogical contribution through their effect on student motivation, rather than making a cognitive contribution to learning biology. While on this view, the detail is not entirely gratuitous, there is reason to think that it plays a more substantive role in learning biology.

The value of the detail involved in pictorial representation lies in relating detailed information about biological individuals to important conceptual themes. For example, viewing the electron micrograph in relation to a diagram has the potential to do more than convey a sense of immediacy: these images are a key source of evidence for cell structure, and understanding the relation between two different kinds of images involves learning not just to relate the form of an electron micrograph to concepts like "mitochondrion" and other forms of representation of cell structure, but to perform the visual abstraction from the micrograph that is a key to understanding the micrograph *as* evidence for the structural claim.

Lynch's discussion of figures that pair an electron micrograph with a diagram suggests how this works. The diagram has a relatively simple visible form, compared to the electron micrograph. The pairing between a pictorial representation and a diagram with a similar form helps the reader identify which visible features of the micrograph are important through a visual comparison of similar parts of the two figures.[21] This type of comparison can be further facilitated by lines that connect areas in the micrograph with those in the diagram (figure 10.1). The pairing not only makes the pictorial representation more comprehensible, it also relates the content of the diagram to an image that is the result of a mechanical detection process. Understanding the connection between the theoretical content of the diagram and the evidence for those claims as presented in the electron micrograph is essential for understanding biology as a science. Presenting detailed pictorial representations allows students to understand that connection through learning how to make the perceptual links between detailed pictures and abstract diagrams. This is a significant perceptual and cognitive achievement. In a prior paper, Lynch shows that the diagram is not merely a simplified version of the micrograph; relative to the pictorial representation, corresponding parts of the diagram are altered in different ways.[22] These include making some parts of the diagram look more similar to each other than do corresponding areas in the micrograph, and increasing the contrast between other parts of the diagram (relative to the corresponding areas in the micrograph.)[23] The use of detailed pictorial representations thus provides an important resource through which students learn to "see for themselves" how evidence relates to theory.

Another example of how highly detailed images can play a substantive pedagogical role is in the use of replete pictorial representations—especially photographs—to teach students about the diversity of living systems, often a key theme in an introductory biology course. While the details of photos can impede the recognition of salient details in a single photograph, clustering multiple photographs offers an easy way to communicate about differences. Photographs' capacity for detail can be exploited; the overall morphological differences are reinforced by the differences in color, textures, and so forth. So, for example, a section on plants might include a clustered figure of several photographs, chosen for distinctive differences in form.[24] The student can enjoy the aesthetic appeal of the image and soak in the details while being in no danger of missing the main point about how these different organisms relate to one another: they share an ancestor but have significant differences in their traits.

Schematic Diagrams

Photographs, electron micrographs, and naturalistic drawings all involve different relations between their visible forms and their contents, but all share the general characteristic that detailed visible features convey information about detailed properties of the subject of the image. Diagrammatic representations, on the other hand, involve significant differences in how their forms relate to their contents. Diagrams all share a low relative repleteness. That is, relative to the examples discussed in the previous section, few visible features of the diagram are interpreted as representing features of the subject matter. Diagrams can be sorted into significantly different types, however, and it is the more specific representational features of each type that explain its potential to play a pedagogical role. Two types of diagrams are especially common in biology textbooks.[25]

The most common type of diagram in general biology textbooks is characterized by the fact that generic visible features, rather than exact visible details, are interpreted as conveying information, and the information they convey pertains to generic, rather than specific, properties. For example, in the diagram in the right panel of figure 10.1 the visible features that convey information about the structure of the endoplasmic reticulum include lines, shading, and black dots. But it does not represent the endoplasmic reticulum as having ribosomes in exact numbers or in locations corresponding to the locations of the black dots in the diagram; instead, the figure represents generic structural features of the organelle, including representing it as having some number of ribosomes attached. Similarly, it does not represent all endoplasmic reticula as having the specific shape that corresponds to that of the curved lines in the diagram. It is not the exact shape of the curves that conveys information, but more generic properties of the image that convey information, like the curved, contiguous nature of the boundary. Those in turn are interpreted to refer to correspondingly more generic properties, so that this line represents the endoplasmic reticulum as bounded by a continuous membrane. Such diagrams have the capacity to represent features that are shared among many individuals, even though those individuals differ in the ways those features are instantiated.

Above I claimed that schematic diagrams could resolve the pedagogical problems inherent in representing biological structure with pictorial representations. Schematic diagrams are effective in this role for two reasons: first, they are less replete, so there is a reduction in detail and corresponding focus of attention on significant content. This solves the problem of gratuitous detail.

Second, schematic diagrams offer more than mere reduction of detail: they convey a different kind of content than pictorial representations. The nature of their form-content relations provides a way to represent biological classes in cases when the individual members of those classes vary. Images whose form-content relationship involves relating relatively generic visible features to relatively generic properties are especially well suited to representing biological features that vary in how they are instantiated. This offers a solution to the problem raised by Daston and Galison, in which an individual is depicted in detail, which then represents a class whose members vary in those specific properties. Instead, schematic diagrams like this offer a way to communicate about the shared features of a class of objects, even when the individuals of that class vary in how they instantiate those shared features. For this reason, they are very effective means to express generalizations about biological structure.

Schematic diagrams are also very effective means for representing the components of biological systems. This is important explanatory content in biology. One of the key aims of introductory courses is to generate an understanding of biological systems in terms of their material composition, and that is explained in large part by identifying the significant parts of a biological system at a particular level of organization. Diagrams relating components at one level of organization to the next are ubiquitous: they are used to represent organelles as the key components of cells, to show that tissues are composed of cells, how organs relate to physiological systems, and so on. Diagrammatic representation involves an important limitation when it comes to communicating about this key theme. While they are very effective means of representing the components of one level of organization, they are not effective means of communicating about relations among multiple levels of organization, due to limits in space and human visual acuity. While it's easy to make out the component parts of a cell diagram, representation of the component parts of the organelles would make their overall structure less visually prominent. Not only would the structure of, say, a mitochondrion be more difficult to pick out, but in addition, the details of mitochondrial components—the structure

Form and Function · 247

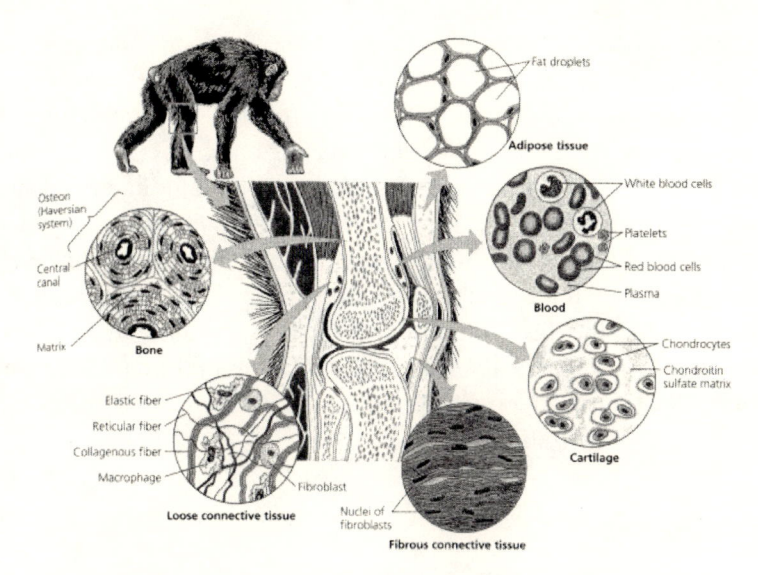

10.2. Linked schematic diagram. Fig. 40.2, p. 837, *Biology* 6th ed. by Neil A. Campbell and Jane B. Reece. Copyright © by Pearson Education, Inc. Used by Permission.

of the ATP synthase, for example—would be difficult to see. The size and scaling of the diagram in how it represents the highest level of biological organization imposes limits on the lowest level that can be included and still be discriminated by human visual perception.

Textbook images rarely push these limits. Instead, textbooks mitigate this limitation in the pedagogical usefulness of schematic diagrams by visually linking multiple diagrams, each of which represents only two levels of organization. A typical kind of figure has a telescoping structure, with a chain of diagrams linking one level of organization to another, which collectively relate multiple levels of organization, from organelle to ecosystem for example.[26] The solution has its own limitations: a telescoping diagram can only trace out a trajectory from one type of cell through one type of organ system, and so on. This technique cannot provide global information about how all the different lower-level structures relate to any one higher-level feature, nor can it show one lower-level structure, like a generic cell, in relation to multiple different higher-level structures. It can, however, relate one kind of biological entity, such as an individual organism, to its component parts, at successively lower levels of organization.

A similar technique can be used to show the structures of several different things at one level of organization, using arrows to indicate that they are components of a wider system. Consider figure 10.2. The arrows in such diagrams are completely arbitrary: unlike the individual structural diagrams at the periphery, understanding the arrows does not involve relating the shape of the arrow to the shape of a biological structure. Instead, the arrow functions as a label, indicating an abstract concept: that one diagram is related to another. Comprehension of this figure depends on understanding the difference in interpretation applied to the arrow compared to the different structure diagrams, including the difference in scaling between the central and peripheral diagrams. Setting aside the question of how students know how to interpret such a diagram, it offers a concise way to summarize the many different kinds of structural relations at one level of organization—that of tissues—and to relate them all to a higher level of organization.

Compositional Diagrams

There is another form of diagram that appears in all contemporary introductory biology textbooks. Compositional diagrams use a more precise type of relationship between diagrammatic form and content than that which characterizes schematic diagrams. Familiar examples of compositional diagrams include chemical diagrams, electrical circuit diagrams, and some diagrams of biological models, such as diagrams of the Krebs cycle. These are less common in textbooks than schematic diagrams, but compositional diagrams are well suited to convey a different kind of content from schematic diagrams. Figure 10.3 is a typical example. Note that at first glance, it doesn't *appear* to be a different kind of diagram from a schematic diagram. However, the difference is not a matter of the visual appearance of the two kinds of figures, but the ways their visible features are related to their referents. Compositional diagrams are composed of discrete visible elements—atomic characters, like arrows, lines, and other shapes—which are used to refer to things in the same way that names refer: they function as labels. Atomic characters are assigned referents by stipulation, and often have no resemblance to their referents at all. Note the shapes used to represent DNA bases in figure 10.3: those shapes have no similarity relation to the shapes of the bases. The important thing about compositional diagrams is that the spatial arrangement of atomic characters in space is significant: spatial relations among the atomic characters are used to represent relations among the things the atomic characters refer to. In figure 10.3, contiguity between the C, A, T, G shapes on the one hand, and one of the ribbons, on the other, is used to represent co-

Form and Function · 249

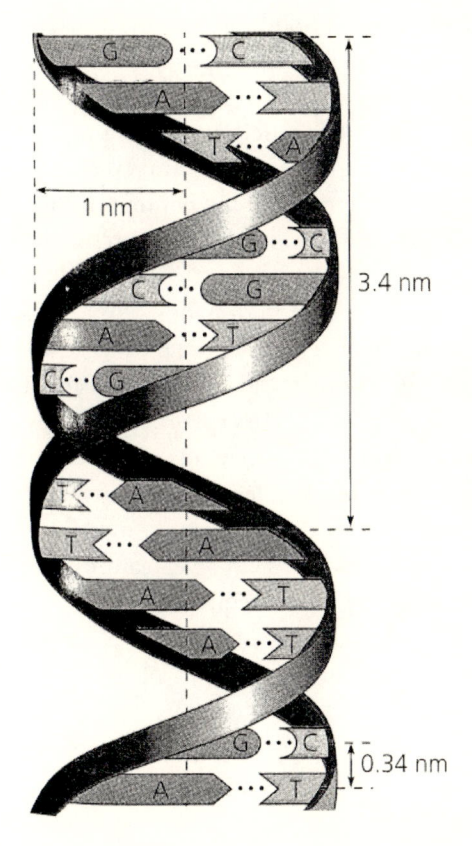

10.3 Compositional diagram. Campbell and Reese, 291. Figure 16.5a. Caption: "The double helix. (a) Key features of the double helix. The 'ribbons' in this diagram represent the sugar-phosphate backbones of the two DNA strands. The Helix is 'right-handed,' curving up to the right. The two strands are held together by hydrogen bonds (dotted lines) between the nitrogenous bases, which are paired in the interior of the double helix." From Campbell, Neil A.; Reece, Jane B.; *Biology*, 6th ed., (c) 2002, p. 291. Reprinted by permission of Pearson Education, Inc., Upper Saddle River, New Jersey.

valent bonding of a specific base to the sugar-phosphate chains. Three dots between a C and a G, or a T and A pair, are used to represent hydrogen bonding between bases. Because the length of the "ribbons" is used to refer to the length of the sugar-phosphate chains, the vertical placement of the C/G and T/A shapes along those lengths represents the base pairs as stacked internally between the sugar-phosphate chains.

This feature is quite different from schematic diagrams, which do not use spatial relations in the figure to represent properties of the subject matter in the same precise, systematic way. Recall that they do not represent precise details of system components; a schematic diagram of a cell interior does not represent precise numbers and locations of mitochondria, for example. Rather, mitochondria are represented as inside the cytoplasm—a more generic feature—rather than at specific distances from the cell's nucleus: the spatial relation between the area representing the mitochondrion and the area representing the nucleus is not interpreted as representing the distance between a mitochondrion and the nucleus inside every cell. It is possible to represent details of how organelles might be packed together. To appreciate both the information that is left out in schematic diagrams and the visual difference made by representing the spatial relations among cell components, see David Goodsell's representations of cell interiors.[27] While experts can appreciate the content of such images, the visual crowding takes the visual focus off of the features of the component parts and distinctions among the system parts.

The use of spatial relations to convey specific information about relations among the referents of the atomic characters, however, is what makes compositional diagrams especially useful for conveying explanatory content in biology. Often properties of a biological system are explained by (a) the components of the system and (b) how those components interrelate.[28] Compositional diagrams use atomic characters to refer to the components of the system, and then spatial relations among those characters model the relations among system components that explain system-level features, like why a macromolecule has a particular shape, or the sequence of reactions that relate the different compounds involved in the Krebs cycle.[29] Because they represent very few details about the component parts of biological systems, the focus is on how system components relate to one another. In figure 10.3, for example, there is no information about the structure of the sugar-phosphate chain.

In compositional diagrams, the visible forms of the atomic characters can be chosen on pragmatic grounds, for visual clarity of relations *among* the atomic characters, easy differentiation between different atomic characters, and easy association of each atomic character with its referent. These choices can be made completely independently of how spatial relations among the atomic characters are interpreted, leaving figure designers with a great deal of flexibility in whether there will be any kind of perceived resemblance between atomic characters or not. As a result, compositional diagrams can be used to represent very abstract content because there is no need for the visible features of the atomic characters to resemble any property of their referent.[30] In addition, because the forms of the atomic characters need not represent any of the properties of the system components, compositional diagrams offer a representational format that emphasizes relations among the components of a biological system—in contrast to schematic diagrams, which are most effective at representing characteristics of the components of a biological system. Schematic diagrams also represent generic part/whole relations, but they are most useful for communicating about generic properties of system components and convey little specific information about how those components relate to one another within a biological system.

The capacity to represent explanatory relations among system components, as well as their capacity to convey abstract content, also explains why textbooks frequently use compositional diagrams to convey information about biological processes. For some diagrams, spatial relations refer to nonspatial features, such as the biochemical transitions between two particular states involved in a biological process. For example, diagrams of the Krebs (or citric acid) cycle involve representations of individual carbon compounds connected in a circular form by arrows. Such variation in the *type* of form-content relations involved in a single figure can result in significant confusion for students. While the diagrams of the carbon compounds *do* use spatial features to represent spatial relations (the lines represent bonds between atoms), the circular pattern does not represent a shape or change in location, but rather a sequence of transitions. This has been demonstrated to cause confusion among undergraduates, who interpreted the circular form to indicate that the reactions occur in a circular area.[31] Nevertheless, such a diagram is standard; in spite of its potential for confusion, it highlights the important relations among the component parts of the cycle: the sequential relations between different carbon compounds, and the enzymes that catalyze the successive transitions.

Conclusion

The study investigated three different kinds of figures in biology textbooks, and analysis of their form-content relations has clarified and explained the pedagogical advantages and limitations of using each of these types of figures in introductory textbooks.

The figures discussed in this paper show that images can provide very effective ways to communicate about biological concepts.[32] In spite of their pedagogical liabilities, replete pictorial representations are common, and we now know that images like photographs and electron micrographs do more than merely convey information about the individual depicted. They play an important function in fostering the inferential move from a detailed representation of the properties of a particular individual—which include many specific properties that will vary from individual to individual—to conclusions about the higher-order structural features that are shared. Schematic diagrams were shown to be extremely effective tools for representing relations between two successive levels of structural organization, and they have the capacity to do so without representing fine-grained details of structure. For this reason, schematic diagrams can communicate the generic features that hold for all individuals at a level without having to choose to represent a class via depiction of a particular individual, with detailed properties that aren't shared. Compositional diagrams use spatial relations to represent functionally explanatory relations among parts of a biological system (figure 10.3). They are excellent tools for representing the relations *among* components at one level of organization that account for structural features or biological processes. In short, images offer significant cognitive value to life science students. The proliferation of images in textbooks is not just a matter of enticing students to engage with the text; rather, the images offer important pedagogical advantages.

Notes

1. Lorande Loss Woodruff, *Foundations of Biology*, 2nd ed. (New York: Macmillan, 1926).

2. George Gaylord Simpson, Colin S. Pittendrigh, and Lewis H. Tiffany, *Life: An Introduction to Biology* (New York: Harcourt, Brace 1957).

3. This study presents an investigation of the visual representations in the textbook itself; all the textbooks surveyed in this study offer students access to a Web page in which they can view animations, and participate in online activities.

4. Neil A. Campbell and Jane B. Reece, *Biology* 6th ed. (San Francisco: Benjamin Cummings, 2002); William K. Purves, David Sadava, Gordon H. Orians, and H. Craig Heller, *Life: The Science of Biology*, 7th ed. (Sunderland, MA: Sinauer / W.H. Freeman, 2004); Peter H. Raven and George B. Johnson, *Biology*, 6th ed. (Boston: McGraw-Hill 2002). There are changes in image styles over different editions, including a trend among current editions toward more integration of text and figures. I chose to focus on books published from 2002 to 2004 in order to focus on a well-defined, yet recent, time period.

5. Michael Lynch, "The Externalized Retina: Selection and Mathematization in the Visual Documentation of Objects in the Life Sciences," *Human Studies* 11 (1988): 210; Laura Perini, "Explanation in Two Dimensions: Diagrams and Biological Explanation," *Biology and Philosophy* 20 (2005): 265–67.

6. Greg Myers, "Words and Pictures in a Biology Textbook," *The Journal of TESOL France* 2 (1995): 118–20.

7. Laura Perini, "Convention, Resemblance and Isomorphism: Understanding Scientific Visual Representations," in *Multidisciplinary Approaches to Visual Representations and Interpretations*, ed. Grant Malcolm (Amsterdam: Elsevier 2004): 39–42.

8. Myers, "Words and Pictures," 120.

9. Perini, "Convention, Resemblance and Isomorphism," 38–39, 46.

10. Ibid. 43.

11. These are relatively replete pictorial representations; most visible features matter for the identity of the picture. Some images involve a detailed match between a limited subset of visible features, on the one hand, and properties of the referent, on the other, and are less replete. In line graphs, for example, the exact curve of the line represents exact features of the relation the line represents, but line width and color are not informative—those visible details are not meaningful. See Laura Perini, "Diagrams in Biology," *Knowledge Engineering Review* (forthcoming).

12. Laura Perini, "Visual Representations and Confirmation," *Philosophy of Science* 72 (2005): 920–21.

13. Greg Myers, "Every Picture Tells a Story: Illustrations in E. O. Wilson's Sociobiology," *Human Studies* 11 (1988): 240–41.

14. John Law and Michael Lynch, "Lists, Field Guides, and the Descriptive Organization of Seeing: Birdwatching as an Exemplary Observational Activity," *Human Studies* 11 (1988): 286–87.

15. Michael Lynch, "Science in the Age of Mechanical Reproduction: Moral and Epistemic Relations between Diagrams and Photographs," *Biology and Philosophy* 6 (1991): 214.

16. Lorraine Daston and Peter Galison, "The Image of Objectivity," *Representations* 40 (1992): 87–96.

17. Ibid. Because those images were handmade, the option to depict an individual that never existed—like an idealized human skeleton—is open. This is not an option for images produced through mechanized imaging techniques (assuming no subsequent manipulation of the image).

18. Myers, "Every Picture Tells a Story," 254.

19. Françoise Bastide, "Iconography of Scientific Images: Principles of Analysis," trans. Greg Myers, in *Representation in Scientific Practice*, ed. Michael Lynch and Steve Woolgar (Cambridge, MA: MIT Press 1990): 196–97.

20. Myers, "Every Picture Tells a Story," 242. This impression is not due to the fact that photos are produced by mechanized processes; Myers notes that naturalistic drawings, due to their "detail and particularity," also have this function.

21. Lynch, "Science in the Age of Mechanical Reproduction," 217.

22. Lynch, "The Externalized Retina," 209.

23. Lynch, "Science in the Age of Mechanical Reproduction," 218. He claims that the diagram performs "gestalt functions," such as those grounding the shift involved from seeing a drawing as a duck to seeing it as a rabbit. I think that this overstates the perceptual shift involved, because the diagram seems to aid more in focusing, allowing for visual abstraction from the micrograph rather than a gestalt perceptual shift. This is a concern because Lynch's formulation may obscure the pedagogical gain involved: If students learn to extrapolate—how to perform their own visual abstraction on new pictorial representations—then they have acquired a cognitively significant perceptual skill.

24. See, for example, Campbell and Reece, *Biology*, 603, figure 30.7.

25. Some pictorial visual representations are non-replete, and thus qualify as diagrams (as I've characterized diagrams.) These images are pictorial due to the very precise correlation of detailed visible features of the figure with specific properties of the referent, but they are also diagrammatic in virtue of relative non-repleteness. Common examples include line graphs in which the exact position of the line is used to represent an exact relationship between two properties, and topographical maps. In diagrammatic forms of pictorial representations, only a small number of visible properties are used to convey information: for pictorial line graphs, only line position, and not width or color, conveys information about the property. This form of representation is used in introductory biology textbooks, but is much less common than the other kinds of diagrams, so I will not provide a detailed discussion of this type of diagram. See Perini, "Diagrams in Biology," for more on diagrams in biology.

26. See for example Purves et al., *Life*, figure 47.7.

27. David Goodsell, Scripps Research Institute, "Molecules in Living Cells," http://mgl.scripps.edu/people/goodsell/illustration/cell (accessed October 24, 2007).

28. William Bechtel and Robert C. Richardson, "Emergent Phenomena and Complex Systems," in *Emergence or Reduction? Essays on the Prospects of Non-reductive Physicalism*, ed. A. Beckermann, H. Flohr, and J. Kim (Berlin: Walter de Gruyter Verlag, 1992): 266–78.

29. Perini, "Explanation in Two Dimensions," 266.

30. The arrows connecting schematic diagrams in figure 10.3 function in this way.

31. T. L. Hull, "Students' Use of Diagrams for the Visualization of Biochemical Processes" (M.Sc. thesis, University of KwaZulu-Natal, South Africa, 2003), which is cited along with other references to works documenting student difficulties with interpreting diagrams in Konrad Schönborn and Trevor Anderson, "The Importance of Visual Literacy in the Education of Biochemists," *Biochemistry and Molecular Biology Education* 34 (2006): 97–98.

32. Indeed, David Sadava, one of the authors of *Life* believes that figures are the *only* effective means of communicating about many of the important concepts in biology (personal communication).

chapter eleven

Neuroimages, Pedagogy, and Society

Adina L. Roskies

Introduction

Many lament that the United States is losing its edge in science and technology. A 2007 study reported that 52 percent of Americans believe that the United States is not performing well in math and science relative to other countries, and 64 percent think that the average American is not scientifically well informed.[1] These popular views are bolstered by data from recent studies. The National Science Board's 2006 Science and Engineering Indicators report, which was based on science and mathematics literacy tests administered to high school seniors in 29 developed countries, showed that even the best U.S. students perform near the bottom internationally.[2] There is a general fear that if current trends continue, America will lose its place on the world stage as a scientific and economic powerhouse.[3]

It is not clear to what American decline in performance should be attributed. The American public has not lost interest in science. On the contrary, most Americans claim that science is interesting and important and maintain that it is important that we remain a world leader in scientific development and research.[4] Some of the decline in performance may be attributable to changes in interest driven by cultural shifts, including greater emphasis on material wealth and the cult of celebrity, but most explanations instead focus on the failure of our educational system to adequately prepare students to engage with scientific and technological discourse.[5] The National Science Board's study concludes, "We know— and this report demonstrates—that there is a need to make drastic changes within the Nation's science and mathematics classrooms. If not, our Nation risks raising generations of students and citizens who do not know how to think critically and make informed decisions based on technical and scientific information."[6]

255

If education, not interest, is the problem, then one way to begin to improve our nation's scientific literacy is to fix upon scientific issues that capture the public's imagination and interest and use these as vehicles for training the public to engage in more sophisticated scientific thinking and discourse. An area that has obvious popular appeal is the study of the human brain. We are perennially fascinated with questions about our own minds. What makes us tick? What makes us cognitively different from other animals? In what ways are we similar? How do we see, think, plan, talk, and move? What is the biological nature of various mental dysfunctions? More generally, how do minds arise from brain function?

The New Sciences of the Brain

Until recently, our scientific understanding of the human brain was rather limited. Our knowledge, such as it was, was based on inferences derived from the study of nervous systems of nonhuman animals, or it was acquired from the study of brains of the deceased or from psychological observations of the damaged brain. Previously unthinkable opportunities for knowledge have now been opened by the development of noninvasive techniques for probing brain function in normal, intact, cognizing human beings. At the center of these techniques are brain investigation methods that provide us with images of the functioning brain. These neuroimaging techniques allow researchers to associate the performance of particular cognitive tasks with measurements reflecting brain activity. By correlating functional components of a task performed while a subject is being scanned with measured differences in signal from which brain activity levels are inferred, we can gain a deeper understanding of how cognitive function is subserved by the human brain. In this chapter I will focus upon functional magnetic resonance imaging (fMRI), a now widely used method for studying the brain bases of cognition.[7]

The development of fMRI over the last decade and a half has spawned a burgeoning new subfield of neuroscience. Although fMRI accounted for just a trickle of papers during its first years of development, now several hundred academic journal articles employing fMRI as a research method are published each month. Not only has the number of studies increased dramatically in the last decade, but the nature of the studies undertaken has changed too.[8] A survey of the topics of study over time reveals a gradual progression in the kinds of the tasks and functions studied, from simple sensory and motor functions to more and more high-level cognitive tasks and sophisticated behaviors.[9] Recent studies have moved toward understanding social interactions and interpersonal relations and have focused on phenomena as diverse as moral judgment, trust, love, retribution, and empathy. Thus, imaging is being used to investigate functions that many have come to think of as uniquely human. Via imaging, we are beginning to put a biological face on aspects of our lives that have heretofore been outside the scope of the natural sciences.

The increase in neuroimaging's profile in science has been paralleled by an increased visibility in popular outlets. Many of these studies are impacting popular culture, brought to the public's attention through the media and lay publications.[10] The number of reports in the popular media both reflects and encourages the enthusiasm of the public for the deliverances and potential of these novel imaging methods. Given worries about America's waning influence and expertise in science, neuroimaging's ability to engage the public's interest in basic science is laudable.

Neuroimages are visual representations of scientific data that allow the layperson to see beyond the familiar barrier of the human skull to the internal mystery, the hidden frontier of the human brain. The technical results from neuroimaging studies are typically rendered in the clearly recognizable, visually striking, and easily digestible format of the brain image. In addition to being an important way of evaluating, representing, and synthesizing results in the field of neuroimaging itself, neuroimages are the primary vehicle used for conveying the results of a study to the public. In these images, the relative degree and extent of neural activity is indicated by localized contours of pseudocolored regions superimposed upon a grayscale image of a whole brain or brain section. The images appear to be visual representations of a concrete object, and the colorful overlay representing brain activity is visually interesting and arresting. In the popular media, these images are usually accompanied by a short caption or explanation of the function they are meant to represent; the implication of this brevity is that the caption suffices to capture the representational value or meaning of the image. The take-home message thus seems simple and straightforward. Adding to the sense of familiarity imparted by this way of representing imaging results is the neuroimage's visual similarity to other types of scientific images, such as the weather maps that appear regularly on the evening news, and more mundane images such as photographs. Unlike many more abstract representations of physiological data, such as graphs, phase-spaces, equations, or the like, the brain image seems concrete, accessible, and easily graspable.

Thus, in imaging we have an advanced technology that is potentially widely harnessed, suited to investigating complex biological processes that are intimately tied to our highest cognitive abilities, and able to deliver results that superficially appear quite straightforward to a largely unsophisticated and enamored populace. This profile brings with it certain dangers, for the potential for misunderstanding and misuse of neuroimaging results is enormous. To illustrate, there are several companies engaged in studying the potential of neuroimaging for lie detection. Although the data these companies report may seem compelling, properly understood, they are not. Given an understanding of the technological limitations, the nature of the data upon which the reports are based, and the rates of false positives and negatives, these techniques would pose serious ethical problems if admitted in court.[11] However, one can well imagine a scenario in which an "expert" could convince a jury that a defendant lied if he backed his case with an image clearly showing activation in, for example, "the deception area" (see figure 11.1).[12] Misuse of imaging is not a potential problem restricted to lie detection, or even the legal system. Similar sorts of scenarios can be easily imagined involving racial prejudice, negative emotions, and so on. In addition, one can imagine instances in which images might be conceived of as screening devices for future behavior, job fitness, health insurance and the like. They could readily be used as vehicles to justify differential treatment of some person or group, especially if not responsibly evaluated. The public simply lacks the degree of understanding necessary for evaluating the reasonableness of claims or assessing the quality of evidence derived from an imaging study.

At this point, the worries expressed above are largely conjecture, but there is evidence that neuroimages lend the impression of evidential support even when they are irrelevant to an explanation. In a series of psychological experiments, Deena Skolnick Weisberg and colleagues provided naive subjects with adequate or poor explanations for a variety of psychological phenomena, and in some instances supplemented these with irrelevant claims about neural activity as shown by imaging. The subjects were asked to evaluate the quality of the explanations. These same explanations given to experts confirmed that the claims about regions of brain activity did not add to the quality of the explanation—indeed, experts rated the explanations with the irrelevant information as less satisfactory than those that only contained relevant information. However, naive subjects consistently rated the explanations with the irrelevant details from neuroimaging as better explanations. This confirms that laypeople tend to

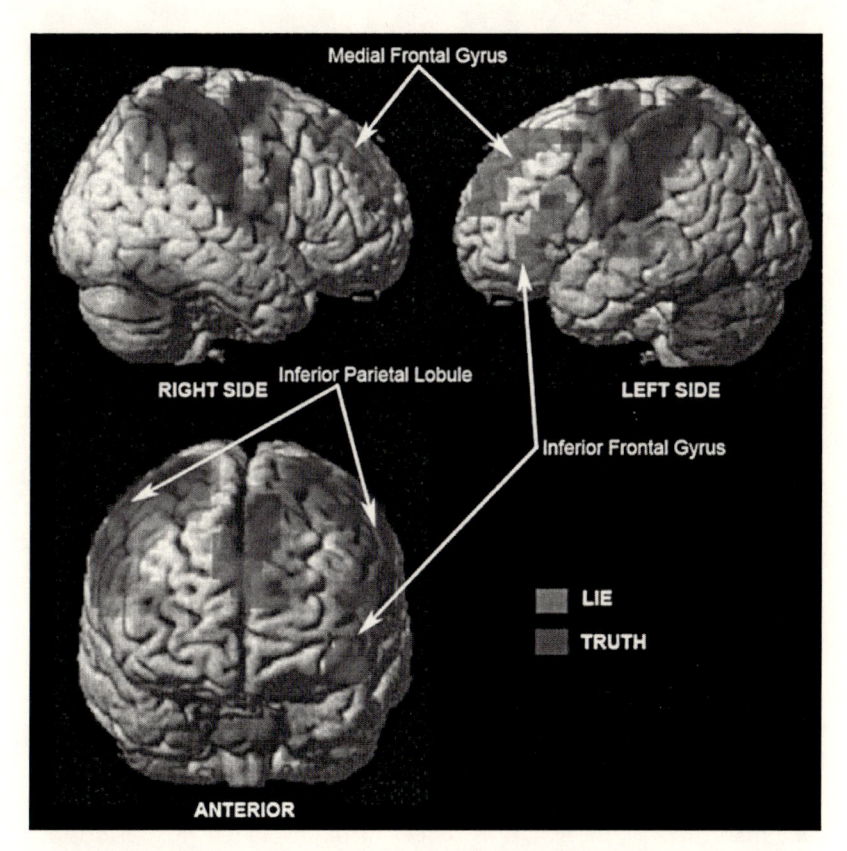

11.1 fMRI image showing average brain activation in 22 individuals performing a modified Guilty Knowledge Test (Langleben et al., "Telling Truth from Lie in Individual Subjects with Fast Event-Related fMRI," *Human Brain Mapping* 26 [2005]: 262). The highlighted areas in the medial and inferior frontal gyri are more active during the lie responses, and areas in the parietal cortex are more active during truthful responses. Image courtesy Kosha Ruparel and Daniel Langleben, University of Pennsylvania

impart to imaging data an evidential status that is unrelated to their actual evidential role. Further empirical studies need to be done to determine exactly how the layperson interprets images and what factors influence their attributions of evidential weight. For instance, useful insight could be gained from understanding how information presented as a brain image is apprehended differently from the same information otherwise presented. If these worries and others that have been expressed in the literature are even approximately correct, it is imperative that we as a society become cognizant of the dangers inherent in naive public consumption of brain images.[13] Otherwise, we may end up hostage to our own technologies.

Part of what is involved in understanding science, being able to think critically about it, and being able to apply scientific knowledge in practical and policy decision making involves being aware of the limitations of science, the epistemic status of scientific findings, and associated ethical issues. A recent survey of articles on neuroimaging in the popular media has assessed them on a number of dimensions relevant to these concerns, including degree to which they are critical of the techniques and the implications of the scientific results, their descriptiveness, and the clarity with which they discuss scientific limitations. The study revealed that, overall, media reports of imaging studies tend to be simplistic and uncritical, and few mention the ethical issues potentially engendered by these novel methods.[14] It is not going out on a limb to claim that media exposure of this sort does not serve to responsibly inform or educate the public; rather, it is a manifestation of the general problem with science education.

This paper will attempt to begin to redress these shortcomings by discussing the epistemological issues involved in neuroimaging. Images are widely used in conveying information about the life sciences because they are cognitively approachable tools. And indeed they are: the seeming accessibility of neuroimages, and the grip they have on the scientific and public imagination makes them important conduits of information about the progress of neuroscience. However, in thinking about the role of images in scientific pedagogy, we typically consider the image as unproblematic tool, rather than as object of discovery. In this discussion, however, we recognize the tool's more problematic aspects, and position the image itself as the object of scientific interest. The simple visual format of the image belies its complexity and provides no indication of its provenance or the wealth of theory that must be employed in order to properly understand the information that it represents. The experimental design, data collection, data analysis, and interpretation of neuroimaging experiments are complex and theory laden, and so is the image generated by these experiments. Failure to appreciate this has the potential to lead to misinterpretation and misuse of neuroimaging data.[15] In what follows, I will consider the source of the popular appeal of neuroimaging, and some potential concerns that accompany the development of such powerful, complex, and yet widely accessible techniques. In particular, I will focus upon a mismatch between the intuitive grasp of neuroimaging results made possible by the presentation of the data in imagistic form, and the interpretational difficulties attending neuroimaging. I suggest that neuroimages are naively taken to be akin to photographs of brain activity.

Photographic images enjoy a certain epistemic status, and if neuroimages are analogous to photographs, it follows that they will be taken to enjoy a similar status. I consider here whether the analogy is warranted, and argue that in important ways it is not. This realization is important in order for the evidential status of neuroimages to be properly appreciated. I will end with a discussion of the role of pedagogy in mediating a responsible public consumption of the yields of these new methods.

Vision as a Privileged Way of Knowing

To appreciate the role of the image in representing and conveying information, we might first look to the modality it harnesses. Images are apprehended by, and to at least some degree interpreted by, our visual faculties. Humans, along with many other primates, are visual animals. Visual cues have been highly salient for survival in our evolutionary history, and we have evolved to be extremely adept processors of visual information. An enormous percentage of our visual cortex is dedicated to visual processing, and even more of it responds in some way to visual signals. In other words, vision is one of the primary ways we have of knowing about the world.

The connection between vision and knowledge is evident both at the folk and the philosophical level. Not only do we see objects, such as cups, computers, and cars. We see properties, such as color, texture, size, and motion, and we see changes in those properties. Some argue that we even see emotions, or abstract objects such as facts or propositions: I see your anger, or I see that you are too tired to drive. The idea that seeing grounds knowledge is implicit in our speech and our practices. For example, commonly used idioms like "seeing is believing," "you have to see it for yourself," "I saw it with my own eyes," or even the simple phrase, "I see" to express understanding illustrate the connection that is drawn between seeing something and presumptive knowledge. The importance of eyewitness testimony in trials is a social manifestation of the role vision plays in generating knowledge, and its privileged evidential status is indicative of the privileged role vision has relative to many other potential sources of knowledge. In philosophy, too, vision is granted a special status over other sensory modalities. Philosophy of perception is dominated by concerns about the relation of visual experience to belief and knowledge, while other sensory modalities are usually viewed as subsidiary.

Vision also enjoys a certain privileged status in the acquisition of scientific knowledge. Through various devices or inferential processes, the power of vision has been extended to lay open to us much of the world that has previously been invisible. For example, we see by means of various intermediary devices such as microscopes or telescopes, objects too small, or too far to be seen with the naked eye.[16] Many devices convert nonvisual quantities into representations in the visual realm: for example, voltage is a nonvisual property that is visually represented by an oscilloscope. We commonly accept that we see what are, at some level, theoretical entities. Because of other ways we have of seeing what is not ordinarily visible, the notion that we can see objects and properties that are not open to the naked eye no longer troubles any but philosophers. The layperson accepts as unproblematic that we see galaxies through telescopes, the sea floor through sonar, and cellular organelles through electron microscopes. It is no great leap for her to accept that we see brains through scanners.

Given the common acceptance that we can see brains, and the commonsense notion that what we see is, in a sense, the objective truth, the presumption is that scanners allow us to see objective features of brain activity. While this is undeniably true in some sense, it can be misleading when taken at face value. The goal here is to explicate the way in which it is true, and ways in which it is not.

The Photography Analogy

My supposition is that the layperson apprehends a neuroimage much as he does a photograph, whether taken from a regular camera, or taken through an optical microscope or a telescope. The neuroimage is also produced by the operation of a mechanical instrument. The image it generates has the look of a snapshot of brain activity, where the brain—a concrete object made up of gray and white matter—is visually apparent in the grayscale background image, and the activity is clearly perceived in the well-defined, precisely located color contours. Brain activity, so represented, seems unmistakably present and localizable. When labeled with a caption, such as "Trust in the Brain," interpretation of these images seems to require no deep theoretical knowledge: trust occurs in the brain in the locations that correspond spatially to where the brightly colored spots are on the brain image. It also seems apparent that this pattern of activity has certain objectivity: the activity is obvious upon cursory visual inspection, and any observer will see the same thing. In many respects, the neuroimage resembles a photograph, but a photograph of the brain in operation, rather than of a scene, a galaxy, or a cell. The objectivity conveyed by the image appears to give the results a certain unimpeachable epistemic status—it seems as if the image leaves no room for disputes.

This is what seems to be the case, and we may suppose that if people take neuroimages to be like photographs of brain activity, then they will think the images have an epistemic status akin to that held by photographs. Thus, we may be able to understand the epistemic status naively accorded to brain images derivatively, by coming to understand the epistemic status of photographs. The epistemic status of photography has been a topic of some discussion in aesthetics, so I will rely upon others' work to set the stage.[17] In particular, I discuss three features that have been identified as contributing to the evidential status of photographs: Transparency, counterfactual dependence, and theory independence. My concern here is not to defend these properties as constitutive of the evidential weight ascribed to photographs, but rather to evaluate the aptness of the analogy between photography and neuroimaging by analyzing the extent to which neuroimages share the above characteristics. In other words, do neuroimages actually have the same epistemic status as do photographs? No. In important ways neuroimaging differs from photography. Moreover, the ways in which it differs are not apparent upon visual inspection of the images. Considerable theoretical sophistication is required to appreciate the ways in which they diverge. This is one reason that neuroimages are apt to mislead.

Exploring the Photography Analogy

Transparency

Kendall Walton has famously claimed that photographs are *transparent*: that we "see through the photograph to the object being photographed."[18] Transparency attributes certain objectivity to photography, because it claims that a photograph literally presents us with the object it is of, rather than a facsimile or a figure that merely bears a likeness to the object. Walton's thesis of transparency is quite radical, but a less radical transparency thesis is both more defensible and a sufficient illustration of the evidential relationship a photograph bears to its object. According to the view of transparency I prefer, a photograph is transparent with respect to a certain type of feature of an object if the photograph instantiates that feature, or shares it.[19] Of course, it is not sufficient that a feature is merely instantiated in the photograph—the feature must be instantiated in such a way that it preserves certain relations among features of the object. In particular, photographs instantiate two-dimensional projections of visual features such as color, shape, shading, and texture, as well as preserving projections of the spatial relations between them.[20] Both Walton and Cohen and Meskin locate a great deal of the evidential import of photographs in the fact that many of the surface visual properties of the photograph are visual properties of its object.[21] Because photographs carry information about many of the visual properties of their objects by instantiating those very properties, in seeing the photograph we can imagine we are seeing the object.[22]

Can neuroimaging support the transparency claim? It may appear to, for it may seem that in seeing a brain image we are seeing the brain at work, and indeed that is what the lay person is apt to think. However, the pseudo-colored image does not instantiate the visual features of an active brain, for brain activity does not actually manifest itself in visible changes, let alone changes that are accurately reflected by the rough pixelated color profiles characteristic of brain images. The properties relevant to neural activity are changes in electrical properties such as voltage and current, and their underlying physical substrates such as ion fluxes, protein conformation changes, and neurochemical diffusion. While this already casts doubt on the transparency claim, it is subject to even more severe problems. For not only are these properties of neural activity not visible properties in the classic sense, but they are not the properties that fMRI measures. Instead, what fMRI measures are changes in magnetic properties of blood and tissue that result from changes in blood flow and volume concomitant with neural activity.[23] These changes in magnetic properties also do not manifest as changes in visual properties. The link between what fMRI measures and neural activity is mediated by multiple theoretical inferences, involving, for example, such theoretical quantities as changes in decay times of coherently precessing particles and magnetic field inhomogeneities. Thus, fMRI measures one type of factor in order to allow the inference of changes in quite different factors. If one traces the inferential steps between what is measured by the technique and what is claimed to be portrayed in the image, it becomes clear that one does not see the properties of the brain instantiated in the image.

There are a number of other ways in which neuroimages fail to be transparent, but here I'll mention just one other. Different neuroimages variously represent the relation of the color spectrum to neural activity.[24] Not much sophistication is required to recognize that the colors exhibited are not meant to indicate colors in the brain, but precisely what they are meant to represent is not always apparent. Importantly, colors are variously used in different studies to represent quite different parameters related to neural activity. Sometimes they are used to represent the percentage signal change during the performance of some task relative to another; other times they are used to reflect the statistical significance of the signal change, and other times the proportion or number of cases that show statistically significant changes in that relative location. Thus, the very same visual image could reflect a very different scientific datum, depending upon the intended interpretation. In many cases, visual inspection of the image does not suffice to specify the intended interpretation. Sometimes the correct interpretation is indicated in the figure legend. However, occasionally the only way the correct interpretation can be discerned is by a reasonably careful perusal of the original scientific paper. In the popular media the proper interpretation of the image data is rarely made explicit, and moreover, the layperson is rarely aware that alternative interpretations are possible.

Neuroimages give the impression of simplicity, but neuroimaging is in fact a very complex and unintuitive technique. The failure of transparency is not evident in the image itself, and most people are completely unaware of the scientific basis of neuroimaging techniques and the sorts of inferential steps necessary in order to draw conclusions about brain activity from a brain image.[25] Were people to acquire even this meta-scientific knowledge, they might be prompted to be more wary of accepting at face value their naive interpretation of a brain image.

Causal and Counterfactual Dependence

There are many different types of images, and some have greater epistemic import than others. Paintings, for instance, can be virtually indistinguishable from photographs, and thus as realistic as them. However, paintings carry less epistemic weight than photographs: from paintings one cannot reliably infer the existence of the subject or its properties, even though they can faithfully represent these qualities. Photographs, on the other hand, seem to be more closely tied to reality. For instance, if we evaluate the suitability of different types of images for serving as legal evidence, it is clear that photography is superior to painting: a photograph of a suspect engaging in a crime is far more compelling than a visually indistinguishable painting would be. This shows that there are aspects of photography that contribute to its evidential status, but do not derive purely from the visual properties of the photograph. As a number of philosophers have argued, these derive from factors related to the technology and conventions employed.[26]

In particular, photographs are suitable as evidence because of the informational relationship they bear to their subjects. Photographs are causally and counterfactually dependent on their subjects in a way that is different from the way paintings or drawings are.[27] Had the subjects of a photograph been differently arranged, differently illuminated, had they had different visual properties, and so on, the resulting photograph would have been correspondingly different. Note that this counterfactual dependence is not thoroughgoing. Certain properties of that which is photographed could be altered without altering the photograph: the hidden aspects of objects could change with no change to the photograph; clever wax duplicates of objects could be substituted without a resulting change in the visual properties of the photograph, and so on. However, the types of changes the subject of the photograph can sustain without a concomitant change in the resulting photograph are limited, and are dependent on the technology, and because we understand both the subjects and the technology, we have a good intuitive grasp of what those limits are.[28] Cohen and Meskin stress the importance of people's knowledge and background beliefs about photography in accounting for its evidentiary status.[29]

Like photography, neuroimages also bear reliable causal and counterfactual relations to the phenomena they represent. It is the reliability of these relations that makes neuroimaging an appropriate technique for scientific investigations of brain function. However, the phenomena they directly represent are not the ones we are most interested in scientifically (see above). Insofar as neuroimages represent neural activity, imaging differs from photography in that the nature of the underlying counterfactual and causal relations is relatively underspecified. While we are beginning to understand how neural activity patterns correlate with fMRI data, we don't have robust explicit or even intuitive theories about what sorts of counterfactual relations they bear to each other.[30] This is due in part to the limited spatial and temporal resolution of the technique, and its relatively low signal-to-noise ratio. It is also due to a lack of understanding of the precise relationships between neural activity and changes in blood flow, volume, and oxygenation. Because it is unclear what range of different activity patterns can lead to the same MR signal, what our inferences about changes in activity should be given a particular pattern is underconstrained by the imaging results. The problem is compounded for the layperson, who is inclined to accept at face value the image as incontrovertible and to assume that there is a one-to-one correspondence between functional image and the underlying activity. If this were correct, the absence of significant activation in a subtraction image would imply that there was no change in underlying neural activity. However, that inference is demonstrably false: different patterns of neural activity can give rise to the same fMRI profile.

As in the case of transparency, the image itself carries no clues as to the limits of interpretability, due to inadequate understanding of counterfactual relations. It is likely that people think that interpretation in imaging is as unproblematic as it is in photography.

Theory Independence

Walton claims that one of the primary reasons that photographs are taken to be objective forms of evidence is that they are belief independent.[31] Whether this is an apt characterization of photography is dubious. Clearly, the beliefs or the desires of the photographer affect what she photographs, how she frames it, what lenses she uses, or how she lights it or exposes it, all of which affect the photograph produced. However, what is correct is that given that she does take a photograph from a certain place, with certain parameters, and so on, the resulting image is independent of her beliefs about her subject and about photography. The resulting image is dependent solely on the scene the camera is pointed at, the film, and the light reaching the film.[32] We might say that photography is theory independent. The nature of the image does not depend upon the photographer's acceptance or denial of particular theories, and typically apprehension of the photograph is similarly theory independent.

The technology of the fMRI, in contrast, is highly theory dependent. While the mechanical process of magnetic resonance imaging is a causal process just as photography is, the image one ends up with is dependent upon a host of theoretically based decisions that the experimenter makes, and many of these must be relied upon in order to accurately interpret the image. For example, the pulse-sequences one uses in obtaining an image can have a dramatic effect on what image is obtained. While the temptation may be to liken this to the photographer's choice of film, the degrees of freedom here are much more extensive and interpretation must be based upon a theoretical understanding of the technique. This makes it substantially different from, for instance, merely noting that the photographic film used is color or infrared. As another example, the statistical analysis one employs in order to process the image data also significantly affects the pattern of activity that an image represents. Merely changing the statistical threshold of the pseudocolored regions can result in different patterns and extents of "activated" regions; changing the color scheme can affect the visual effect it has on the viewer. Neuroimaging and photography occupy two extreme ends on a spectrum of theory dependence for technologies.[33]

More importantly but perhaps less obviously, the experimental paradigm itself is highly theory dependent. Interpretation of images cannot be dissociated from an understanding of the tasks employed. Neuroimaging studies typically contrast a task of interest with either a baseline task or some comparison task. Because of this, what an image depicts is not a signal correlated with absolute levels of neural activity, but with differences in neural activity between task conditions. The profile of activity for the very same task of interest could look entirely different given two different contrast tasks. For example, regions that are active in a task can appear as deactivated regions if they are more activated in the baseline task.[34] The dependence of the image on the details of the tasks employed is an ineliminable aspect of neuroimaging studies, but because the task-dependence is not apparent in the image itself, it is apt to be overlooked by naive consumers of neuroimaging. In general, a significant amount of background knowledge is necessary in order to interpret the results of any neuroimaging study. While such details are often available in the technical papers in which neuroimaging results are primarily reported to the scientific community, the requisite background information is rarely included in any popular discussion of the study, and insufficient information is presented to allow a proper understanding of the nature of the study itself, or an adequate assessment of the results.

Here I have identified three characteristics that are important to the epistemic status of photographs that are not characteristics that apply (or apply in the same way) to neuroimages. However, one cannot look to the brain images themselves in order to appreciate the differences between photography and neuroimaging. Rather, one must develop a more sophisticated understanding of the science and technology in order to adequately comprehend the ways in which imaging differs from photography, and the ways in which the deliverances of the two fail to be on par epistemically. Failure to do so can lead people to come to unwarranted conclusions about the meaning of a neuroimage or its status as evidence for or against some scientific or social claim.

The Importance of Pedagogy

Neuroimaging is not like photography, but the epistemic power of the photograph may lend the superficially similar neuroimage added epistemic force. The potential for neuroimaging to mislead may also be based in part on the public's poor understanding of how science works in general. The degree to which scientific progress is contingent, the fact that scientific knowledge is subject to revision, and arguments against foundationalism and for holism tend to be lost on the average citizen. In addition, many feel that science is best left to the experts, and they tend to suppose that more is understood or is on firmer footing than may be the case. This naïveté about the pursuit of science may also contribute to people's readiness to accept brain images as unproblematic, objective data.

Given the foreseeable dangers that come with a public fascination with powerful scientific techniques that are deceptive in their accessibility, we have a spectrum of options. At one extreme, we can stick our heads in the sand, cognizant of the potential of public misinterpretation of such data to lead to ill effects, but unwilling or unable to take preemptive measures to avoid such scenarios. In such circumstances we merely hope for the best, resigning ourselves to cleaning up problems after they arise. While this is clearly the easiest route to take in the short term, it is likely not the best. The old adage about an ounce of prevention is apt here. Moreover, the problems that do arise are likely to arise in a variety of circumstances and each in its own guise. The piecemeal presentation of problems will likely engender a piecemeal response, and although it is not a foregone conclusion that such a response would be suboptimal, it does highlight the potential for fragmentation, incoherence and conflict.

At another extreme, we could limit the access nonspecialists have to neuroimaging data, and in this way try to ensure that the capacity the data have to mislead is not realized. However, this approach is plainly inimical to the American ideal of an open, egalitarian society, and blatantly in conflict with scientific practice and the transparency that ensures many of its virtues. It is the accessibility of scientific data, and the answerability of scientists to scrutiny by peers and public, that helps make the American scientific community the productive, inventive, and responsive enterprise it is. To limit the dissemination of data would only hobble scientific pursuit. And, as the recent introduction of fMRI evidence into the courts and the rise of neuro-marketing suggest, it is too late. The future is already here.

It will probably come as no surprise, then, that my preferred method of addressing the dangers mentioned above focuses on pedagogy. In this I am in agreement with the National Science Board's recommendations for science education, although my focus is on neuroimaging as an entry point for stimulating interest and promoting scientific literacy. The goals of educating the public about neuroimaging should be severalfold. First, efforts should be made to disseminate information about the basics of neuroimaging techniques, as well as the basics of brain function. Perhaps the best way to do this would be to institute a segment on brain science in the high school biology curriculum, and with it, a section on neuroimaging. Targeting high school students will eventually result in a citizenry with a common basic understanding of brain and imaging techniques, much as today the populace shares a basic knowledge of American history. However, since only those of high school age or younger will be exposed to this material, it will take many years for such knowledge to become truly widespread among the segments of the population most likely to be called upon to engage in policy decisions that could be affected by imaging results. For this reason, it is also worthwhile to pursue other avenues of education to reach older segments of the population as well. The 1996 Science and Engineering Indicators study reported that American's get the majority of their information about current science from the popular media; the more recent 2006 study stressed the rise of the Internet for disseminating information about science, especially among younger Americans.[35] Thus, articles in popular magazines, features on television shows, educational websites, and public lectures are all potential candidate vehicles for education.

Another avenue that has already garnered some success is to promote brain literacy with science competitions at the high school level, such as the Society for Neuroscience's Brain Bee.[36] The Society promotes these regional, state, and international competitions for high school students. The questions in the Brain Bee are based upon information available in SFNs free educational materials, and prizes are awarded to the top participants. The Brain Bee has enjoyed success in parts of the United States and Australia. More effort could be devoted to informing participants about the complexities involved in interpreting neuroimaging studies, as well as to providing basic information about the technologies. Efforts to educate judges and lawyers about brain imaging and its use in legal contexts are being undertaken by the MacArthur Project in Law and Neuroscience in conjunction with the Gruter Institute for Law and Behavioral Research.[37]

Skepticism here is not unwarranted. Weisberg et al.'s study on the misleading epistemic effect of neuroimages provides cause for concern.[38] They found virtually the same pattern of data for subjects who had had a semester of college-level cognitive science instruction as they did for naive subjects. This suggests that a moderate level of expertise is insufficient to combat the epistemic force of imaging; only experts in the field were able to reliably discern the irrelevance of the neuroscientific information. If this is correct, the prospect of addressing the epistemic worries with a cursory educational program are dim. However, one wonders whether such a high level of expertise is really necessary, or whether the instruction given to the students in cognitive science was not geared to teaching the critical skills necessary for distinguishing relevant from irrelevant neuroscientific data or for understanding the epistemic value of imaging. If this is the case, then education may well be the best solution, but it may be that the curriculum of neuroimaging basics is something that will require considerable thought and effort to develop adequately. Perhaps more important than imparting to people basic technical understanding is sketching for them the variety of complexities inherent in neuroimaging. Because one of the greatest dangers involves underestimating the interpretational difficulties of imaging, making explicit the extent of the factors relevant to interpreting neuroimaging studies will help provide people with a sense of what they do not know. Instilling a sense of personal modesty with regard to interpretational ability is one of the most immediate pedagogical goals; balancing this modesty with a more sophisticated understanding is one of the most challenging ones. Finally, it is also important to try to impart to people a sense of the limits of science. First, there are limitations imposed by the technologies themselves: limitations in what they can show us about the brain because of limits to spatial or temporal resolution or because of the nature of what the techniques measure. Second, scientific data are always interpreted in the context of current theory. Interpretations are often dependent upon auxiliary information or hypotheses about brain function. To the extent that such information is speculative or hypothetical, so will our interpretations be. Thus, our ability to make inferences about the data is limited by current knowledge and technical considerations. The public is woefully ignorant about these matters, often imparting more authority to scientific reports than is warranted, and attributing far more power to the reach of scientific (or pseudoscientific technologies) than is reasonable. The discussion about imaging could profitably be placed in the context of a more general discussion about scientific progress. Images from other sciences could be employed to illustrate how scientific conceptions change over time, thereby emphasizing the evolving nature of our scientific understanding.

These are tricky waters to navigate, and doing so successfully will require the cooperation of scientists and the media. There is likely to be resistance in both corners. Scientists want their work to be properly understood, but neuroimagers also tangibly benefit from the popularity and hype surrounding neuroimaging. For example, many have reported that in certain fields the probability of getting a grant application funded is much higher for studies that include neuroimaging than for those that do not.[39] It may be in the interests of neuroimagers to fan the flames of neuroimaging ardor rather than do anything to dampen them. A related but more selfless worry might lead scientists to balk at the prospect of trying to educate the public about the intricacies of neuroimaging: namely, the risk that in trying to impart a proper understanding of the epistemic import of neuroimages, people might be pushed instead toward an extreme skepticism about what imaging can deliver, or even about science in general. This could conceivably have serious impacts on funding, maybe ranging far beyond imaging. For this reason it is important to stress neuroimaging's scientific value, and the value of science, and not just interpretational and methodological difficulties. It should be possible to instill respect for the difficulties inherent in the quest for knowledge without undermining the value of the endeavor.

The media, too, may be resistant to cooperating with an effort to educate the public. They benefit from the dazzle of the image and the sound bite. Good scientific reporting is difficult, and requires much more than a pretty picture. It often involves more detail than the media are prepared to provide, and the nuanced result sometimes lacks the pizzazz of the simple story. Incentives in this case are also difficult to come by. However, efforts should be made to make science journalists recognize public education as both a duty and an invaluable civic service. To be sure, balancing competing interests with responsible scientific education and reporting will be an ongoing challenge.

Summary

Vision is one of the main ways we have of forming beliefs about the world. One explanation for the compelling nature of the data thus presented is that they provide the sense that we, through imaging, are seeing the brain in action, much as through photographs we see the things that are photographed. Neuroimaging harnesses this perceptual system, and purports to allow us to see brain activity. While interpreting many kinds of images is relatively unproblematic, this is not so with functional images of the brain. Characteristics of brain images differ in important ways from the characteristics of photographs. While these deviations do not undermine neuroimaging's usefulness as a scientific method, they do suggest that the analogy between the two techniques is potentially misleading, and that their differences should afford neuroimaging a much different epistemic status than nonspecialists are inclined to attribute it. The way to achieve a proper understanding is through an improved, targeted scientific education.

Acknowledgments

This work was supported in part by a grant from the Leslie Humanities Center at Dartmouth College and in part by a fellowship from the Australian Research Council and the University of Sydney.

Notes

1. "Americans Fear Decline in U.S. Performance in Math and Science," *Eurek-Alert* (Eins Communication, 2007), http://www.eurekalert.org/pub_releases/2007-02/ec-afd013107.php; Research!America, "Americans Support Bridging the Sciences" (Charlton Research Co., 2007) http://www.researchamerica.org/uploads/btspollreport.pdf

2. National Science Board, *Science and Engineering Indicators 2004* (Washington DC: National Science Board, 2004); National Science Board, *Science and Engineering Indicators 2006* (Washington DC: National Science Board, 2006). Analysis indicates that average performance of U.S. students has remained relatively stable over the years, but student performance in other countries has risen.

3. Committee on Prospering in the Global Economy of the 21st Century, National Academy of Sciences, National Academy of Engineering, Institute of Medicine, *Rising above the Gathering Storm: Energizing and Employing America for a Brighter Economic Future* (Washington, DC: The National Academies Press, 2007); Chris Mooney and Sheril Kirshenbaum, *Unscientific America: How Scientific Illiteracy Threatens our Future* (New York: Basic Books, 2009).

4. Research!America, "Americans Support."

5. Mooney and Kirshenbaum, *Unscientific America*.

6. National Science Foundation, "America's Pressing Challenge—Building a Stronger Foundation," in *A Companion to Science and Engineering Indicators* (Arlington, VA: National Science Foundation, 2006).

7. Much of what I will have to say applies as well to positron emission tomography, or PET, which preceded fMRI and generates similar images.

8. Judy Illes, Matthew Kirschen, and J. D. E. Gabrieli, "From Neuroimaging to Neuroethics," *Nature Neurosciences* 6 (2003), 205.

9. Judy Illes, Eric Racine, and Matthew Kirschen, "A Picture Is Worth 1,000 Words, but Which 1,000?" in *Neuroethics: Defining the Issues in Theory, Practice, and Policy*, ed. Judy Illes (Oxford: Oxford University Press, 2005).

10. Eric Racine, Ofek Bar-Ilan, Judy Illes, "FMRI in the Public Eye," *Nature Reviews Neuroscience* 6 (2005): 9–14; Eric Racine, Ofek Bar-Ilan, Judy Illes, "Brain Imaging: A Decade of Coverage in the Print Media," *Science Communications* 28 (2006): 122–42.

11. Scott T. Grafton, Walter P. Sinnott-Armstrong, Suzanne I. Gazzaniga, Michael S. Gazzaniga, "Brain Scans Go Legal," *Scientific American Mind* 17 (December 2006): 30–37.

12. In fact, the defense in a recent court case aimed to use functional neuroimaging as "truth verification," but the evidence was ruled inadmissible by the judge because of current reliability of fMRI for lie detection. See Alexis Madrigal, "Brain Scan Lie-Detection Deemed Far from Ready for Courtroom," *Wired Science* (2010). Future use was not ruled out. For example, see the discussion in Mark Harris, "Liar," *IEEE Spectrum* 47, no. 8 (August 2010): 40–53.

13. Judy Illes, ed., *Neuroethics: Defining the Issues in Theory, Practice, and Policy* (Oxford: Oxford University Press, 2005); Racine, Bar-Ilan, and Illes, "FMRI in the Public Eye."

14. Racine, Bar-Ilan, and Illes, "Brain Imaging."

15. Neuroscientists involved in neuroimaging experiments are not apt to misapprehend the evidential status of brain images in the ways I suggest here, and are quite aware that their enterprise involves interpretational challenges. But because these images are disseminated far beyond the neuroimaging community, these arguments become particularly relevant.

16. Jutta Shickore, *The Microscope and the Eye: A History of Reflections, 1740–1870* (Chicago: University of Chicago Press, 2007).

17. Kendall Walton, "Transparent Pictures: On the Nature of Photographic Realism," *Critical Inquiry* 11 (1984): 246–76; Jonathan Cohen and Aaron Meskin, "On the Epistemic Value of Photographs," *Journal of Aesthetics and Art Criticism* 62 (2004): 197–210.

18. Walton, "Transparent Pictures."

19. Cohen and Meskin, "On the Epistemic Value of Photographs"; Jonathan Cohen and Aaron Meskin, "Photographs as Evidence," in *Photography and Philosophy*, ed., S. Walden (New York: Blackwell, in press).

20. See John Kulvicki, *On Images: Their Structure and Content* (New York: Oxford University Press, 2006).

21. Walton, "Transparent Pictures;" Cohen and Meskin, "On the Epistemic Value of Photographs;" Cohen and Meskin, "Photographs as Evidence."

22. Walton, "Transparent Pictures."

23. See Richard B. Buxton, *Introduction to Functional Magnetic Resonance Imaging: Principles and Techniques* (Cambridge: Cambridge University Press,

2002); Adina Roskies, "Is Neuroimaging Like Photography?" *Philosophy of Science* (forthcoming).

24. Joseph Dumit, *Picturing Personhood: Brain Scans and Biomedical Identity* (Princeton, NJ: Princeton University Press, 2003).

25. For a more in-depth treatment of this issue see Roskies, "Is Neuroimaging Like Photography?"

26. Walton, "Transparent Pictures;" Cohen and Meskin, "On the Epistemic Value of Photographs."

27. Walton, "Transparent Pictures;" Cohen and Meskin, "On the Epistemic Value of Photographs."

28. Our increasingly sophisticated abilities to digitally manipulate photographs has clearly changed the character of the counterfactual dependence of photographs on their subjects.

29. Cohen and Meskin, "On the Epistemic Value of Photographs."

30. Nikos K. Logothetis, "The Underpinnings of the BOLD Functional Magnetic Resonance Imaging Signal, " *Journal of Neuroscience* 23 (2003): 3963–71; Nikos K. Logothetis and Brian A. Wandell, "Interpreting the BOLD signal," *Annual Review of Physiology* 66 (2004): 735–69; Marcus E. Raichle and Mark A. Mintim, "Brain Work and Brain Imaging," *Annual Review of Neuroscience* 29 (2006): 449–76.

31. Walton, "Transparent Pictures."

32. Again, this characterization of photography downplays the role of the photographer too much. Not only do the photographer's beliefs have important effects on the resulting negative, they also play an important role in the processing and developing stages, and thus of the final print. Nonetheless, the more simplified characterization above captures an important feature of photography, and accounts for why photographs are viewed to be better evidence of the nature of reality than are, for instance, paintings.

33. I should stress here that this is overall a simplification, and that I am talking about standard photography, and people's appreciation of it. Even photography is more theory laden than most people tend to acknowledge, and its status as evidence is affected by the ubiquity and alterability of digital photography.

34. Deborah A. Gusnard and Marcus E. Raichle, "Searching for a Baseline: Functional Imaging and the Resting Human Brain," *Nature Reviews Neuroscience* 2 (2001): 685–94; Raichle and Mintim, "Brain Work and Brain Imaging."

35. National Science Board, *Science and Engineering Indicators 1996* (Washington DC: National Science Foundation, 1996); National Science Board, *Science and Engineering Indicators 2006*.

36. See Society for Neuroscience, "Brain Bees," available at http://www.sfn.org/index.aspx?pagename=baw_brain_bee, and http://www.internationalbrainbee.com/.

37. See Law and Neuroscience Project, http://www.lawneuro.org/; Gruter Institute, http://www.gruterinstitute.org/Home.html

38. Deena S. Weisberg, Frank C. Keil, Joshua Goodstein, Elizabeth Rawson,

and Jeremy R. Gray, "The Seductive Allure of Neuroscience Explanations," *Journal of Cognitive Neuroscience* (forthcoming).

39. Paul Bloom, "Seduced by the Flickering Lights of the Brain," *Seed* (June 2006).

chapter twelve

The Anatomy of a Surgical Simulation
The Mutual Articulation of Bodies in and through the Machine

Rachel Prentice

> All the energy we spend on motion
> All the circuitry and time
> Is there any way to feel a body
> Through fibre-optic lines?
> —CASSANDRA WILSON

Surgical learning traditionally has included intensive and structured training of a surgical resident's skills of seeing, interpreting, and intervening manually in a patient's body. Residents now receive most of their training in the operating room, working on actual patients under the close supervision of an attending surgeon. In the last decade, however, changes in hospital economics have squeezed operating room time. Medical students and beginning residents often are relegated to roles as observers,[1] even as a growing body of medical research indicates that constant practice is critical to surgical success rates.[2] In response, researchers in several universities and private companies have begun to develop virtual reality training systems, modeled on flight simulators, that might one day train medical students outside the operating room, potentially freeing staff surgeons' time and giving students a higher level of aptitude before they work on patients. Surgical simulators also could be used to train experienced surgeons' skills with emerging visualization technologies and minimally invasive surgical techniques.[3] The ideal virtual reality simulator would provide visual and physical experiences similar to minimally invasive surgery, teaching the fine motor movements needed to clamp, cut, or suture virtual tissues, and giving students and surgeons opportunities to practice their skills *in silico* before trying them in vivo.

Medical technology researchers are building two types of computerized simulator: physical simulators, in which a mannequin with sensors represents a human patient's body, and virtual reality simulators, in which a graphic creation existing entirely in the computer models the patient's body.[4] Mannequin-based simulators are useful for teaching physical skills, such as palpation, particularly when the student or physician cannot see structures to be palpated, as with pelvic and prostate exams. Virtual reality simulators hold promise for teaching skills, such as cutting, that would rapidly destroy a mannequin.[5] Virtual reality simulators are most commonly developed for minimally invasive procedures. There are three reasons for this: because of the preexisting relationship of instrument to screen, because minimally invasive procedures are harder to learn than open procedures, and because students and residents can practice many skills for open surgery on ordinary objects. Most virtual reality simulators are prototypes, whose expense and technological difficulties make their future uncertain. The technological challenges are significant and simulator makers often say their creations do not "feel right."

Building virtual reality simulators for teaching surgical skill and other medical procedures has become an active research area among computer experts, engineers, and physicians interested in medical informatics. To build virtual reality simulators, researchers have had to break down and reformulate knowledge about patients' bodies and surgeons' actions in ways that are technologically compatible with digital computers. The computer as a surgical teaching tool thus becomes a crucial nonhuman actor in this research arena.[6] In this chapter, I dissect the research that went into creation of a surgical simulator developed by an interdisciplinary medical informatics laboratory at Stanford University School of Medicine to teach minimally invasive gynecological procedures, such as the removal of an ovary.

Surgery is embodied action, action that creates particular physical relationships between patients and surgeons.[7] The surgeon must understand the patient body's materiality—its specificity, its pathologies, its interactions with other bodies. The very origin of the word *surgery* in the Greek *cheir* ("hand") and *ergon* ("work") suggests that surgical learning must include training of a surgeon's hands, though this neglects the extent to which the surgeon's entire body participates in surgery.[8] Surgeons and anatomists repeatedly told me that a trainee's physical experience of dissection is a critical component of anatomical learning.[9] My experience confirms this. After months of observing dissections and handling tissues, I picked up a scalpel and—under the careful supervision of a hand surgeon—performed a mock "surgical" procedure on a cadaver arm, the transposition of an ulnar nerve, a procedure typically done to relieve pain associated with a pinched nerve in the elbow. I began to understand how much easier distinguishing tissues and remembering names and spatial relations becomes when tactile sensation and visual knowledge come together. Differences among tissues become palpable. Skin slightly resists a scalpel, giving a feel for the skin's fibrousness. The same scalpel slides easily through fat. Scissors, used in "reverse," to spread tissues rather than cut them, puncture and widen incisions in fascia only with some difficulty. Nerves are hard and slippery. One surgeon likened nerves to pasta cooked until it's *al dente*, soft on the outside with a harder core. Blood vessels, which are hollow tubes that in cadavers often look like nerves, give the sensation of two slippery layers gliding against each other when rubbed between gloved fingers. The vivid tactile and kinesthetic experience of dissecting strengthened my knowledge of anatomical structure. Medical students begin to make visual and tactile distinctions among anatomical features while dissecting cadavers, but the distinctions grow finer for those who elect surgery as a career.[10]

Trainees spend years practicing under the supervision of attending surgeons, developing skill they can generalize from one procedure to another and from one body to another. Through this extended apprenticeship, they acquire a "muscular gestalt" that, combined with knowledge of anatomy, pathology, and problem-solving skills, leads to "the power to respond with a certain type of solution to situations of a certain general form."[11] Broad surgical skill can be divided into tacit and explicit knowledges. Explicit knowledge can be codified in textbooks, procedural scripts, and verbal instructions. Tacit knowledge cannot be taught solely by verbal means.[12] Tacit knowledge has two forms: physical skill and unspoken social knowledge. A surgical simulator addresses only the physical aspects of surgical skill, though some simulators also incorporate explicit lessons about surgical procedure. Harry Collins and others call this type of physical knowledge "mimeomorphic," meaning skills that can be taught without complex socialization in a group.[13] For a simulator to represent the experience of surgery, the user must see the body on the screen and feel its responses to surgical actions. The computer must facilitate a visual and kinesthetic interaction between the surgeon-user's body and the virtual patient's body, representing the user's actions and the model body's reactions as graphic and "haptic" feedback.[14] Haptics is tactile and kinesthetic feedback. In computer device research, haptically enabled instruments provide physical feedback from a virtual object to the user, creating the sensation of interacting with a material object. Adding haptics to a simulator creates a tight link between sensation and action, a significant research challenge for simulator makers that is neatly captured by taking literally singer Cassandra Wilson's question, "Is there any way to feel a body through fibre-optic lines?"[15]

Stanford's simulator incorporates haptic feedback. The construction of haptically enabled surgical simulators involves three distinct but related research areas: graphic modeling, haptic interface design, and studies of haptic cognition. Each research area requires surgeons, computer experts, engineers, and others to develop new understandings of the model patient's body and the user's body and to incorporate these understandings into computer software and interface devices. Surgical simulator makers must parse the physical components of surgical skill. Looking at technical practice in medicine can illuminate the construction of bodies in medical work in new ways.[16] Studying haptically enabled simulators as they emerge provides an opportunity to examine surgical practice and the construction of surgical knowledge by following how researchers construct a digitally and mechanically mediated relationship between hands and patient. This chapter shows how studying the construction of a medical teaching technology can reveal facets of surgical practice that are not as readily apparent when observing traditional operating room instruction. The process of simulator construction reveals, I argue, the shaping of the patient's body by the surgeon and, reciprocally, of the surgeon's body by the patient. It also reveals how bodies and machines are mutually constructed during simulator design. I call these processes mutual articulation.

Mutual Articulation in Surgery and Simulation

Mutual articulation emerges from the concept of articulation and provides a means of studying the acquisition of surgical skill and the design of surgical simulators. My examination of a surgeons' knowledge at the interface of the surgeon's hands with the objects of surgical action—instruments and bodies—fits well with recent studies of medical practice[17] and with critiques of observers' tendencies to interpret their observational perspective as equivalent to actors' perspectives.[18] Examining the role of hands follows a recent trend in science studies toward an emphasis on the objects of medical knowledge as they are brought into being through practice, "Instead of the observer's eyes, the practitioner's hands become the focus point of theorizing."[19] (Studying the relationship between hands and objects in surgery and surgical simulation moves the focus of the observation away from visual and cognitive models toward a focus on what happens at the interface of hands and instruments. Although anatomy and surgery are undeniably visual, the role of physical interaction in the development of surgical knowledge remains underexplored.

"Mutual articulation" follows from Bruno Latour's concept of "articulation," which describes how bodies come into being through sensory interactions with the world.[20] Latour acknowledges the difficulty of describing what a body is. He suggests that the body is most usefully imagined as an interface that becomes increasingly differentiated as it interacts with more elements in the world. Bodies and body parts come into being through the process of learning to articulate differences. Following this approach, attending to the body means focusing on what the body becomes aware of. The sensing body becomes increasingly articulate as the senses learn to register and differentiate objects. Latour cites the example of a kit the perfume industry employs to teach future perfume makers the art of smelling. Using this kit, students learn how to differentiate extremely dissimilar smells and then to make progressively finer distinctions. Skilled perfume experts become known as "noses." The metonym reveals how sniffing skill and body part become synonymous, how they come into being together. The play on multiple meanings of the word *articulate* as "jointed," in the sense of a body having joints, and "speaking intelligibly," suggests that bodies and knowledge come into being together. Viewed from this perspective, much of medical education is a process of articulating two bodies—the patient's body and the physician's body. Latour says scientific and technological instruments extend the senses and the process of instrumental discovery mirrors sensory learning. Acquiring knowledge, whether through the senses or through the mediation of instruments, becomes a process of articulating differences in the world.

The concept of articulation works well when, as occurs in the case of the perfume kit, the teaching tool is standardized and stable. In surgery, a surgeon must create the surgical site, sculpting flesh, with all its variations, into an approximation of an anatomical model. What happens when, as occurs when the human body undergoes surgery, knowledge of the object is embodied in the surgeon at the same time that the surgeon brings that object into being? I argue that patient and surgeon shape each other through a process of mutual articulation. The physician crafts the anatomical body from the indistinct tissues of the patient's body[21] even as practice defines and reinforces the surgeon's skill. Mutual articulation is particularly important when creating models from complex objects, such as human bodies. With each surgery, the surgeon creates a version of the model from a body's broad anatomical variations and fleshy opacity.

Medical anthropologists and historians describe sight as the privileged sense in biomedicine. The visual is critical to the concept of the physician's abstract "gaze" that collects information from the eyes, ears, and fingers and translates it into an image that could be seen if the living patient could be opened up and viewed with the same clarity as at autopsy.[22] The difficulty with the language of mental models is that we imagine the physician having an anatomy atlas, complete with labels, residing somewhere in his or her brain.[23] The idea that humans have an internal, visual representation of the world—a mind's eye—is an ancient one.[24] And contemporary brain imaging indicates that internal visualization activates the visual cortex in ways and at intensities very similar to actual viewing.[25] Anatomists and physicians have described anatomical learning to me as the process of creating internalized, "mental models" of structures in three dimensions and learning to connect this spatial knowledge with anatomical language.[26]

But what are mental models? Oliver Sacks, in an article on mental "imagery," describes wide variations in states of internal representation among people who went blind as adults. These range from absolute darkness and an accompanying atrophy of visual concepts, such as "in front of," to powerful mental images augmented either by rich imaginings or through cautious checking against real-world referents. Even among those with unimpaired sight, how the world outside is "seen" internally varies from precise, three-dimensional visual images to complete darkness. Sacks describes at one extreme his mother, a surgeon and comparative anatomist, who once studied a lizard skeleton for just a moment, then drew a series of lizard skeletons, each rotated 30 degrees from the last, without glancing at it again. He contrasts her extraordinary visualization skill to a vascular surgeon who, evidently genetically, lacked internal, visual models. Understanding of human structure clearly was deeply embedded somewhere in the surgeon's body, but it was not visual knowledge; that is, he did not see human structure in his "mind's eye." Sacks suggests that the mind may have its own language that is not visual or linguistic or auditory or tactile, but is all these things, and then some.[27] His tales of variations in visual imagining point to the notion that the idea of skill when conceived of as mental imagery may mislead. Further, it may neglect the rest of the body's role in learning. Looking at the interface—at how surgeons and anatomists use their hands to articulate bodies in practice—may be a more effective way to measure their abilities.

A focus on hands encourages reexamination of some classic ethnographic work on medical learning. Medical anthropologist Byron Good describes medical students' first explorations of human bodies in the anatomy laboratory as primarily visual training. The ability to distinguish among the reds and whites of different tissues develops with weeks of experience and practice. Good describes how anatomy students become aware of the internal dimensions of bodies and how they experience perceptual shifts that change the ways they look at bodies. Bodies become something different from the bodies all of us look at every day. A medical student told Good, "I'll find myself in conversation . . . I'll all of a sudden start to think about, you know, if I took the scalpel and made a cut [on you] right here, what would that look like."[28] As Good notes, many medical students describe their difficulty separating everyday bodies from biological bodies at the beginning of their medical training. But bringing greater attention to hands in this story brings forth another aspect: the student describes this process of looking as initiated by the scalpel. This student's knowledge of the body's insides develops while he dissects. Seeing is inextricably bound up with sculpting. The process of learning anatomy is fully embodied, not merely visual.

To take another example, ethnographer Stefan Hirschauer says physicians acquire two bodies. They learn an "abstract body," which is the body as it is represented in anatomy texts and plastic models. They also acquire their own bodies as experienced practitioners. He describes anatomical knowledge and surgical experience as being engaged in a "permanent cross-fading of experience and representation."[29] Hirschauer describes how surgeons sculpt the body, reproducing the abstract body of anatomical representation in the patient's body. He says knowledge and skill develop together, combining "the anatomical *knowing that* of the visible, and the anatomical *knowing how* of making something visible."[30] And he says anatomical images reflect physical means—usually dissection—of their production. But, although he connects abstract anatomical knowledge as contained in atlases to the skills needed to produce those images, he says, "the body of the anatomic atlas, with its clear-cut divisions, different colors, numbered and labeled structures, is present in the surgeon's mind."[31] Hirschauer creates a separation between the skilled work of hands and the visual knowledge of the atlas, which he says resides in the mind. Anatomical knowledge, according to this view, resides in the mind, separate from anatomical skill. But considering anatomical knowledge as it is practiced in the act of sculpting the anatomical body—studying practice at the interface of a surgeon's hands and a patient's body—eliminates worries about the completeness or accuracy of mental models and about the surgeon's ability to translate mental knowledge into physical action. Hands, eyes, and mind are no longer considered separately. Practicing on patient bodies teaches young doctors how to make the fine visual and visceral distinctions among tissues that they will need as surgeons. Attending surgeons use real bodies, and the contrasts between them, to teach students to see and to feel differences among tissues. Students learn these distinctions through the process of operating upon bodies. The abstract anatomical body depicted in atlases and models does not exist in the flesh until it is created by an anatomist or surgeon. The anatomical body comes into being through practice. Practitioner and body mutually articulate each other as the student learns to create the abstract anatomical body from the undifferentiated, unarticulated patient's body.

Treating the mutual articulation of bodies in surgical learning is a useful way to focus attention on what occurs at the interface of a surgeon's hands and a patient's body. In simulator design, model bodies and user bodies must be articulated for the computer. The process of construction of a surgical simulator reveals how surgical skill must be articulated for the computer and, ultimately, for its users. The objectification of the relationship of hands, instruments, and anatomies breaks the process of surgery into many components, forcing surgeons and programmers to make explicit elements of the tactile experience of surgery, such as the elasticity of a uterus or the delicacy of an ovary, that often remain tacit. This, too, is a process of mutual articulation: the construction of a surgical simulator makes explicit the two-way movement of mutual articulation. Engineers and programmers must literally build the relationship between hands and machines by decomposing the action of hands into two components: action and sensation. Hands learn while they do. The eyes and other senses also learn while they do, but the connection is much less direct. Simulator researchers explore the body as an interface in precisely the way Latour describes to understand the elements of information required to pass from hands to object and back. This reveals how researchers articulate the physical connection between hands and model for the computer. And studying simulator research can provide a wealth of new questions for observational studies of surgery, particularly about such areas as haptic knowing and the social aspects of surgery that cannot be taught with a simulator.[32]

The Ethnographic Setting: Merging Disciplines

The Stanford University Medical Media and Information Technologies (SUMMIT) laboratory occupies half a floor of a burnt-sienna stucco office building at the northwest corner of the Stanford University School of Medicine. I did ten months of participant observation at SUMMIT, starting in late 2001, just after Silicon Valley's "dot-com" boom had "dot-bombed." The laboratory shares a floor and a loose affiliation with Stanford's Medical Informatics group. The twelve-year-old laboratory has twin roles: information technology research and service to the medical school. Much of SUMMIT's work falls within the emerging field of medical informatics, which seeks to apply computer science and technologies to medicine.[33] The laboratory looks at first glance like a small Silicon Valley cubicle farm, containing offices, a computer laboratory, a server room, and a small conference room. A closer look reveals the presence of its other major cultures: medicine and medical education. In an open hallway and waiting area, copies of the *Journal of the American Medical Association* occupy shelves next to *Internet Week*; *Syllabus*; *Academic Medicine*; and, bridging the cultures of medicine and computing, the *Journal of the American Medical Informatics Association*. In computer rooms and individual offices, atlases of anatomy and histology occupy space on bookshelves next to handbooks on programming and designing with C++, Perl, and Director. These objects reveal the heterogeneous disciplines that SUMMIT researchers draw from when building technologies.

The laboratory employs about thirty people: a director, seven or eight researchers, web designers, project managers, students, and support staff. Researchers in the lab include four surgeons, mechanical and electrical engineers, several educational technologies experts, and a haptics researcher. Collaborators, including computer programming and networking experts, work from other laboratories at Stanford and in other universities, connecting with the group via telephone, e-mail, and video-conferencing systems. Researchers come to SUMMIT from medicine, computing, education research, and engineering. The group employs roughly equal numbers of men and women at all levels and has an extraordinary diversity of races, ages, and backgrounds of U.S. and non-U.S. origins, reflecting a cultural pattern among Silicon Valley residents that values "dense networks of skilled, mobile, and 'diverse' professional workers."[34] The group's cultural and professional diversity seems to ease cross-disciplinary research: the lab contains such a broad mix of disciplines that no one domain dominates. Rather, the collaborative work of building computer technologies and the creation of programs, graphics, and medical content strongly shape the laboratory's identity.

Researchers at SUMMIT tend to fall into one of two groups that can be loosely described using terms borrowed from information theory: The physicians and educators, "content" people, develop the pedagogical contents of applications and ensure their accuracy and validity as teaching tools. The "information" researchers, mostly programmers and engineers, study ways to transmit those contents — information — to users, doing networking research, device building, and programming. Though the cultures of medicine and computing are distinct, a danger exists in describing laboratory members as rigidly bound to one or the other. The physicians and others I have described as occupying the "content" side of SUMMIT's research work are all highly computer literate. They participate, at various levels, in computing, the high-tech world of Silicon Valley, and "a computer culture that in one way or another touches us all."[35] Three of four surgeons working in the group have studied programming, hardware wiring, or Web design, and the fourth has done extensive work with digitized medical images and models. Conversely, many engineers and programmers, the "information" people, have spent years creating medical devices and applications. However, as the group director pointed out to me, all lab members received the bulk of their training within one discipline and very few, if any, can create both contents and the information structures to deliver them to users. Similarly, surgical simulator design requires, at a minimum, software writing skills, mechanical and electrical engineering, anatomy knowledge, and surgical skill, clearly more than any one person can master. The diverse abilities of laboratory workers are reflected in the physical space they inhabit and in the technological objects they build.

Body Objects: Emerging Technologies

SUMMIT's projects bring researchers' diverse knowledges together in interactions around objects: hardware, software, terminology. These "object worlds"[36] become focal points for negotiations about bodies and machines, medical and engineering practices, and how they interact.[37] For example, I watched the arrival of a second-generation haptic interface for the virtual reality pelvic simulator. The device was designed to mimic the look and feel of handles used in laparoscopies, a set of minimally invasive abdominal procedures. A surgeon, who has retired from his gynecological post in the medical school and now is a full-time simulator researcher, examined the device together with several members of the lab, including collaborators remotely linked to the group via a video-conferencing system. The gynecologist fiddled with the device for a few minutes, feeling its handles, their weight and movement, then said: "This is a significant advance. . . . It's lighter weight and it doesn't feel so resistant in your hand. . . . These [handles] are lighter weight. They feel less metallic. They are less metallic because they're plastic and it gives a better sensation. They're still wide. They're still heavier, but they've got to accommodate a lot of stuff. And the rotation works smoothly, just the way it ought to."

A bit later, the gynecologist introduced the device to one of the remote collaborators, calling it the "Number three interface for the surgery workbench" and describing its "five degrees of freedom and force feedback." When comparing the device's weight and spacing to analogous surgical instruments, the gynecologist was discussing the device as a surgeon, comparing it to more traditional tools. When describing the device as an interface, and discussing its degrees of freedom and force feedback capability, he was thinking about the device in engineering terms, as a component in a surgical simulation system. Researchers in this multidisciplinary space tend to "cycle through" various professional identities—as physicians or engineers, as medical technologies researchers, as computer users—while they work with people from other disciplines to build these technologies.[38] Scientists reconfigure natural objects into laboratory objects, and laboratory objects reconfigure the social worlds of scientists.[39] This is true of SUMMIT's technologies, including the simulator interface. They are products of negotiations among physicians, engineers, and computer programmers who must absorb knowledge from scientific cultures outside their own—physicians learn about programming and engineering, programmers and engineers learn about medicine—to create these objects. The fields represented at SUMMIT—surgery, engineering, computer science, and education—are not merging in this new disciplinary space merely because researchers inhabit a shared space, but because they work together to build these hybrid objects.[40] This work of negotiation and construction is the work of developing a new field and of building objects that are the field's "signatures."[41] Bringing computers into the teaching of surgery means bringing engineers and computer experts into the study of surgery. At SUMMIT, a new discipline is forming around modeling and other skills needed to build objects.

The heterogeneous groups of researchers at SUMMIT create multiply articulated objects—objects designed to be used by both computers and users. I call these objects "body objects," representations of human bodies as they have been engineered to inhabit computers. Body objects are teaching tools, diagrams, and models that reflect SUMMIT's character as a computer research laboratory for creating medical teaching tools.[42] On a shelf in the director's office sits a cardboard model of a child's skull, an artifact from an early project. The director and other researchers created the model by programming a computer to calculate the curves of a real skull from the outlines depicted on a series of CT cross-sections of a skull. Those calculated curves then became outlines of cross-sections of skull, which they cut out of cardboard. Stacking sequential cardboard cutouts created the three-dimensional skull model, an early cardboard proof-of-concept of graphic models now common in medical modeling. In another office is a whiteboard drawing of a finger overlaid with a schematic intended to show the physics of finger motion and what happens mechanically when it fractures. The drawing became a computer animation of a broken finger driven by a mathematical description of its motion. The gynecologist's office contains a small pink and white foam model of a uterus. The uterus served as the model for a CAD (computer assisted design) model of a uterus, a prototype for a virtual reality surgical simulator. The project, a commercial venture, failed and the gynecologist began building models that originated with real cadavers. Each object is a body part as it has been built or defined in relation to a particular technology: the skull is described as computed cross-sections, the finger as force vectors that allow the computer to model the movements of a broken finger, and the uterus as a foam model that would have to be resolved into its most elementary shapes and reformed in the computer as a CAD model.[43] Regardless of their purpose or success, such objects reveal how the combined engineering, computational, and medical knowledges of the group come together in these computational body objects.

Body objects are hybrids: the skull, finger drawing, and foam uterus are medical and computational or engineering objects. They are models and representations of bodies, all originating in medicine, that have become intertwined, visually and semiotically, with knowledges culled from engineering and computer science and, physically, with sensors, wires, and processors. Body objects also are narrow: because the computer requires specific mathematical descriptions to calculate a line or determine a trajectory, body objects cannot be loosely described in ways humans understand intuitively. Body objects are representations of bodies articulated graphically and haptically, so humans can understand them, and mathematically, so computers can understand them. Applying new disciplinary knowledges to surgery articulates bodies in new ways. Both the surgeon's body and the model-patient's body must be articulated for the computer in ways that previously were unnecessary for surgical teaching.

To build a virtual reality simulator, researchers must create body objects that are incorporated in the computer. This requires a crucial epistemic move: the body must become mathematical, described using equations the computer can interpret. Actions and sensations surgeons usually experience physically must be calculated, just as the finger's motion and the skull's curvature must be calculated in the examples cited above. In the world of surgical simulation, a virtual body must interact with both computer and user as a mathematical and a visual-physical entity. The laboratory director describes the mathematics of modeling bodies: "The only way the computer can understand things is, in this case, through geometry. It needs geometry. It needs to know how to compute a sequence of forces with equations, which previously, in a sense, [surgeons] did in their heads. You knew how to predict what was going to happen. You didn't solve an equation to do that, it was just part of the experience. So it's the computer that forces you to put that mathematical construct on."

As the director indicates, surgeons predict the consequences of their actions based on their experience as it becomes incorporated in their bodies and others' experience distilled in papers, procedural scripts, and apprentice-style teaching. In contrast, computers must "understand" bodies and their actions mathematically. The computer requires that each step in a body's motion be modeled as a discrete mathematical state acted upon by the movements—forces—of tools wielded by the surgeon. The feel of surgery, which surgeons' bodies typically experience phenomenologically—as practice—must be parsed, calculated, incorporated into the computer's programming, and ultimately, fed back to the human user, who then will experience the sensations of performing a surgical procedure phenomenologically.

Materializing the Virtual Patient

Traditional methods of practicing surgical technique outside the operating room include suturing bananas and other natural objects, practicing on rubber and plastic mannequins, and performing procedures on cadavers. All are used in practice, but they have limitations. Bananas bear some tactile resemblance to skin, but the analogy to surgery ends there. Mannequins are expensive and wear out quickly. Cadavers require the presence of an anatomy laboratory and staff, who must maintain a willed-body donation program, an expense many medical-school administrators want to reduce.[44] A procedure can only be performed on a cadaver or a rubber model a few times before it falls apart. Researchers also see several potential positive reasons for adopting virtual reality models rather than physical models to virtual reality. Unlike a cadaver, a simulation is reversible—the computer can be reset—so students can practice as often as needed to acquire a skill. The computer also can track student progress and, ideally, suggest corrective measures, which researchers hope will help the student master a procedure correctly and reduce operating room errors.[45] Simulator makers also are discussing their technologies with specialty certification boards, which might eventually adopt simulated exams as a means of ensuring student competence. Enrolling specialty boards in simulator researchers' networks also might ensure simulators' adoption by medical schools, their use by students and, ultimately, their success.

Simulator research at Stanford, and most virtual reality simulator research elsewhere, focuses on minimally invasive procedures. To perform a minimally invasive procedure, a surgeon inserts a camera and instruments through small incisions in the body and performs the procedure while looking at a monitor that shows surgical action taking place inside the body's interior, a move one surgeon at SUMMIT describes as "operating on images, not on patients." Because minimally invasive surgery already occurs "on-screen," the move to simulate these procedures is easier than with open surgery. Although efforts exist to simulate open surgery, surgeons often use their hands directly inside the body when doing open surgery, a practice that would be more difficult to simulate than surgery with instruments. And minimally invasive surgery involves more kinesthetic than tactile sense, making the provision of haptic feedback easier. Simulating surgery also takes advantage of a feature of all surgeries: the operating field is separated from the rest of the patient's body, which usually is covered with sterile drapes.[46] A simulated patient represented as a fragmented body part on a computer monitor may resemble the surgeon's visual experience of the operating field more than might be apparent.

The system requires a user, graphic models of patient body and surgical tools, an interactive device designed to look and act like the surgeon's end of an instrument, a computer to manage the haptic device, and a separate computer to run the simulation. Making the system work requires definition of how these components work together. Materializing tools and bodies in cyberspace requires what are, in effect, three feedback loops that make up the interaction between user and model. The first—or virtual—feedback loop defines the interaction between instrument tips and model body as the model responds to the instruments and, in turn, provides haptic feedback to the user. This is the domain of computer modeling. Researchers— programmers and surgeons—wrestle with the question: how can we create a graphic and physical model that accurately represents the body interacting with the instrument? The second—or mechanical—loop describes the interaction between the user's hand and the instruments as the instruments respond to user and model. This is the domain of mechanical engineering research, which aims to answer the question: how can we ensure that our device works properly—feeding correct haptic information to the virtual world and back to the user's hands? The third—or cognitive—loop connects the user's mind, his or her intent, to the user's hands, while hands and device interact. The cognitive loop represents the domain of haptics research, and this question predominates: how does a body learn and what mental models do our tactile and kinesthetic actions help us create? Each of these loops represents a research area among simulation experts. Each requires descriptions of the virtual patient's and the material user's bodies as they interact with the simulation. Though I describe these loops as independent entities and, at SUMMIT, they represent somewhat independent research projects, researchers want to build a simulator from this complex assemblage of hardware, software, and expert knowledge that can represent a visual and physical experience similar enough to performing surgery to help the student learn. Although each component of the simulator defines the relationship between model and user slightly differently, the components attempt to give the user a seamless experience of surgery.

Modeling: Constructing the Model Patient's Body

SUMMIT's laparoscopic simulator contains a model of the female pelvis made from ninety-five digitized photographs of pelvic cross-sections. The sections came from an anonymous thirty-two-year-old woman who willed her body to Stanford before she died. Anatomists at Stanford froze the pelvis in an upright position, then ground layers off at roughly 2-millimeter intervals. After removing each layer, they took a photograph of the newly exposed cross-section. The retired gynecologist used the collection of cross-section photographs as the foundation for Stanford's virtual reality simulator. He named the collection the Stanford Visible Female, linking it to the National Library of Medicine's better-known Visible Human Male and Visible Human Female, which were created using similar techniques.[47] He scanned the ninety-five cross-section images into a computer. He then spent more than a year tracing the structures he wanted to model into files using an early version of PhotoShop, a commercial image-manipulation application: one file for each structure on each cross-sectional image. He describes this process, called segmentation, simply as "drawing circles" around each structure he wanted to model and saving the contents of each "circle" as a "mask" with its own computer file: "I would make a mask and I would put it in the muscle file. And I'd make a mask and I'd put it in the bone file. And then I'd go to the next slice, put the bone in the bone file. Next slice. And so I ended up with all of these files that had individual masks and then we took the software . . . and made models from those masks."

The gynecologist initially segmented only the reproductive system, leaving the six pelvic bones and many muscles as undifferentiated aggregates labeled "bone" and "muscle" respectively. Subsequent iterations differentiated pelvic bones and muscles and added less critical features, such as fat. The gynecologist segmented the reproductive organs and a collaborating orthopedist segmented the bones and muscles. They produced 2,200 masks from ninety-five cross-section slices encompassing the female reproductive system and the surrounding musculo-skeletal system. The division of labor occurred because each physician had a slightly different area of anatomical knowledge. Segmentation includes several of the "transformative practices" Michael Lynch identifies in relation to model making, including "upgrading" the images by making strong borders between tissue types and "defining" the images by sharpening contrasts.[48] But segmentation is not done to make the cross sections readable by human eyes. Rather, anatomists segment cross-sectional images to create outlines that are integrated into computer-modeling programs, that is, to articulate cross sections for computers.

The orthopedist compares the difference in the two surgeons' anatomical expertise to geographical knowledge of highways and interstates in the Bay Area:

> I think it is a question of with what granularity you look at [the body], with what amount of detail. To try to give an analogy, it's like . . . a map. If you look at a map, say you're looking at the map of the Bay Area and it's an overall map, and you say, there's [U.S.] 880 and there's [U.S.] 101, and you have a fair idea of the map and that's your basic anatomy. But now if you want to know about Palo Alto, then you need to zoom down. Oh, there's El Camino and there's this and there's this. So now you know a little bit more detail there. It hasn't changed your 880 and 101 knowledge, which is over the Bay Area. . . . So, if you're talking about the radial nerve, if someone doesn't have to deal with the radial nerve surgically, they have an idea, OK, the radial nerve comes from there and goes there. But the finer bends and curves only somebody who is dealing with it would know.

The analogy between anatomical knowledge and a map is quite common, but ignores several complexities inherent in anatomical segmentation. First, cross-sectional images of the body have no labels to guide the surgeons as they segment. Second, although some radiological images, notably CT, are cross-sections, surgeons rarely see actual bodies in cross-section, so interpreting cross-sectional images requires a mental extrapolation in three dimensions from one angle of approach to another, the mental equivalent, perhaps, of trying to read a map of the Bay Area from a diagram of its geological strata.[49] The level of anatomical knowledge required to segment one female pelvis also speaks to the extreme specialization of surgical-anatomical knowledges and to the difficulty of producing a comprehensive model body.[50] The anatomical body, even in a partial area, such as the pelvis, required digital articulation by specialists from two surgical disciplines.[51] This is an example of Annemarie Mol's "body multiple."[52] The female pelvis is a single, albeit complex, anatomical region that is, in practice, a gynecological pelvis, an orthopedic pelvis, and more. The orthopedist's term, "granularity," can be thought of as the multiplicity of practices that bring different areas of anatomical region into being.

Up to this point, medical experts—the two surgeons—did the work of delineating body parts. The next modeling steps multiplied the body in another realm of practice: the world of computer modeling, a subspecialty of medical informatics. A computer-modeling student took the segmented masks and computationally stacked them, creating models of organs, muscles, bones and other features (as stacked slices of bread create a loaf). To connect cross-sections into a surface model, the student transformed stacked outlines into a "mesh," a digital, mathematically generated net that mapped the model's surface. Modeling using this technique takes advantage of a digital photograph's resolution into pixels. Once gynecologist and orthopedist outlined the structures to be modeled on the two-dimensional cross-sections, the modeling student wrote computer algorithms to connect the outlined pixels across adjacent cross-sections, creating a geometry the computer could understand. These connected pixels formed a mesh conforming to each structure's surface. Because this model is made of both graphic pixels and the mathematical mesh, the model body is simultaneously a graphic and a mathematical representation of a body—a representation that can be viewed and manipulated by a human user in ways the computer can calculate. These graphic models are "silicon second natures," digital artifacts that mirror natural objects, but also offer to replace them as resources for medical learning and research.[53]

The gynecologist, who spent eighteen months doing the first segmentation of the Stanford female pelvis, described the first time he saw the reconstructed uterus made from the masks he drew:

> And so when I saw that uterus the first time, the thing that blew me away was not what I expected to see, but what I hadn't expected to see and that is where the utero-sacral ligaments attach to the cervix and support the uterus in the pelvis. There are a couple of little bumps, little sharp points there where those take off that I could see [on the model]. And, of course, that relates a lot to my surgery, which is on those ligaments where endometriosis occurs. So many laparoscopies I did finding endometriosis on those ligaments and in the region of the pelvis that I was so drawn to the image. There they are. And I could see them.

The gynecologist described the process of drawing outlines of structures on cross-sectional photos as a process of abstracting the human body's complexity and specificity. But when the model came together, the resemblance of the model uterus in the computer to an actual uterus gave the gynecologist a sense of wonder, pleasure, and reassurance that tedious months of drawing circles produced a model that looks like a uterus. Hirschauer describes anatomical exposition in surgery as sculptural practice. This is a process of carving a body resembling an anatomical model out of messy, indistinct flesh. The cross-sectional photographs the gynecologist began with are themselves representations that neither computers nor inexperienced medical students can use to distinguish anatomical features. By drawing outlines of anatomical structure that could then be computationally stacked, the gynecologist and the programmer performed a sculptural process analogous to surgical exposition. The surgeon's experience confirms the computational procedure's success: The utero-sacral ligaments depicted on the model looked like ligaments he had operated upon. Modeling transforms the photographic cross sections into a neat, three-dimensional model of a uterus that has already had fat dissected away, in other words, a graphic model that resembles an anatomical site that an anatomist has already exposed. The surgeon physically—using a computerized drawing pen instead of a scalpel—articulated a model that represented his experience. The model body then affirms for the gynecologist that this computational procedure worked and has produced a tool he considers adequate for teaching surgical practice to simulation users.

The gynecologist named the newly modeled reproductive systems Lucy 2.0, describing it as the "digital daughter" of the famous hominid bones found by Stanford researchers in Africa in 1974.[54] This model human body is a laboratory object: it is the image of the original object (in this case a human body), detached from its natural environment, and no longer beholden to the original's temporality.[55] Unlike a living or dead human body, the model body can travel through a computer network, can be pulled apart and put back together, or modified to reflect pathologies, all without causing it harm. The model body becomes an "immutable mobile," a digital reconstruction of the original with the advantage of "mobility, stability, and combinability."[56] But the model in this state is useful primarily for teaching anatomical structures.[57] It is visual, but it cannot yet interact with the user as a material body would. It is not yet a patient and it is not yet prepared for surgery because surgery, at its most basic, physical level, involves interactions of bodies and instruments. Before the model pelvis could become what one Stanford researcher calls a "patient-on-demand," it had to become responsive to surgical action. In Latourian terms, it had to become articulate, or able to be "moved, put into motion by other entities, humans or non-humans."[58] To make the pelvis deformable, a programmer added algorithms to the model that describe how tissues stretch, separate, or come together—that is, how tissue deforms—when pulled, cut, or sutured. The programmer began with the mesh structure of the surface model and defined the lines connecting points on the mesh as springs. Pulling on any point of the virtual mesh causes the surrounding virtual springs to stretch, "deforming" the model according to well-defined physics equations that describe the resistance of springs. Spring-based deformations are useful for small, relatively slow movements of tissue, as are common in surgery. Stiffer springs lead to tougher-feeling tissues.

To set values for spring stiffness, the gynecologist and the programmer developed heuristics describing the feel of pelvic tissues. These mathematical descriptions are constructions based on the gynecologist's physical memories—what he calls "haptic memories"—of the feel of performing surgery on various tissues. The gynecologist expressed his haptic memories in terms both of his sense of differences among tissues and his sense of the feel of a particular tissue.[59] To develop the haptic program, gynecologist and programmer created algorithms that attempt to represent the surgeon's physical experience in a form the computer can use. To do this, the programmer had to learn something about surgery. He learned the physical differences between structures in a woman's reproductive system. He also learned some terminology of anatomy and surgery. Most importantly, he found a way to physically describe the gynecologist's embodied actions. He said he created a description of "how the world works" at a deeper level than typical surgical instructions to cut, clamp, or suture. In effect, the engineer developed a physical model of the movements behind each of those verbs.

Traditionally, tissue stiffness is known only through surgeons' bodies and might be communicated to a student as a general warning about the potential to harm delicate tissue, such as a warning that damaging or cutting a nerve during surgery could be a "million dollar [malpractice] injury." The surgeon's understanding of tissue feel comes from years of practice. Constructing a quantitative model of a patient body's physical response to surgery only becomes necessary when haptic feel moves from body to computer. The redefinition of a patient's body from the body experienced by the physician to the body defined for the computer is an important new articulation of bodies. Moments where these reconstructions become evident can be both revealing and amusing. During a demonstration, the programmer runs into a technical glitch and tries to describe to the gynecologist how the uterus feels: "Hey, do you want me to reset your uterus there? . . . Do you want me to bump up the stiffness so it behaves like muscle? Now it's behaving like a thin skin. I think that's something I learned from you [the gynecologist]: that the uterus is basically like a tough muscle. Now it's behaving like a thin skin."

The idea of "resetting" a uterus comes from computer science and shows how the conceptual vocabulary from that discipline contributes to defining the body in the world of virtual anatomical modeling and surgical simulation. Verbally, the programmer describes what he has learned from the surgeon about tissue feel. Mathematically, he attempts to approximate the surgeon's bodily experience, translating a surgeon's experience of a body's feel, which usually remains tacit, into equations describing the stiffness of springs. The virtual model body is put into motion as a function of the movement of springs. This is the type of "mathematical construct" the group director refers to when she says knowledge that once was primarily experiential must become mathematical when translated into a computational idiom. The feel of the model body's movements becomes articulated in relation to the gynecologist's experience as it gets translated into algorithms. In turn, the differences in tissue feel incorporated into the model will help articulate the students' bodies; that is, these differences will help students learn the feel of model bodies, feel that, if the simulation succeeds, will allow the transfer of the surgeon's skills from simulated to material bodies. Tissue feel can be described, but only using relative terms, such as "delicate" and "tough."[60] Students can use these descriptions to guide them while relative differences in tissue feel become embodied knowledge. But the computer requires experiential haptic knowledge of difference to be articulated as mathematical values. The surgeon constructs differential values from his experience and the programmer translates these into mathematical descriptions of tissue feel. The model's deformability does not, cannot exist apart from the thing it interacts with, in this case, the surgeon's body as mediated by instruments. Deformability is a quality of model bodies defined exclusively at their interface with other bodies. Values of tissue feel used in deformable models are products of the mutual articulation of bodies.

Interacting: Characterizing the User's Body

By making the virtual model deformable, programmers had built the possibility of movement into the model body, but it could not yet be put into motion by a user. The next key step in making the surgical simulator was to create an instrument to act upon the body. Because the user activates the instrument, which then acts upon the model body, the instrument becomes, in effect, a bridge from a body in the real world to a model body in the virtual world.[61] A bridge can take the form of several types of devices, but ones I have seen share this feature: they all exist both on and off the screen. This existence in both worlds resembles many gaming devices, but medical researchers pay more attention to giving users a realistic feel for surgical interaction. The coupling of haptic action and reaction is tighter and more rigorously defined and is itself a unique research area. SUMMIT's gynecology simulator uses the two-handed, or "bi-manual," device described earlier, which was designed to mimic the feel and motion of instruments used in laparoscopic surgeries. SUMMIT developed the device jointly with Immersion Corp., a San Jose medical device manufacturer. The device is a heavy, metal box with two protruding handles. Each handle has a scissor-like mechanism at the end that allows the user to manipulate virtual instrument tips.

When a user turns the simulator on, graphic representations of surgical instrument tips—the patient ends—appear on the computer screen in the same space as the body model. A multiprocessor graphics computer runs the simulation. The computer uses a method known as "collision detection," which tells the instrument tips and model body to react when they enter each others' coordinate space; that is, when they touch. Outside the computer, the surgeon's ends of the instruments resemble surgical instruments whose virtual tips move as the user moves the handles, giving the illusion that real handles and virtual tips are continuous. Closing the metal, scissor-like handle in the real world clamps the virtual instrument tips in the virtual world. When the user pulls the handle, virtual tip and tissue move with it, allowing what the gynecologist calls "tool-tissue interactions." The instrument acts in two directions. The bi-manual device allows the user to perform actions on the handles that translate into action at the tips; the tips, in turn, act on the model body. The device also transmits back to the user's hands—in real time—the effects of those actions on instrument and model, providing haptic feedback.[62] When I clamped the instrument onto a virtual ovary, for example, I felt a distinct snap as the instrument locked onto it and resistance when I pulled the virtual tissues. In reality, all I pulled was the physical interface handle; on the screen, the instrument tip retracted, pulling the ovary with it.

Within the context of the mechanical feedback loop, the user's body emerged in relation to the haptic device as engineers designed the device and began to study how it operates in practice. A mechanical engineer said engineers and surgeons had lengthy conversations during the design process to resolve such details as distance between the handles and the range of movement the device should have: "There was considerable debate from engineers like me who wanted to simplify things by removing some degrees of freedom, but surgeons argued you needed it." Each new

Anatomy of a Surgical Simulation · 299

capability makes the device more difficult to manage mechanically and computationally, but surgeons demanded fidelity to surgical experience. Realism requires that the device faithfully mimic not only the feedback of interacting with patients' bodies, a software design challenge, but also the spatial and tactile feel of instruments themselves, a hardware design challenge. Designing a device that correctly interprets the signals it receives from the human user and correctly feeds the haptic response back to the user gives rise to a fascinating problem: characterizing the human user's effect on the system. During an eight-hour meeting of laboratory researchers with an external reviewer, who is an expert in educational technologies, participants tackled the question of how to consider the user's body as it interacts with the device:

> Mechanical engineer: We will have to do a study that accounts for variability among subjects.
>
> Laboratory director: When [our collaborator in Texas] uses Immersion stuff, she's always complaining that she's not getting the kind of frequency response they claim it should have.
>
> Mechanical engineer: The dynamic response slows if a human hand is holding the device.
>
> Laboratory director: It's like having a sloppy, wet mass holding the thing.

Human bodies, viewed here as research objects, create several difficult problems for investigators. Bodies are variable; that is, not all bodies affect the device the same way. And users' bodies slow the device down, compromising its ability to faithfully transmit the sensations of interacting with the model. The research question becomes how to manage the effects of this "sloppy, wet mass" (or many, varied sloppy, wet masses) on the device's response. In surgery, the surgeon's body and tools, when they're performing well, are the unproblematic agents of surgical action.[63] This is the essence of embodied skill: with years of practice, surgeons learn to use tools as extensions of their bodies. Technique becomes fully embodied and, therefore, largely unconscious, when all proceeds smoothly.[64] But the effect of the surgeon's—or user's—body on the bi-manual device and the virtual simulation must be characterized mechanically and compensated for by the simulator, so the interaction of cyberbody and material body feels like an interaction between two material bodies. The user's and the model body's ability to mutually articulate each other depends on the ability of programmers, surgeons, and instrument makers to create a good enough representation of the feel of performing surgery on a live body. This requires articulating the user's body for the instrument and for the programs that control the instrument. Researchers must account for the sloppiness and variability of users' bodies so the user can properly articulate the model body and receive useful physical feedback. This is another example of mutual articulation: the user's body must be articulated for the instrument, so it can articulate the feel of doing surgery for the user.

Embodied Cognition: Integrating and Translating Skill

The cognitive feedback loop—the work that happens between hand and mind—takes up the question of what we learn through our bodies and how what's transmitted to the body gets interpreted and learned. A physicist turned cognitive scientist does haptics research at Stanford. She has conducted a series of experiments intended to elucidate poorly understood haptic concepts, such as the delineation of edges, which we use to understand our world through tactile and kinesthetic sense. She also is investigating how many times a particular pattern in space must be repeated before the body learns the pattern. She wants to better understand the role of physical learning in surgery and to help develop more effective devices, including surgical simulators. She sums up the research project as the attempt to characterize "somato-conceptual" intelligence: "Haptic sensations are personal. I cannot tell you exactly what I feel. It's personal. It's felt by the touching person only. It's determined by the touching forces. Each person exerts different forces. There's a different coefficient of forces for muscle, so we experience different things."

In this researcher's study of haptic cognition, material bodies become bodies that exert forces on objects and receive forces from those objects. But bodies vary. And varying bodies exert different forces on objects, so experience also varies. According to this concept of haptic learning, physical experience is reduced to a set of forces exerted upon and received by muscle, so experience and learning are determined by the interaction of muscular forces with an object. The contribution of other types of experience—of cognitive memory, knowledge culled from procedural scripts, and explicit instruction, gets bracketed.

Studying the path from physical force to learning presents enormous problems for researchers, so the problem gets redefined in terms of the force transmitted to the hands and the user's interpretation of that force.

During the same external review cited above, researchers tackled the problem of how to understand what's happening inside the user's body:

> Haptics researcher: How do you make it so everybody feels the same thing?
> Reviewer: It gets metaphysical very quickly. If we all touch the table, do we all feel the same thing?
> Haptics researcher: It's a bad question because you can't answer it.
> Reviewer: It's a good question; it just shows you're not a philosopher.
> Haptics researcher: Yes, but as a physicist, I understand the question.
> Reviewer: That's because physics and philosophy are close together.
> . . .
> Surgeon: What is felt by the user? What is the force? What is the interpretation of force by the user? Is it possible to measure?
> Haptics researcher: Different surgeons would make the same interpretation when they feel the same lump.
> Reviewer: That's as far as you can go. If everybody says it's a ring, you're in good shape.
> Mechanical engineer: Or 85 percent of them.
> Reviewer: But if you want to get to their subjective experience, then it's the metaphysical problem. . . . You could frame it as a signal-to-noise problem. You can't guarantee the same experience for everybody. But if you can build enough signal into it so most people give you the same interpretation. . . .
> Laboratory director: There may be various sources of signal: how do you know what they're telling you?
> Surgeon: What in the brain it is, you can't measure it.
> Reviewer: You know right where they are and you know what they're interpreting.

This conversation reveals a process of defining the surgeon-user's body in a way researchers can manage. They do this by defining the user's body in relation to the device. They begin with a broad question: How can they ensure that simulator users all have the same physical experience? They recognize that if they try to answer the question in terms of subjective experience, it becomes a philosophical issue, not a question amenable to medical, cognitive, or engineering research. This is the haptic equivalent of Sacks's struggles with internal, visual states: we can image brains, but not minds or experiences. What a user senses through his or her body— whether studied as forces on muscles or descriptions of experience—is inaccessible to scientific research. If haptic knowledge consists of forces exerted on users' bodies and the interpretation of those forces, then studying the connection between force and interpretation becomes very difficult. The researchers then reformulate the user's subjective experience as a question of consistency of interpretation or, in more scientific terms, reproducible results. They realize they cannot know what bodies experience directly or whether two people experience the same sensations when touching the same object. They cannot know whether many users' internal experiences of touching an object, such as a lump, are identical, but they know that many surgeons would give the same interpretation of that object. As the reviewer suggests, shifting the definition of haptic experience away from metaphysical questions about internal experience—away from the body's physical and subjective insides—and toward the body's interface with an object might allow researchers to elicit consistent interpretations of that experience.

Defined as a body that palpates and interprets a lump, researchers can study what the body knows. As scientists, however, they can go one step further. They can augment the signal from the object to encourage more consistency among interpretations. By defining haptic cognition as a relation of signal to noise, they can ensure that the device sends a strong enough message to the user's body that most users give the same response. By observing where on the model the user is working, they begin to understand what signals are strong enough to provide a consistent interpretation. The pathway between the user's body and his or her understanding—the mind-body connection—becomes, in effect, black boxed. It cannot be characterized the way a device might be, or mathematicized the way a model patient's body might be. Rather, the user's body in haptics research gets defined in terms of the signal the rest of the system sends to the user's body and the accuracy with which the user interprets that signal. The question is no longer what the body is, but how the body interprets action; the ontological body becomes the interpreting body. The challenge shifts from trying to interpret what happens inside the user's mind and body toward understanding how to create a model body that surgeons can be sensitive to in identical—or mostly identical—ways. Augmenting the model's signal helps make the interpretations of experience more articulate. The model articulates what the user's body knows, which helps the user articulate what the model is.

Discussion: Vision, Touch, Embodiment

The simulator is an assemblage of hardware and software, shaped by knowledges from multiple disciplines. Simulator research falls into three areas —modeling and deformation, interactive device making, and studies of haptic cognition. Research into each of these areas requires definitions of the model patient's body, the user's body, and how they interact in simulated surgeries. Within each research area, the physical connection between user and model must be delineated. Simulator makers must make mathematical models of surgical actions that usually remain tacit, such as the movements a surgeon makes when clamping, cutting, or suturing, and the response of tissues to those movements. I have laid out how each of the three research areas articulates the user's body in relation to the simulated model body and vice versa. What remains to be done is to consider the implications of mutual articulation for studying the teaching of manual skill.

The deformable model's utility as a teaching tool is limited without values representing haptic feel. The representation of the gynecologist's physical experience that gets incorporated into the model shapes how the model will react to the user and how the model will shape the user's experience.[65] The model body's resistance to surgical instruments is defined in relation to the gynecologist's embodied memories and the resulting algorithms describing the model's resistance will, in turn, shape the user's body. The haptic interface must compensate for the fleshiness of the user's body well enough that the mutual shaping of model and user will provide a meaningful learning experience for beginning surgeons. To do this, researchers will study many bodies, so they can incorporate a model of their variations into the device. And haptics research attempts to define what parts of physical interaction are meaningful for learning by studying what happens at the interface of body and model. Among other methods, this can be done by altering signals the model sends to the user to elicit particular interpretations. The model's ability to articulate the user's body will be measured in terms of users' interpretations. At each stage of this research, the user's body is articulated in relation to the simulation system and vice versa.

Haptics—researching and designing an interface that feeds sensory information to the user's hands—makes the mutual articulation of the user's and the model's bodies apparent because the connection between the hands and the model must be carefully constructed. Technologically and physiologically, the link between the object's effects on the user and the resulting action is much tighter with touch than vision. A haptics researcher best describes how touch differs from other senses: "Touch and force sensations convey information about the environment by that enabling action. Successful bodily acting requires "touch and feel" information from the environment simply because, unlike any other sense, haptics (touch and kinesthetics) is not only a sensory channel to receive information, but also a channel for expressiveness through actions. The hands are both sensors and actuators, using sensory information to control their acts."[66]

This dual nature of hands connects actor to object much more directly than vision, smell, or hearing. Hands simultaneously perceive an object and act directly on it. The effects of touch can be measured as effects on the object. Simulator researchers at Stanford realize this: they know that a poorly designed model of tissue feel or a poorly designed interface may fail to provide the kind of "muscular gestalt" that Dreyfus describes. Conversely, they can boost the signal sent to the hands to make interpretation easier. With a simulated model body, researchers can study directly what forces users exert when dissecting tissues. Researchers also know they can observe exactly what part of the model reacts to the body's actions, making the study of the connection between model and learning more direct. Because hands themselves contain the means of both sensation and action, they embody mutual articulation in a way that forces researchers to place tight constraints on the connection between sensing and acting. The reviewer in the dialogue cited above makes the critical point about touch and knowing, "You know right where they are and you know what they're interpreting." The hand, as a perceptual instrument that senses while it acts, can make studying the interpretations that result from these perceptions and actions easier to study than other senses. Simulator researchers, if they can make haptically enabled simulators work effectively, can guide and enhance the student's tactile learning. Guiding surgical experience is vital to the development of a surgeon's multisensory medical gaze; that is, the incorporation of bodily skills that creates the surgeon's body. With hands, how sensation, action, and interpretation intertwine can be constructed at the interface with an object, as the ability of the user to articulate the model body through anatomical sculpting and the ability of the model to articulate the user's body in terms of surgical skill.

The concept of mutual articulation for understanding surgical simulation addresses a problem that arises when discussing simulation. Latour's concept of articulation specifically attempts to avoid a world of subjects and objects in which the subject houses an internal representation of the object whose accuracy must be verified.[67] The notion of abstract anatomical knowledge and the surgeon's ability to sculpt the body to resemble an anatomical model tends to reproduce this concept of a representation of human anatomy housed somewhere inside the surgeon (typically imagined as inside his or her mind). Considering the creation of anatomical understanding as the development of physical skill that comes with years of practice allows one to consider not the accuracy of an internal visual model, such as may or may not exist, but simply the surgeon's ability to create such anatomy in the patient's body. Thus, surgical skill can be thought of as the interface between a surgeon's hands and a patient's body, as it exists in practice. Whether taught by a simulator or by another surgeon, the surgeon's skill becomes his or her ability to sculpt the operative site from highly variable patient bodies. Simulation reveals that the patient's body plays a role in that shaping.

With a simulated "patient-on-demand," students may have many more opportunities to practice surgical procedures when they want, as often as they want, and on as many types of pathologies as can be programmed into the simulator. Haptics will change the nature of the interactions from viewing and perhaps acting upon the body with a mouse to *feeling* the cyberbody react. The incarnation of bodies in cyberspace that can provide haptic feedback will make these interactions bodily in ways unlike earlier computer technologies, undoubtedly with implications for other fields in which haptic interactions are important. Haptics research, as a field that studies how hands learn, can reveal how bodies mutually shape each other. Additionally, information gathered from research into modeling, deformation, mechanical haptic interfaces, and haptic cognition will contribute not only to simulator research, but also to the development of future medical and surgical technologies, such as radiological modeling, surgical planning, remote surgery, and surgical robotics.

At each point in the creation of the surgical simulation described here, researchers pooled various disciplinary knowledges of anatomy, surgery, computation, education, cognition, and engineering to develop an object (a model, a software program, a device) that has a particular relationship to the user's body. At each point, then, researchers are working to create interpretations of what human bodies are in relation to these objects, that is, to articulate the body in new ways. As I argue, these technological ways of knowing human bodies are multiple, but not unconstrained. The simulator must be relevant for the medical student. It must work as a teaching tool. The simulator must not only articulate patients' and users' bodies as they relate in surgery, it must also help incorporate knowledge of those relations — surgical skill — into the student's body.

Notes

1. Kenneth Ludmerer, *Time to Heal: American Medical Education from the Turn of the Century to the Era of Managed Care* (New York: Oxford, 1999), 49.

2. Atul Gawande, *Complications: A Surgeon's Notes on an Imperfect Science* (New York: Henry Holt, 2002); Raja Mishra, "Study Cites Risks of Low-Volume Surgeries," *Boston Globe*, March 2003: A1.

3. Howard Rheingold, *Virtual Reality* (New York: Simon & Schuster, 1991); Pearl Katz, *The Scalpel's Edge: The Culture of Surgeons* (Boston: Allyn and Bacon, 1999).

4. A third type of medical procedural simulator, called "augmented reality," seeks to put virtual structures and actual hands and tools in the same space, usually through the use of special screens and/or glasses. Augmented reality systems are not part of this discussion. The premise of surgical simulation most closely resembles that of flight simulation, in which students practice physical and cognitive skills, sometimes following simulated scenarios. Other simulations, of economic processes for example, are primarily mathematical constructs that sometimes represent numbers graphically, but lack physical feedback.

5. The social challenges of incorporating simulators into traditional medical school curricula may be a challenge as great or greater than the technological challenges of building simulators; these social challenges include such questions as how to restructure curricula and students' time to accommodate simulation exercises.

6. Bruno Latour, *We Have Never Been Modern* (Cambridge, MA: Harvard University Press, 1993); Donna Haraway, *Modest_Witness@Second_Millenium.FemaleMan_Meets_OncoMouse: Feminism and Technoscience* (New York: Routledge, 1997).

7. I do not wish to suggest that vision is disembodied. Rather, vision is sensory and, therefore, prior to action, even when it is as profoundly part of that action as the kind of hand-eye coordination a surgeon employs.

8. For a discussion of bodily learning in surgery, see Rachel Prentice, *Bodies of Information: Reinventing Bodies and Practice in Medical Education* (Cambridge, MA: MIT Press, 2004).

9. The importance of dissection in medical education is hotly debated. Medical school administrators in the United States and Canada have been cutting back teaching time and resources for gross anatomy for several decades. The justifications for this move are many: maintaining a willed body donation program is expensive, gross anatomy is time consuming, only surgeons really need the in-depth anatomical knowledge provided by gross anatomy, medical students need actual clinical experience earlier. Anatomists are fighting this threat to their discipline

with many arguments, including the importance of the training in physical skills and three-dimensional visualization that gross anatomy provides.

10. Human gross anatomy is sufficiently complex that an anatomist practicing for five decades described learning new structural features with each dissection. A surgeon practicing for two decades described reviewing the anatomy of regions where she rarely operates. Surgeons and residents in teaching hospitals also constantly review and reinforce their anatomical knowledge when a surgeon quizzes a resident, during surgical planning, and while operating.

11. Hubert Dreyfus, *What Computers Still Can't Do: A Critique of Artificial Reason* (Cambridge, MA: MIT Press, 1992), 248–49.

12. Donald MacKenzie, *Knowing Machines: Essays on Technical Change* (Cambridge, MA: MIT Press, 1996). For more on tacit knowledge, see Michael Polanyi, (1966) *The Tacit Dimension* (Garden City, NY: Doubleday, 1966); H. M. Collins, *Changing Order: Replication and Induction in Scientific Practice* (London: Sage, 1985).

13. H. M. Collins, et al., "Ways of Going On: An analysis of Skill Applied to Medical Practice," *Science, Technology & Human Values* 22 (1997): 267–85.

14. Stanford's simulator requires at least three pieces of hardware: a graphics computer to run the simulation, an interface device, and another computer connecting the interface device with the graphics computer. I use "computer" and "simulator" interchangeably throughout this essay.

15. Cassandra Wilson, *Right Here, Right Now*. Capitol Records, 1999.

16. Monica J. Casper and Marc Berg, "Constructivist Perspectives on Medical Work: Medical Practice and Science and Technology Studies," *Science, Technology & Human Values* 20 (1995): 395–407.

17. Annemarie Mol and Marc Berg, "Differences in Medicine: An Introduction," in *Differences in Medicine: Unraveling Practices, Techniques, and Bodies*, ed. M. Berg and A. Mol, (Durham, NC: Duke University Press, 1998), 1–12; Annemarie Mol, *The Body Multiple: Ontology in Medical Practice* (Durham, NC: Duke University Press, 2002).

18. Pierre Bourdieu, *Outline of a Theory of Practice* (Cambridge: Cambridge University, 1977); H. M. Collins, "Dissecting Surgery: Forms of Life Depersonalized," *Social Studies of Science* 24 (1994): 311–33.

19. Mol, *The Body Multiple*, 152.

20. See also Francisco Varela, "The Reenchantment of the Concrete," in *Incorporations*, ed. J. Crary and S. Kwinter (New York: Zone, 1992), 320–38; Bruno Latour, "How to Talk about the Body? The Normative Dimension of Science Studies," *Body & Society* 10 (2004): 205–29.

21. Stefan Hirschauer, "The Manufacture of Bodies in Surgery," *Social Studies of Science* 21 (1991): 279–19.

22. Michel Foucault, *The Birth of the Clinic: An Archaeology of Medical Perception* (New York: Vintage Books, 1973).

23. Hirschauer, "The Manufacture of Bodies in Surgery," 310.

24. Ibid., 310.

25. Oliver Sacks, "The Mind's Eye: A Neurologist's Notebook," *New Yorker* 79 (July 28, 2003), 48.
26. Prentice, *Bodies of Information*.
27. Sacks, "The Mind's Eye."
28. Byron J. Good, *Medicine, Rationality, and Experience* (Cambridge: Cambridge University Press, 1994), 73.
29. Hirschauer, "The Manufacture of Bodies in Surgery," 310.
30. Ibid, 310, italics in original.
31. Ibid., 310.
32. Collins et al., "Ways of Going On"; Prentice, *Bodies of Information*.
33. Diana E. Forsythe, *Studying Those Who Study Us: An Anthropologist in the World of Artificial Intelligence* (Palo Alto, CA: Stanford University), 3.
34. J. A. English-Lueck, *Cultures@Silicon Valley* (Palo Alto, CA: Stanford University Press, 2002), 20.
35. Sherry Turkle, *The Second Self: Computers and the Human Spirit* (New York: Simon & Schuster, 1984), 18.
36. Louis Bucciarelli, *Designing Engineers* (Cambridge, MA: MIT Press, 1994), 62.
37. Peter Galison, *Image and Logic: A Material Culture of Microphysics* (Chicago: University of Chicago Press, 1997). Galison borrows the term *trading zones* from anthropology to describe spaces where physics experimenters and theorists interact. SUMMIT definitely is a medical and engineering trading zone, but the trading that occurs seems to most significantly revolve around objects (from simulators to grant proposals).
38. Sherry Turkle, *Life on the Screen: Identity in the Age of the Internet* (New York: Simon & Schuster, 1995), 12.
39. Karin Knorr Cetina, *Epistemic Cultures: How the Sciences Make Knowledge* (Cambridge, MA: Harvard University Press, 2000), 28–32.
40. See Galison, *Image and Logic*.
41. Sharon Traweek, *Beamtimes and Lifetimes: The World of High Energy Physicists* (Cambridge, MA: Harvard University Press, 1988), 49.
42. Body objects might be considered a type of boundary object in that they have different meanings and research purposes for different social groups. But I treat body objects as more narrowly defined than boundary objects. These objects are specifically bodies and body parts that have been engineered to inhabit computers. My purpose in making the distinction is to highlight the reengineering of representations of bodies that occurs in the world of medical informatics.
43. Gary Lee Downey, *The Machine in Me: An Anthropologists Sits among Computer Engineers* (New York: Routledge, 1998).
44. Anatomy programs also face competition from cadaver brokers, who provide bodies for various types of continuing medical education seminars. This lesser-known and sometimes questionably legal use of cadavers has led to charges of illegal body sales against some medical schools and their employees. Annie Cheney, "The Resurrection Men: Scenes from the Cadaver Trade," *Harper's* 308

(2004), 45–54; Alan Zarembo and Jessica Garrison, "Profit Drives Illegal Trade in Body Parts," *Los Angeles Times* (March 7, 2004), A1.

45. Simulator researchers have picked up a 1999 report on errors in medicine by the National Institute of Medicine as a strong justification for the repetitive procedural training a simulator can provide. Institute of Medicine, Linda J. Kohn, Janet Corrigan, et al., eds., *To Err Is Human* (Washington, DC: National Academies Press, 2000). See also Gawande, *Complications*.

46. Hirschauer, "The Manufacture of Bodies in Surgery," 299.

47. Visible Human Project information and images can be viewed at www.nlm.nih.gov/research/visible/visible_human.html. See also Lisa Cartwright, "The Visible Man: The Male Criminal Subject as Biomedical Norm," in *Processed Lives: Gender and Technology in Everyday Life*, ed. J. Terry and M. Calvert (New York: Routledge, 1997), 123–37; Lisa Cartwright, "A Cultural Anatomy of the Visible Human Project," in *The Visible Woman: Imaging Technologies, Gender, and Science*, ed. P. Treichler, L. Cartwright, and C. Penley (New York: Routledge, 1998), 21–43; Catherine Waldby, *The Visible Human Project: Informatic Bodies and Posthuman Medicine* (New York: Routledge, 2000); Thomas J. Csordas, "Computerized Cadavers: Shades of Representation and Being in Virtual Reality," in *Biotechnology and Culture: Bodies, Anxieties, Ethics*, ed. P. Brodwin (Bloomington: Indiana University Press, 2001), 173–92. Lynda Birke, *Feminism and the Biological Body* (New Brunswick, NJ: Rutgers University Press, 1999) briefly discusses both the Visible Human Project and the Stanford Visible Female.

48. Michael Lynch and Steve Woolgar, *Representations in Scientific Practice* (Cambridge, MA: MIT Press, 1988), 160–61.

49. One anatomist at Stanford teaches students to check their knowledge of anatomy by attempting to label structures on cross sections. He says the ability to "rotate" a two-dimensional image by 90 degrees and then label its structures indicates that the student has begun to understand anatomical terminology and the body's three-dimensional structure.

50. An anatomist at the University of Washington who works on computer applications for teaching anatomy told me that research funding also stands in the way of creating comprehensive anatomical applications. Funding agencies will pay for new applications, usually limited to one area of the body, but claim that applying new computer technologies to an entire body is production work, not research, and ought to be done by the private sector. However, this anatomist claims, and others confirm, most companies have found the labor of creating a comprehensive computer body model not worth the cost.

51. Hirschauer, "The Manufacture of Bodies in Surgery."

52. Mol, *The Body Multiple*.

53. Stefan Helmreich, *Silicon Second Nature: Culturing Artificial Life in a Digital World* (Berkeley: University of California, 1998), 11–12.

54. Stanford Visible Female, "Fun," http://summit.stanford.edu/ourwork/PROJECTS/LUCY/lucywebsite/fun.html. Accessed: March 1, 2003. Donna Haraway argues that we must pay attention to the material and the semiotic natures

of objects. Donna Haraway, *Simians, Cyborgs, and Women: The Reinvention of Nature* (New York: Routledge, 1991), 200. By naming this model "Lucy 2.0," the researchers who created the model brought it into narratives of evolution and reproduction, narratives in which the female often is associated with matter, while the male is associated with form (see Judith Butler, *Bodies That Matter: On the Discursive Limits of "Sex"* (New York: Routledge, 1993). Csordas, "Computerized Cadavers," describes how the male and female bodies from the National Library of Medicine's Visible Human Project have been described as a digital Adam and Eve. The Stanford model similarly has been baptized with an origin story, but a more evolutionary and Stanford-specific story (Cartwright, "The Visible Man;" Cartwright, "A Cultural Anatomy;" Birke, *Feminism and the Biological Body*; Waldby, *The Visible Human Project*.

55. Knorr Cetina, *Epistemic Cultures*, 27.

56. Bruno Latour, *Science in Action* (Cambridge, MA: Harvard University Press, 198), 7.

57. Stanford Visible Female, "Our Work," http://summit.stanford.edu/ourwork/PROJECTS/LUCY/lucywebsite/infofr.html (accessed March 1, 2003).

58. Latour, "How to Talk about the Body," 1.

59. The model is an ideal body: it does not take into account variations among patient bodies or in sense of feel experienced by different surgeons, though these are additions that simulator makers say they will incorporate into future iterations.

60. Trevor Pinch, et al., "Inside Knowledge: Second Order Measures of Skill," *Science, Technology, and Human Values* 21 (1996), 163–86.

61. I do not use the obvious word *interface* here, though it is technically correct, because it has visual implications that I want to avoid.

62. Some experiments have been done with haptic interaction between two users in remote locations, but technically this creates a problem separating signals that are feeding forward from users' bodies from signals that are simultaneously feeding back to users' bodies. Human nervous systems have no trouble with this kind of "signal processing," but it still is a challenge for machines.

63. See Heidegger, quoted in Lucy Suchman, *Plans and Situated Actions: The Problem of Human Machine Communication* (Cambridge: Cambridge University Press, 1987), 53–54.

64. Polanyi, *The Tacit Dimension*.

65. The gynecologist plans to incorporate values for haptic feel based on the experiences of many surgeons in a future iteration of the simulator.

66. Miriam Reiner, "The Role of Haptics in Immersive Telecommunication Environments," (unpublished manuscript, no date).

67. Latour, "How to Talk about the Body."

Contributors

NANCY ANDERSON is assistant professor of visual studies at the State University of New York at Buffalo. Her research interests are nineteenth- and twentieth-century scientific and medical imaging technologies and the use of images and physical models in biological research. She has published on such topics as the early use of the image intensifier in cell biology and the study of bioluminescence as well as on the use of Buckminster Fuller's geodesic dome as a scientific model for virus studies. She is currently writing a book entitled *The Scale of the Event: Science Image Model, 1945–1965*, which will explore the problem of scale in the emerging field of molecular biology and its imaging and modeling practices.

SABINE BRAUCKMANN is senior researcher at the Estonian Institute of Humanities of Tallinn University (Estonia). Her primary interest is in studying the epistemic history of embryos and cells in text and images from around 1800 to the 1960s. She coedited *Graphing Genes, Cells and Embryos* (2009) and worked with Scott F. Gilbert on embryonic cells in Gustav Klimt's paintings (*Fertilization Narratives in the Art of Gustav Klimt, Diego Rivera, and Frieda Kahlo*, 2011). Most recently she initiated a trans-disciplinary network between University of Tartu and the Estonian Academy of Sciences, focusing on environmental humanities and, in particular, on the cultural and visual history of "plant-spaces" in the nineteenth century.

T. HUGH CRAWFORD is an associate professor at the Georgia Institute of Technology, where he teaches in the Science, Technology and Culture program. He is the author of *Modernism, Medicine and William Carlos Williams*, past-president of the Society for Literature, Science and the Arts, and former editor of *Configurations*. His current research in is culture, cognition, and tool use, and he is writing a book tentatively entitled *Treeontology*.

SCOTT CURTIS is associate professor of Radio/Television/Film at Northwestern University, where he teaches film theory and history. He is primarily interested in the institutional appropriation of motion pictures, such as educational filmmaking or the use of moving image technology as a scientific research tool or diagnostic instrument.

A former medical photographer, he has published numerous articles and chapters on medical cinematography and the scientific use of motion pictures.

MICHAEL R. DIETRICH is professor of history and philosophy of biology in the Department of Biological Sciences at Dartmouth College. His research takes an interdisciplinary approach to scientific controversies in twentieth-century biology. He has authored numerous articles on the history of evolutionary genetics and is currently completing a biography of the exiled German biologist Richard Goldschmidt. He also recently coedited *Rebels, Mavericks, and Heretics in Biology* (Yale University Press, 2008) with Oren Harman. He is also an adjunct senior scientist at the Marine Biological Laboratory at Woods Hole, Massachusetts.

MICHAEL J. GOLEC is associate professor of the history of design at The School of the Art Institute of Chicago and is the 2011 Anschutz Distinguished Fellow in American Studies at Princeton University. Golec's scholarship focuses on twentieth-century design in the United States as it intersects with the history of art, the history of technology and science, and philosophical aesthetics. While his interests range across and touch on all manner of designed objects, Golec's research emphasizes graphic design, visual communications, and print culture. He is the author of *Brillo Box Archive: Aesthetics, Design, and Art* (Dartmouth College Press, 2008) and, along with Aron Vinegar, coedited and contributed to *Relearning from Las Vegas* (University of Minnesota Press, 2009).

RICHARD L. KREMER is associate professor of history at Dartmouth College, where he teaches history of science, especially its material culture, and curates the King Collection of Historic Scientific Instruments. His review of nineteenth- and twentieth-century physiology appeared in *The Cambridge History of Science*, vol. 6 (2009). He has also published articles on the apparatus of high-speed photography and the early development of the American electronics industry, especially the General Radio Company of Cambridge, Massachusetts.

MARA MILLS, assistant professor of media, culture, and communication at New York University, trained in biology (MA) and history of science (PhD) at Harvard University. Working at the intersection of disability studies and media studies, she is currently completing a book on the historical relationship between the telephone system, deafness, and signal processing. A former high school biology teacher, she also has a particular interest in visual culture and pedagogy in the life sciences.

KIRSTEN OSTHERR is associate professor of English at Rice University. She is the author of *Cinematic Prophylaxis: Globalization and Contagion in the Discourse of World Health* (2005), and has published articles on David Lynch, science fiction, surgical training films, and biocultures. She is currently completing *Medical Visions: Producing the Patient through Film, Television, and Imaging Technologies*, forthcoming from Oxford University Press.

LAURA PERINI is assistant professor in the Department of Philosophy at Pomona College. Her research focuses on the diverse array of visual representations scientists use, both as part of the research process and in presenting and defending new ideas in publications. She aims to clarify what visual representations contribute to scientific reasoning and to explain how they do so. She has published several articles and book chapters examining the nature of scientific visual representations, how they function as evidence, and their capacity to convey explanatory content.

RACHEL PRENTICE is an associate professor in the Science and Technology Studies Department at Cornell University. Her work examines embodiment and technology in biomedicine. Her forthcoming book is titled *Bodies of Information: An Ethnography of Anatomy and Surgery Education* (Duke University Press). The book examines the embodiment of perception, affect, ethics, and judgment in anatomy and surgery education at a moment when educators have begun to introduce computational tools, such as surgical simulators, into training curricula. She is now working on a project on movement and perception in biomedicine and alternative medicine.

ADINA L. ROSKIES, an associate professor of philosophy at Dartmouth College, has pursued a career in both philosophy and neuroscience. Her research and writing has focused on philosophy of mind, philosophy of science, and ethics, including neuroethics. She received a PhD in neuroscience and cognitive science in 1995 from the University of California, San Diego. She then did a postdoctoral fellowship in cognitive neuroimaging at Washington University in St. Louis, using positron emission tomography and the then newly developing technique of functional MRI. After serving two years as senior editor of the journal *Neuron*, she went on to complete a PhD in philosophy at the Massachusetts Institute of Technology in 2004. Dr. Roskies joined the Dartmouth faculty in the fall of 2004. She has been a visiting fellow in philosophy at the University of Utah, the Australian National University, and the University of Sydney. Her work has been supported by grants and fellowships from National Institutes of Health, the McDonnell-Pew Foundation, the NEH, and the Mellon Foundation. She is a member of the MacArthur Law and Neuroscience Project and holds the Melville and Leila Straus 1960 Faculty Fellowship. Dr. Roskies is the author of some fifty articles published in academic journals, including one on neuroethics for which she was awarded the William James Prize by the Society of Philosophy and Psychology.

HENNING SCHMIDGEN is a senior research scholar at the Max Planck Institute for the History of Science in Berlin, Germany, as well as Professor of Media Aesthetics at Regensburg University. His primary interest is in the history of physiological and psychological time. Fascinated by the possibilities of new media technologies, he has cocreated and continues to develop the "Virtual Laboratory," an archival resource and publication platform for the history of the experimental life sciences (http://vlp.mpiwg-berlin.mpg.de). His most recent book deals with Hermann von Helmholtz and Marcel Proust: *Die Helmholtz-Kurven: Auf der Spur der verlorenen Zeit*, Merve, 2009.

Index

Page numbers in *italics* refer to the illustrations.

aging, 17, 25, 29–30, 33, 34
anatomy, 54, 56–57, 58, 59–60, 96, 101, 283, 284, 292–293
anesthesia, 51–53, 56
animal experimentation, 51–58, 59–62, 95–96. *See also* ethics
animation, 126, 127, 129, 137, 173
Anschauung, 8, 94, 100, 221–222
autopsy, 78

Batten, Barton, Durstine and Osborn, 198–199; for Ethyl Gasoline Company, 198–200
Beale, Lionel, 48–58; microscopic images, *51*, *57*
Bergson, Henri, 32
Bernard, Claude, 59, 62, 68, 69, 78
Bichat, Xavier, 45
Billroth, Theodor, 76, 84
"biological gaze," 17–18, 35
Biskind, Morton, 111–113
Breeder, C. M., 205
Bronowski, Jacob, 164
Bronowski, Judith, 164–165, 167, 176
Buck-Morss, Susan, 52–53
Bunk, Frederick, 14
Butler, Elizabeth, 225

Carrel, Alexis, 31–33, 35
cell: death, 17, 21, 27–30, 28, 33, 35; division; 23–24, 176; growth, 21, 29, 31, 176; schematic diagram of, 240–242, *241*
Czermak, Johann Nepomuk, 94–96, *96–97*, 101, 102, 105; projections, 94–96, *95*, *97*, 113

Daston, Lorraine, 6, 18, 83, 195, 242, 246
Deleuze, Gilles, 47, 48, 62–63
Derrida, Jacques, 48, 158n16
diagram: compositional, 248-251, *249*; schematic, *241*, 245–248, *247*
Dieterle, William, 142, 143–144, 145, 146, 147, 148, 149, 150–151, 154–156; *Dr. Ehrich's Magic Bullet*, 142, 151–155; *The Story of Louis Pasteur*, 142, 143, 146, 147, 150–151
Disney studios, 122, 136. *See also* Rockefeller foundation, *Unhooking the Hookworm*
dissection, 47–48, 49–51, *51*, 3–54
Dr. Ehrich's Magic Bullet (Dieterle), 142, 151–155
Du Nouy, Pierre Lecomte, 31–32

Eames, Charles, 162–163, 165–166, 167, 168–170, 180; *A Communications

Primer; 168–170. See also *Rough Sketch . . .*
Eames Office, 162–163, 165, 167, 170
Edgerton, Harold, 3, 186–189, 191, 192, 193, 194–197, 198, 200–201, 207; career, 202; experiments, 202–204, 204, 205–206; photographs, 186–190, 191–193, 194, 195, 196, 197, 205, 207
Ehrlich, Paul, 151–153, 154–155
Eliot, George (*Middlemarch*), 45, 55
embryo: development/growth, 22, 25–27, 213–215, 217, 218–219, 220, 222; *Image*, 221, 222, 226; staining, 219–220, 220, 221, 223, 224; use in experimentation, 24-25, 27, 58–59, 59, 213, 217, 223, 225, 226, 227
embryology, 14, 24, 25–26, 27, 213, 216, 217, 218, 223–224
ethics, 56, 62–63

"fate maps," 214, 216–217, 218, 222, 228
Flash!, 188, 189, 192, 193–194, 194, 196, 197
Fleck, Ludwick, 75, 79, 80, 141
Foucault, Michel, 1, 17, 69, 79
functional magnetic resonance imaging (fMRI), 256, 259, 264, 266, 267–268

Galison, Peter, 6, 18, 195, 242, 246
Good, Byron, 283
Gray, James, 203
Gräper, Ludwig, 214, 217, 224
growth, 14–15, 22
Guattari, Félix, 47, 62

Hall, G. Stanley, 30
Hamburger, Viktor, 214, 222, 223
haptic feedback, 280, 301–302, 303–304
healing, 31
heredity, 33
Hirschauer, Stefan, 283–284
Hopwood, Nick, 5, 6, 15, 25–27
Huxley, Thomas Henry, 55–56, 57–58

Jacobj, Carl, 97–100, 101–105, 108, 109, 111, 113; projections, 97–100, 102, 103, 104–105, *104*, 109–111, *110*
Jameson, Fredric, 143, 144,
Jendrassik, Andreas Eugen, 105–106, 110

Kaiserling, Carl, 108–109
Keller, Evelyn Fox, 17
Killian, James Rhyne, 189–190, 192–193
Koch, Robert, 56, 73, 75, 152
Kretz, Richard, 81–83

laboratories, 141, 152–153
Landecker, Hannah, 16–17, 35, 37n24
Latour, Bruno, 4, 78, 144, 148, 155–156, 156n3, 164, 166, 172, 281, 284, 304–305; inscription, 150–151; laboratories, 62, 141–143; *Pasteurization of France*, 147, 148–149
Levinas, Emmanuel, 47–48
Lewis, Frederick, 15
Lewontin, R. C., 153
Lobdell, Harold, 189–190
Loeb, Jacques, 19, 33–34,
Luhmann, Niklas, 170–171
Lynch, David (*Blue Velvet*), 180–181
Lynch, Michael, 4, 50, 63, 187, 244, 253n23

Mall, Franklin, 16, 34, 40n77
McLeish, John (*Looking at Chromosomes*), 177–178
McLuhan, Eric, 166, 172–173
"medical gaze," 69, 84
medicine, 44–47, 56; education in, 45–47; studying abroad, 45. *See also* surgical simulators
microphotography, 72–73
microscopy, 20–21, 48, 50, 54, 146
microtome, 24–25, 24
Minot, Charles, 14–16; career, 15–16, 18–22, 34; cell images, 26, 28; experiments, 25–27, 29; lectures, 29; works by, 24–25, 27, 30

Moore, Paul, 205–207
motion pictures, 123, 124, 132, 135–136, 143, 146–147, 150, 163, 171, 173–175; domesticity in, 164, 171; physical aspects of, 166–167, 168; race in 132–134, 136; scientific, 142, 143, 144, 146–147, 148, 149, 163, 171
mutual articulation, 280–281, 282, 284–285, 303–304
Myers, Greg, 237, 241, 243

National Science Board, 255, 270
neuroimaging, 256, 257, 258–261, 259, 262–263, 264–265, 266–267, 272, 273. *See also* functional magnetic resonance imaging (fMRI)
neurology, 256–258, 259

objectivity, 18, 69, 73, 82, 123, 150, 156, 195, 263; history of, 157n13, 242; practical, 144, 157n10
observation, 68–69, 79; medical, 70, 71, 80; scientific, 69, 71, 84
optical constancy, 172–174, 181
organicism, 16

Pasteels, Jean, 214, 225, 226
Pasteur, Louis, 144, 145, 147–148, 151; Pouilly le Fort experiment, 148, 153
patient studies, 73–74, 75–76
Pauwel, Luc, 187
Pearl, Raymond, 34
pedagogy, 4–5, 61, 134–135, 143, 154–155, 204, 236–237, 242–243, 261, 269–272
Pestalozzi, Johann Heinrich, 100–101
photography, 68, 71, 72, 84–85, 143, 238–239, 242, 244; causal and counterfactual dependence, 265–267; clinical portrait, 75–78; "faithful to nature," 82–83; high speed, 186–189, 190–191, 192, 195, 198, 200, 201; medical, 71–72, 73–74, 74, 76, 77, 80; series, 79–80, 81;

theory independence, 267–268; transparency, 263–265
Powers of Ten (Eames), 163–164
projection, 94–101, 95, 102, 103–104, 106–107, 107, 112, 113–115; apparatus, 109–111, 110; episcopic, 107–109, 113–114
Purkinje, Jan Evangelista, 101–102

Quicker'n a Wink (Smith), 200–201, 205

Ranvier, Louis-Antoine, 59, 60–61
Rawes, Mary, 225, 227; *Image*, 226
Richards, Stewie, 51
Rockefeller Foundation, 121–123, 124, 125, 126, 127, 131, 132, 134, 135–136; reactions to; 125–126, 130, 131, 134; *Unhooking the Hookworm*, 121–124, 125, 126–127, 127–133, 130, 134, 135, 136
Rough Sketch for a Proposed Film Dealing with the Powers of Ten and the Relative Size of the Universe (Eames), 163–164, 165, 167, 168, 170, 171–173, 174–177, 177, 178, 179–180, 181
Rudnik, Dorothy, 214, 223–229, 227
Rutherford, William, 53

Schmiedeberg, Oscar, 96
senescence, 14, 17, 35
Shapin, Steve, 143, 145, 146
Smith, Pete. See *Quicker'n a Wink*
"specification maps," 227–228
Stanford University Medical Media and Information Technologies (SUMMIT) laboratory, 285–288, 291–292, 298
The Story of Louis Pasteur (Dieterle), 142, 143, 146, 147, 150–151
Stricker, Saloman, 107–108
surgery, 278–279, 281–282; using visualization in, 282–284

surgical simulators, 277–278, 279–280, 293, 294–295, 296–297, 299–302, 304–306; physical, 278, 288–289, 290; virtual reality, 278, 279–280, 284–285, 287, 289, 290–291, 293–294, 295–300, 301–302, 303, 304–305. *See also* medicine

Technical Review, 188, 189, 190, 198, 206, 207
textbooks, 46, 235–236; figures in, 2, 237, 238, 245, 247, 248–252, 249; images in, 46–47, 235, 236, 238–239, 240–246, 241, 247, 251–252; pedagogical value of; 236–237, 242–243
tissue cultivation, 16, 17, 31–35, 216–217

Unhooking the Hookworm. See under Rockefeller Foundation

vision, 261–262
Vogt, Walther, 214, 218, 228, 229; education, 215; experimentation, 218–221, 220, 221, 222–223, 229; photomicrographs, 218
Voronoff, Serge, 30–31

Walton, Kendall, 263, 264, 267
Wetzel, Robert, 214, 215, 223, 224–225, 226
Woodruff, Lorande, 236

x-rays, 2, 75